Protein Structure Prediction

The Practical Approach Series

SERIES EDITOR

D. RICKWOOD
Department of Biology, University of Essex
Wivenhoe Park, Colchester, Essex CO4 3SQ, UK

B. D. HAMES
Department of Biochemistry and Molecular Biology
University of Leeds, Leeds LS2 9JT, UK

★ indicates new and forthcoming titles

Affinity Chromatography
Anaerobic Microbiology
Animal Cell Culture (2nd edition)
Animal Virus Pathogenesis
Antibodies I and II
★ Antibody Engineering
Basic Cell Culture
Behavioural Neuroscience
Biochemical Toxicology
Bioenergetics
Biological Data Analysis
Biological Membranes
Biomechanics—Materials
Biomechanics—Structures and Systems
Biosensors
Carbohydrate Analysis (2nd edition)
Cell–Cell Interactions
The Cell Cycle
Cell Growth and Apoptosis
Cellular Calcium
Cellular Interactions in Development
Cellular Neurobiology
Clinical Immunology
Crystallization of Nucleic Acids and Proteins
Cytokines (2nd edition)
The Cytoskeleton
Diagnostic Molecular Pathology I and II
Directed Mutagenesis
★ DNA and Protein Sequence Analysis
DNA Cloning 1: Core Techniques (2nd edition)
DNA Cloning 2: Expression Systems (2nd edition)
★ DNA Cloning 3: Complex Genomes (2nd edition)
★ DNA Cloning 4: Mammalian Systems (2nd edition)
Electron Microscopy in Biology
Electron Microscopy in Molecular Biology
Electrophysiology

Enzyme Assays
★ Epithelial Cell Culture
Essential Developmental Biology
Essential Molecular Biology I and II
Experimental Neuroanatomy
★ Extracellular Matrix
Flow Cytometry (2nd edition)
★ Free Radicals
Gas Chromatography
Gel Electrophoresis of Nucleic Acids (2nd edition)
Gel Electrophoresis of Proteins (2nd edition)
Gene Probes 1 and 2
Gene Targeting
Gene Transcription
Glycobiology
Growth Factors
Haemopoiesis
Histocompatibility Testing
★ HIV Volumes 1 and 2
Human Cytogenetics I and II (2nd edition)
Human Genetic Disease Analysis
Immunocytochemistry
In Situ Hybridization
Iodinated Density Gradient Media
★ Ion Channels
Lipid Analysis
Lipid Modification of Proteins
Lipoprotein Analysis
Liposomes
Mammalian Cell Biotechnology
Medical Bacteriology
Medical Mycology
Medical Parasitology
Medical Virology
Microcomputers in Biology
Molecular Genetic Analysis of Populations
Molecular Genetics of Yeast
Molecular Imaging in Neuroscience
Molecular Neurobiology
Molecular Plant Pathology I and II
Molecular Virology
Monitoring Neuronal Activity
Mutagenicity Testing
★ Neural Cell Culture
Neural Transplantation
★ Neurochemistry (2nd edition)
Neuronal Cell Lines
NMR of Biological Macromolecules
Non-isotopic Methods in Molecular Biology
Nucleic Acid Hybridization
Nucleic Acid and Protein Sequence Analysis
Oligonucleotides and Analogues
Oligonucleotide Synthesis
PCR 1
PCR 2
Peptide Antigens
Photosynthesis: Energy Transduction

Plant Cell Biology
Plant Cell Culture (2nd edition)
Plant Molecular Biology
Plasmids (2nd edition)
★ Platelets
Pollination Ecology
Postimplantation Mammalian Embryos
Preparative Centrifugation
Prostaglandins and Related Substances
Protein Blotting
Protein Engineering
Protein Function
Protein Phosphorylation
Protein Purification Applications
Protein Purification Methods
Protein Sequencing
Protein Structure
★ Protein Structure Prediction
Protein Targeting
Proteolytic Enzymes
★ Pulsed Field Gel Electrophoresis
Radioisotopes in Biology
Receptor Biochemistry
Receptor–Ligand Interactions
RNA Processing I and II
★ Subcellular Fractionation
Signal Transduction
Solid Phase Peptide Synthesis
Transcription Factors
Transcription and Translation
Tumour Immunobiology
Virology
Yeast

Protein Structure Prediction

A Practical Approach

Edited by

MICHAEL J. E. STERNBERG

Biomolecular Modelling Laboratory,
Imperial Cancer Research Fund,
PO Box 123, 44 Lincoln's Inn Fields,
London WC2A 3PX, UK

at
OXFORD UNIVERSITY PRESS
Oxford New York Tokyo

Oxford University Press, Walton Street, Oxford OX2 6DP

Oxford New York
Athens Auckland Bangkok Bombay
Calcutta Cape Town Dar es Salaam Delhi
Florence Hong Kong Istanbul Karachi
Kuala Lumpur Madras Madrid Melbourne
Mexico City Nairobi Paris Singapore
Taipei Tokyo Toronto

and associated companies in
Berlin Ibadan

Oxford is a trade mark of Oxford University Press

Published in the United States
by Oxford University Press Inc., New York

A catalogue record for this book is available from the British Library

Library of Congress Cataloging in Publication Data

Protein structure prediction: a practical approach/edited by
Michael J. E. Sternberg.
(A Practical approach series; 170)
Includes bibliographical references and index.
1. Proteins—Conformation—Computer simulation. I. Sternberg,
Michael J. E. II. Series.
QP551.P69764 1996 547.7′5—20 96–8970 CIP

ISBN 0 19 963497 1 (Hbk)
ISBN 0 19 963496 3 (Pbk)

Typeset by Footnote Graphics, Warminster, Wilts
Printed in Great Britain by Information Press, Ltd, Eynsham, Oxon.

Preface

Prediction of the three-dimensional structure of a protein from its sequence is a problem faced by an increasing number of biological scientists as they strive to utilize genetic information. *Protein Structure Prediction: A Practical Approach* provides a timely description of current approaches and their validity. It describes in detail the various computer-based strategies for translating sequence into structure, highlighting the degree of confidence that can be attributed to different algorithms.

The increasing sizes of the sequence and structural databases, the improvements in computing power, and the deeper understanding of the principles of protein structure have led to major developments in the field over the last few years. To present an integrated review of the practical approach, the book is organized along the same lines as the procedure one follows starting with a sequence. First, the book summarizes the principles underlying prediction and overviews the general procedure. Chapters 2 and 3 focus on sequence analysis to identify homologies and motifs. The next two chapters review secondary-structure and topological predictions for globular and transmembrane proteins. Chapter 6 describes obtaining coordinates from comparative (homology) modelling with the application of this methodology to antibody structure being reported in the following section. Chapter 8 describes the exciting recent developments in fold recognition. The combinatorial approach reported in Chapter 9 spans secondary and tertiary structure prediction. The application of energy calculations to protein modelling is reported in Chapter 10. A major goal of prediction is to assist in the design of novel ligands and the present methodology is then reported. Finally the results of blind trials of the success of prediction are described.

Protein Structure Prediction: A Practical Approach is a unique compendium of facts and advice in this increasingly important field. It will be invaluable both for non-specialists who require guidance in identifying and evaluating appropriate strategies and for experts who require up-to-date information on the latest techniques and approaches.

London M. J. E. S.
August 1996

Contents

Contributors

GEOFFREY J. BARTON
Laboratory of Molecular Biophysics, University of Oxford, Rex Richards Building, South Parks Road, Oxford OX1 3QU, UK.

TOM L. BLUNDELL
Laboratory of Molecular Biology and ICRF Unit of Structural Molecular Biology, Department of Crystallography, Birkbeck College, Malet Street, London WC1E 7HX, UK.
*Present address: Department of Biochemistry, University of Cambridge, Tennis Court Road, Cambridge CB2 1QW, UK.

CHARLES L. BROOKS III
Department of Molecular Biology, The Scripps Research Institute, 10666 North Torrey Pines Road, La Jolla, CA 92037, USA.

FRED E. COHEN
Department of Pharmacology, University of California, San Francisco, California 94143, USA.

K. GURUPRASAD
Laboratory of Molecular Biology and ICRF Unit of Structural Molecular Biology, Department of Crystallography, Birkbeck College, Malet Street, London WC1E 7HX, UK.
*Present address: Department of Biochemistry, University of Cambridge, Tennis Court Road, Cambridge CB2 1QW, UK.

ANDREW H. HENRY
School of Biology and Biochemistry, University of Bath, Claverton Down, Bath BA2 7AY, UK.
*Present address: Oxford Molecular plc, Magdalen Science Park, Oxford, OX4 4GA, UK.

DAVID T. JONES
Department of Biochemistry and Molecular Biology, University College London, Gower Street, London WC1E 6BT, UK.
*Present address: Department of Biological Sciences, University of Warwick, Coventry CV4 7AL, UK.

MARY E. KARPEN
Department of Chemistry, Grand Valley State University, Allendale, MI 49401, USA.

ROSS D. KING
Biomolecular Modelling Laboratory, Imperial Cancer Research Fund, PO Box 123, 44 Lincoln's Inn Fields, London WC2A 3PX, UK.

CHRISTINE A. ORENGO
Department of Biochemistry and Molecular Biology, University College London, Gower Street, London WC1E 6BT, UK.

JAN PEDERSEN
School of Biology and Biochemistry, University of Bath, Claverton Down, Bath BA2 7AY, UK.
*Present address: CARB, 9600 Gudelsky Drive, Rockville, MD 20850, USA.

SCOTT R. PRESNELL
ZymoGenetics Inc., 1201 Eastlake Ave E, Seattle, WA 98102, USA.

ANTHONY R. REES
School of Biology and Biochemistry, University of Bath, Claverton Down, Bath BA2 7AY, UK.

MANSOOR A. S. SAQI
Bioinformatics Group, Glaxo Medicines Research Centre, Gunnels Wood Road, Stevenage, Hertfordshire SG1 2NY, UK.

STEPHEN J. SEARLE
School of Biology and Biochemistry, University of Bath, Claverton Down, Bath BA2 7AY, UK.

BRIAN K. SHOICHET
Institute of Molecular Biology, University of Oregon, Eugene, OR 97403–1229, USA.
*Present address: Dept. of Molecular Pharmacology and Biological Chemistry, Mail Code S215, Northwestern University Medical School, 303 E. Chicago Ave, Chicago, IL 60611-3008, USA.

N. SRINIVASAN
Laboratory of Molecular Biology and ICRF Unit of Structural Molecular Biology, Department of Crystallography, Birkbeck College, Malet Street, London WC1E 7HX, UK.
*Present address: Department of Biochemistry, University of Cambridge, Tennis Court Road, Cambridge CB2 1QW, UK.

MICHAEL J. E. STERNBERG
Biomolecular Modelling Laboratory, Imperial Cancer Research Fund, PO Box 123, 44 Lincoln's Inn Fields, London WC2A 3PX, UK.

JANET M. THORNTON
Department of Biochemistry and Molecular Biology, University College London, Gower Street, London WC1E 6BT, UK.

GUNNAR VON HEIJNE
Department of Biochemistry, Stockholm University, S-106 91 Stockholm, Sweden.

NICHOLAS WHITELEGG
School of Biology and Biochemistry, University of Bath, Claverton Down, Bath BA2 7AY, UK

Abbreviations

AChE	acetylcholinesterase
ALPPS	a language for the prediction of protein substructures
ASA	accessible surface area
ANN	artificial neural network
BLAST	basic local alignment search tool
CD	circular dichroism
CDR	complementary determining region
CSD	Cambridge Structural Database
EDN	eosinophil-derived neurotoxin
EM	excitation maximization
FCD	Fine Chemical Database
FTIR	Fourier transform infrared
Fv	antibody variable region
GOR	Garnier–Osguthorpe–Robson
hGH	human growth hormone
hIL-4	human interleukin 4
ILP	inductive logic programming
indels	insertions and deletions
MD	molecular dynamics
mIL-4	mouse interleukin 4
MSP	maximal segment pair
NMR	nuclear magnetic resonance
ORF	open reading frame
PAM	percentage of acceptable point mutations per 10^8 years
PDB	Protein Data Bank
pdf	probability density function
PLANS	a pattern language for amino and nucleic acid sequences
rms	root mean square
rmsd	root mean square deviation
SCR	structurally conserved regions
SD	standard deviation
SVR	structurally variable region
TS	thymidylate synthase

1

Protein structure prediction—principles and approaches

MICHAEL J. E. STERNBERG

1. Introduction

Anyone who examines the atomic structure of even the smallest globular protein cannot but marvel at the convoluted path of the polypeptide main chain and the complex packing arrangements of main and side chain atoms. Surely then it must be a misguided attempt with today's understanding of physical chemistry and the available computational power to try to predict theoretically protein structure from sequence? It is the aim of this book to give a practical approach towards a limited achievement of this ambitious goal.

The central dogma motivating prediction is that: 'the three-dimensional structure of a protein is determined by its sequence and its environment without the obligatory role of extrinsic factors.'

This hypothesis arises from the classical renaturation studies on ribonuclease by Anfinsen in the sixties (1). Since then, many other proteins have been refolded into their active form (2). However the hypothesis can be challenged. Chaperons and disulfide interchange enzymes have been identified as assisting the folding process. But the experiments support the view that these molecules only assist rather than determine the final native state. There has been sufficient success of predictions to justify the use of the central dogma as a working hypothesis.

Prediction is becoming a pressing problem for many biologists as the discrepancy continues to increase between the number of known protein sequences and the number of experimentally-determined structures (*Figure 1*). At the end of 1994 there were over 40000 protein sequence entries in release 30 of SWISS-PROT (3). This corresponding to about 33800 chains of between 30 and 1500 residues that were non-transmembrane sequence. When homologies of over 30% identity are removed this corresponds to over 9300 chains (4). In contrast, the Brookhaven Data Bank (5) at the end of 1994 had 4045 experimentally-determined protein chains. When homologies of over 35% identity are removed this corresponds to only 510 chains. If

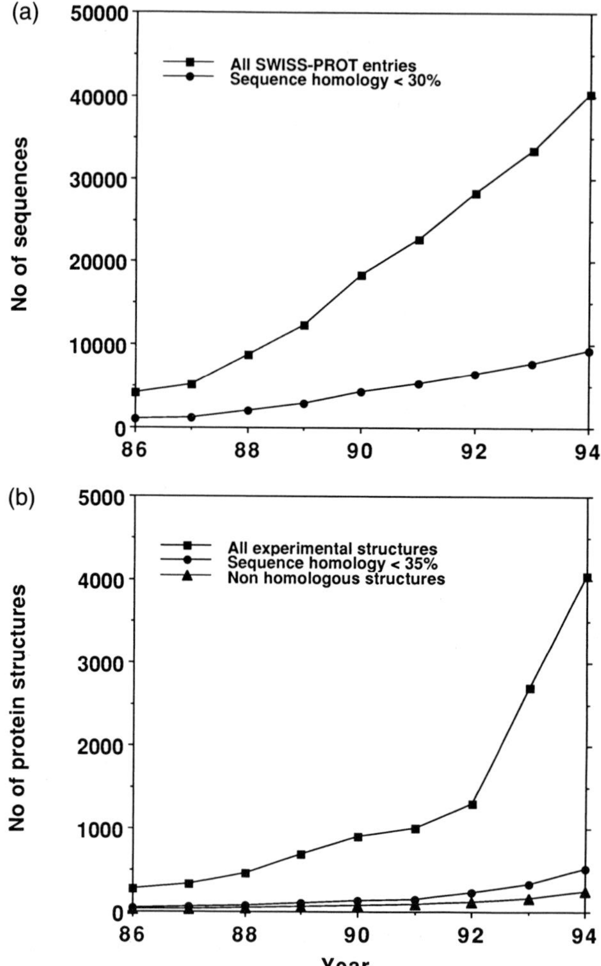

Figure 1. The number of protein sequences and structures. The figure is based on ref. 4. (a) The number of sequence entries in SWISS-PROT and an estimated number of non-homologous, non-transmembrane chains of between 30 and 1500 residues (the estimate is based on the ratio for 1992). (b) The number of polypeptide entries of experimental structures in the Brookhaven Protein Data Bank. From these entries homologies of ≥ 35% identity are removed. Then chains with a structural similarity of a SSAP score of ≥ 80 are further removed.

homologous structures are further identified by structural similarity this yields a representative set of 373 chains (data kindly supplied by Dr Christine Orengo). There is, therefore, roughly a 20-fold discrepancy between the number of representative sequences and structures.

Part A of this chapter will summarize the principles of protein structure as they relate to the prediction problem. Then Part B will overview the approaches that are detailed in the subsequent chapters.

1A—Principles of protein structure

2. Dominant effects in protein folding

In theory, molecular dynamics simulations in solvent with accurate potentials and run over sufficient time would model the folding of a protein (see Chapter 10). Since this is not feasible at present, it is instructive to describe the individual effects that govern the protein/solvent system.

2.1 Net protein stability

The diverse chemical properties of the protein main chain and side chains (*Figure 2*) give rise to an interplay of non-covalent and entropic effects that determine the structure of the molecule. Most globular water soluble proteins have only marginal stability at their physiological conditions: the change in Gibbs free energy from the unfolded to folded state typically is between -5 to -20 kcal mol^{-1}. Understanding and quantifying the thermodynamic effects remains a challenge; for details see refs 6 and 7.

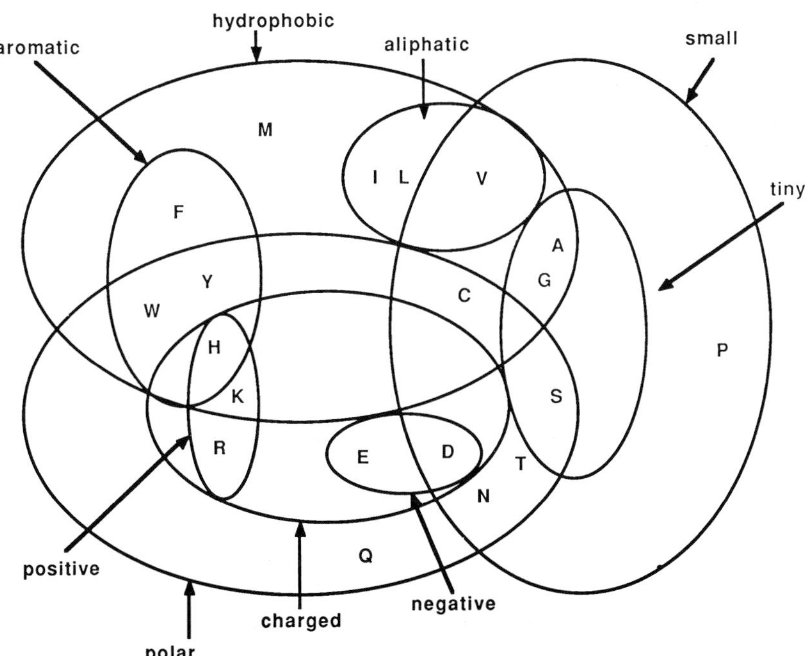

Figure 2. A Venn diagram representation of the properties of the residues.

2.2 The hydrophobic effect

It is widely regarded that protein folding is driven by the hydrophobic effect (8). This describes the energetic preference for non-polar atoms, such as hydrocarbons, to associate and reduce their contact with water—a phenomenon which is readily observable on mixing oil and water (hence the crude oil-drop model for protein folding). At room temperature, the effect is mainly entropic. Bulk water molecules are sampling a variety of conformations and consequently are continually forming and breaking hydrogen bonds with other waters. The presence of a non-polar atom restricts the number of degrees of freedom of the neighbouring water molecules compared to water molecules in the bulk medium.

Experimental measures of the magnitude of hydrophobicity for different side chains come from partition experiments in which the concentrations of compounds modelling side chains are measured in a medium representing the protein core and in water. Fauchère and Pliska (9) used octanol as the medium, a molecule which has a long non-polar aliphatic chain but a polar hydroxyl capable of interacting with a charge. This scale (*Table 1*) is widely used today but many others are tabulated in ref. 10.

The relationship between the hydrophobic effect and the accessible surface area (ASA) of the solute has dominated many aspects of protein modelling (11). ASA is defined as the locus of the centre of a water probe as its rolls around the surface of a molecule (*Figure 3*). Molecular surface is the sum of the area of the solute atoms in contact with this water probe (i.e. contact surface) and the re-entrant surfaces of the water probe. There is an approximate linear relationship (12) between hydrophobicity and the ASA of all non-polar atoms in side chains in their extended conformation. The slope of the data in *Table 1* yields a relationship of 24 cal mol^{-1} per $Å^2$ of accessible area buried in a transition.

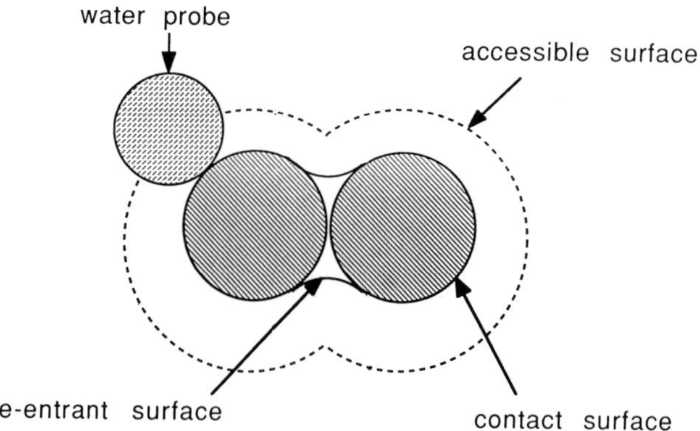

Figure 3. Definition of accessible surface area, contact area, and re-entrant area for two methane atoms in contact.

Table 1. Properties of residues

Residue	Code	Residue ASA ($Å^2$)	Residue ASA–NP ($Å^2$)	ΔG-HYD (kcal/mol)	ΔG-Miller (kcal/mol)	$T\Delta S$ rotamer (kcal/mol)	$T\Delta S$ fusion (kcal/mol)
Ala	A	71	71	−0.42	−0.20	0.00	0.00
Arg	R	206	80	1.37	1.34	−2.03	0.00
Asn	N	107	45	0.82	0.69	−1.57	−1.24
Asp	D	103	50	1.05	0.72	−1.25	−1.37
Cys	C	98	98	−1.34	−0.67	−0.55	−0.65
Gln	Q	143	53	0.30	0.74	−2.11	−1.91
Glu	E	136	61	0.87	1..09	−1.81	−2.04
Gly	G	33	33	0.00	−0.06	0.00	0.00
His	H	148	96	−0.18	−0.04	−0.96	−1.64
Ile	I	140	140	−2.46	−0.74	−0.89	−0.83
Leu	L	143	143	−2.32	−0.65	−0.78	−0.83
Lys	K	166	118	1.35	2.00	−1.94	−3.50
Met	M	159	159	−1.68	−0.71	−1.61	−1.54
Phe	F	164	164	−2.44	−0.67	−0.58	−0.93
Pro	P	120	120	−0.98	−0.44	0.00	0.00
Ser	S	79	48	0.05	0.34	−1.71	−0.89
Thr	T	104	76	−0.35	0.26	−1.63	−0.73
Trp	W	210	186	−3.07	−0.45	−0.97	−1.88
Tyr	Y	177	135	−1.31	0.22	−0.98	−1.35
Val	V	116	116	−1.66	−0.61	−0.51	−0.15

For each residue, the one letter code is given followed by: the accessible surface area (ASA); the non-polar ASA (ASA-NP); the experimental free energy of transfer from Fauchère and Pliska (9) (ΔG-HYD); the empirical hydrophobicity scale of Miller *et al.* (13) (ΔG-Miller); the side chain conformational entropy from a rotamer library (22) and from fusion (23) at 300 K ($T\Delta S$ rotamer and $T\Delta S$ fusion).

An empirical measure of the hydrophobicity of side chains can be derived by enumerating the relative frequencies in a set of proteins of a residue type being exposed and buried. The empirical scale of Miller *et al.* (13), given in *Table 1*, correlates roughly with the experimental scale of Fauchère and Pliska (9). This empirical approach has been refined by Eisenberg and McLachlan (14) who evaluated the area dependence of hydrophobicity separately for each atom type.

Recent mutagenesis studies, for example see ref. 15, measuring the contribution of buried aliphatic side chains to protein stability suggest a much higher relationship of area to hydrophobicity (around 40–50 cal mol^{-1} per $Å^2$). It has been proposed (16) that transfer experiments need to be re-interpreted to consider the relative sizes of the solutes and solvents. However in mutagenesis, deleting part of a side chain can leave an unfilled cavity in the protein. The energetics of cavity formation can be estimated from scaled particle theory (17) and shown to explain this apparent doubling of the hydrophobic effect. Furthermore this theory also suggests that a better model for the hydrophobic effect is a linear relationship with molecular, rather than accessible, surface area.

2.3 Atomic packing

The net effect of attractive and repulsive van der Waals interactions between atoms is to favour close atomic packing. Thus to a first approximation the protein core resembles the solid state. Surface residues and most atoms of the chain in the unfolded state are less ordered and resemble in part the liquid state. Thus for residues that are in the protein core, folding leads to a liquid → solid transition. This transition is primarily enthalpic whose magnitude has been estimated from the freezing of model hydrocarbons as about 0.6 kcal mol^{-1} per CH_2 group favouring the solid state (18).

2.4 Conformational entropy

The formation of the folded structure restricts the dihedral conformational space sampled by the main chain and the buried side chains. This freezing of rotamers is entropically unfavourable.

Early estimates of the effect of backbone conformational entropy ranged from 2–4 kcal mol^{-1} at room temperature (19) and recent simulations are giving similar values (7). Side chain conformational entropy has been estimated as the numbers of rotamers that are frozen during folding and this leads to values of around 0.5 kcal mol^{-1} at room temperature. Recently several groups have re-addressed the problem of estimating this entropic effects (see ref. 20). One approach is that the maximum loss of conformational entropy when a side chain is buried and restricted to a single rotamer can be obtained from Boltzmann equation:

$$S = -R \Sigma p_i \ln p_i$$

where p_i is the probability of being in rotameric state i. Creamer and Rose (21) estimated p_i from Monte Carlo simulations. An empirical scale (*Table 1*) from analysis of the observed distribution of rotamers in X-ray structures was suggested by Pickett and Sternberg (22). Verification of the magnitude of this effect comes from the entropy of melting (fusion) of organic crystals as the solid → liquid transition frees up internal rotation (23). In general values for side chain conformational entropy range from 0–3 kcal mol^{-1} at room temperature.

2.5 Electrostatic effects—ion pairs and hydrogen bonds

The net effect of hydrophobicity, close packing, and conformational entropy would probably lead to a compact protein that lacks a specific architecture. The specificity for the tertiary structure could be considered as residing in the location of the hydrogen bonding and the ion pairing groups. A starting point to consider these electrostatic effects is Coulomb's law which gives the potential energy of two point charges q_i, q_j separated by r as:

$$V_{ij} = \frac{q_i\, q_j}{4\pi\, \varepsilon_o\, \varepsilon_r\, r}$$

where ε_o is the dielectric constant *in vacuo*, and ε_r represents the bulk shielding effect of the intervening medium with suitable values for water and the protein core being around 80 and 3. This suggests that in a protein it will be thermodynamically favourable for all charge groups to form the maximum number of possible electrostatic interactions. However electrostatic effects in the protein/solvent system are complex.

An individual fully charged (or just apolar atom) extending from the protein surface into water will be surrounded by a solvation shell of water molecules. Transferring this charged atom into the protein core is energetically very unfavourable due to removing the solvation shell. Thus, isolated charges are very rarely observed buried within proteins.

The formation of protein–protein electrostatic interactions must compete with the charges interacting with water and thus a charge interaction will be far less favourable energetically compared to the *in vacuo* effect. Despite the disadvantageous effect of partial desolvation, ion pairs on the surface tend to stabilize a protein and on average one-third of charged residues in a protein are involved in salt bridges. However there is an adverse effect of burying an ion pair in a low dielectric environment with only about 20% of such pairs being fully buried (24). Mutagenesis (15) suggests that one solvent exposed salt bridge generally contributes around 0.3–1 kcal mol^{-1} of stability.

There is a competition between protein–protein and protein–solvent hydrogen bonds. Although hydrogen bonds abound in proteins both forming secondary structures and involving side chain/main chain and side chain/side chain interactions (25), it remains unclear whether hydrogen bonds, particularly if buried, actually stabilize a protein. The formation of α-helices and β-sheets is probably the consequence that the periodic hydrogen bonding provides the best method of arranging complementary main chain amide and carboxyl groups within a hydrophobic core. Mutagenesis (15) suggests that the increase in protein stability from an exposed hydrogen bond may only be in the range 0.0–0.5 kcal mol^{-1}.

2.6 Disulfide bridges

The common view is that disulfide cross-links stabilize the folded state by entropically restricting the degrees of freedom of the unfolded state compared to the same chain without cross-links (19). For a single link, the stability increases with the length of the link but for multiple bridges there are complex effects (26). Typically a link will yield a few kcal mol^{-1} of stability. There are small proteins whose stability is considered to be enhanced by the entropic effect of multiple disulfide bridges.

3. Analyses of protein structure

Given the problems in quantifying the interplay of the different effects determining protein conformation, many workers primarily follow a knowledge-based (i.e. empirical) approach to structure prediction. As more protein structures were determined to high resolution, analyses were increasingly able to identify principles of protein architecture. These principles then form the basis for predictive algorithms. Below some features that are particularly relevant to prediction are highlighted. For further accounts of protein architecture see refs 27–30.

3.1 Residue conformation

(a) The main chain backbone torsion angles adopt allowed states conventionally represented as a Ramachandran (ϕ, ψ) plot. Gly adopts a larger and Pro a smaller region of allowed (ϕ, ψ) space (31–33).

(b) Proline is the only residue that adopts a *cis* peptide conformation with a relatively high probability (34).

(c) Side chains adopt distinct conformations that are dependent on backbone conformation. These conformations are conveniently represented as rotamer libraries (35–37).

3.2 Periodic secondary structure

(a) α-Helices can be curved or bent due to interactions with solvent, the presence of proline (38), or an α-aneurism (39).

(b) There are different preferences for residues to occur in the middle of an α-helix, and at the three N- terminal residues (N-cap), just before the N-cap, at the C-cap, and just after the C-cap (40–43).

(c) Nearly all β-sheets have a right-handed twist along the strand direction and consequently a left-handed twist between strands (44).

(d) A common distortion to the β-sheet is the β-bulge (45,46).

(e) Certain residues preferentially occur within β-strands (40).

(f) Right-handed 3.10 helices are relatively common (about 4% of residues). These helices can occur independently (typically less than six residues) or can form a terminal few residues of an α-helix (47).

(g) Left-handed polyproline II helices are relatively common (about 4% of residues) but their recognition was delayed due to the absence of periodic hydrogen bonds (48). These helices tend to be less than six residues long.

3.3 Non-periodic secondary structure

(a) A β-turn refers to four residues that achieve a 180° chain reversal. There are preferred sequence patterns for the different conformational families of β-turns (49,50).

(b) Three residues can achieve a chain reversal via a γ-turn (51).

(c) The β-hairpin linking a sequential pair of antiparallel β-strands can adopt distinct conformations (32,52).

(d) Sequential pairs of α-helices with their short connections can adopt distinct packing geometries and loop conformations (32,53).

(e) The α/β and β/α connections in βαβ units can adopt distinct conformations (32,54).

(f) The N-terminal cap of an α-helix can adopt a distinct structural motif (55).

(g) The N- and C-terminal caps of parallel β-sheets can adopt a distinct structural motif known as a β-breaker (56).

(h) The Ω-loop is formed when the N- and the C-termini of the loop are spatially close (57).

(i) Irregular loops can be classified as according to their linearity, planarity, and location of N- and C-termini (58).

3.4 Residue burial and interactions

(a) Protein cores are close packed (11,59).

(b) The buried protein core consists of non-polar residues and of polar residues satisfying their hydrogen bonding/ion pair requirements (11,59). Unsatisfied charged residues are very rarely fully buried (60).

(c) The exposed protein surface consists of about one-third non-polar residues and the remaining polar atoms interact with one another or with solvent (11,59).

(d) Salt bridges are formed by about one-third of charged residues but only 20% of these bridges are buried (24).

(e) There are geometric tendencies for side chain/side chain packing particularly for charged groups and aromatic rings (61).

3.5 Association of secondary structures

(a) α-Helices generally pack with distinct angular ranges due to the ridges and grooves formed from residues on one helix face intercalating with the ridges and grooves of the other helix face (62,63).

(b) A pair of α-helices can pack via a coiled coil in both fibrous and globular proteins (64).

(c) There are commonly observed packing geometries for β-sheet/β-sheet packing and for α-helix/β-sheet packing (65–69).

(d) The strands in β-sheets exhibit a variety of topological patterns including meanders, Greek keys, and jelly-rolls (29).

(e) The connection between parallel β-strands, particularly βαβ units, are almost invariably right-handed (70–72).

(f) New packing motifs for secondary structures are being identified as the number of known structures increases. For example the split βαβ, the double intertwined βαβ, and the βββα meander (73).

3.6 Folds of protein domains

(a) Some proteins form a single compact unit whilst others fold into spatially distinct regions termed domains. The upper size of a domain generally is about 400 residues (74,75).

(b) Domains, either as the complete tertiary structure or as a substructure, are classified according to the predominant sequential arrangement of secondary structure (76):

- α/α in which the structure is mainly α-helices
- β/β in which the structure is formed mainly from one or more β-sheets
- α/β in which the chain tends to alternate between α-helices and β-sheets forming βαβ units
- α + β in which there are both α-helices and β-sheets but the structures tend to segregate into different regions of the domain
- small proteins (< 100 residues) with little or no regular secondary structure often exhibiting several disulfide bridges and/or metal–cys interactions

(c) β-Sheets can cyclize round to form parallel or antiparallel barrels (69).

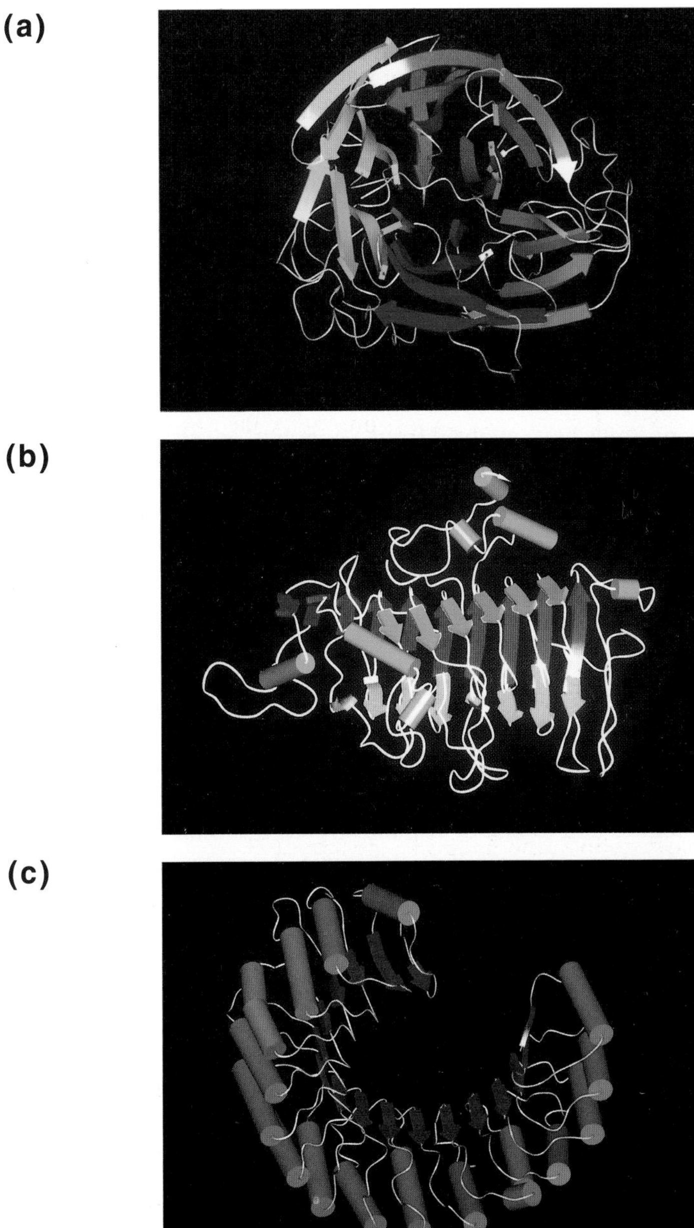

Figure 4. Examples of surprising new folds. (a) The structure of neuraminidase (125) is termed a β-propeller with six sheets each with four strands with the same topology (Brookhaven code: 2bat). (b) The β-helix observed in pectase lyase (126)—there are three sheets and the β-strands along the chain spirals between them (code: 2pec). (c) The 17-fold repeat of an β/α fold from ribonuclease inhibitor (code: 1bnh) (127). *Figures 4* and *5* are produced by the graphics program PREPI developed by Dr Suhail Islam (Imperial Cancer Research Fund) which is available upon request.

(d) Certain folds such as the fourfold α-helical bundle (77), the stacked immunoglobulin sheets (78), and the $(\beta\alpha)_8$-barrel (79) are commonly observed in several proteins (4,80). A classification of protein folds is presented in Chapter 8. Updated classifications are available via the World Wide Web in SCOP (81) and CATH (Thornton and co-workers, see Chapter 8).

(e) New folds are observed that are radically different from arrangements seen before although they still obey the principles of protein architecture (82,83). This has been particularly true for proteins with β-sheets (see *Figure 4* for illustrations of some new folds).

3.7 Evolution of proteins

(a) Homologous proteins have evolved from a common ancestor, often share a related function (e.g. serine proteinase), and adopt similar three-dimensional structures. Sequence identity can be below 20%. The protein core, mainly formed from packed α- and/or β-structures, tends to be more conserved both in sequence and structure than the connecting loops (see *Figure 5a*) (30).

(b) Analogous proteins adopt similar three-dimensional folds but have sequence identity below 20% (typically 10%). There tends to be greater variations in the loops and the packing geometries of secondary structures (see *Figure 5b*). The commonality of fold may be the result of convergence to a favourable architecture (84).

(c) As the sequence identity decreases from closely homologous proteins to distant homologues and to analogues, the structural similarity of the core and the conservation of secondary structures decrease, and there are fewer conserved interresidue side chain contacts (see *Figure 6*).

(d) Recognition of homologous and analogous folds provides a powerful approach for structure prediction (see below).

1B—Approaches to prediction

The approach for prediction is summarized in *Figure 7* and is reflected in the organization of the book. However, the impression gained from such a flow diagram is that there is there is a suite of algorithms that one can use with the output of one program being the input for another. The key point is that today the most successful practical approach for structure prediction involves a combination of the use of the algorithms with human intervention. The objective of obtaining the best prediction, say to direct an experimental study, is distinct from the challenge of developing the most accurate algorithms that yield reproducible results and can be widely used by non-experts. Clearly there is a need to incorporate more of the human expert knowledge and reasoning into the algorithms. But today this has not been achieved to the level that one should shy away from using human intervention.

(a)

(b)

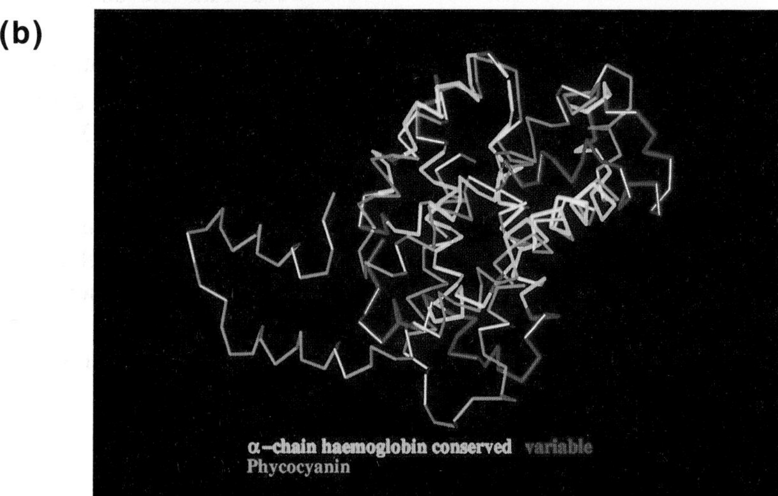

Figure 5. Structural conservation in homologous and analogous structures. (a) C^{α} atoms of erythrocruorin (128) in blue (Brookhaven code: 1eca) superimposed on the α-chain of haemoglobin (129) with conserved residues in yellow and variable in mauve (2hhb). (b) C^{α} atoms of phycocyanin (130) in blue (1cpc) superimposed on the α-chain of haemoglobin.

4. Sequence analysis

4.1 Homology searching

Generally the first step in any prediction is to identify if the target sequence of interest is homologous to other sequences in the database (see Chapter 2). A major difficulty is that many hits give scores in the 'twilight zone'—a pair-

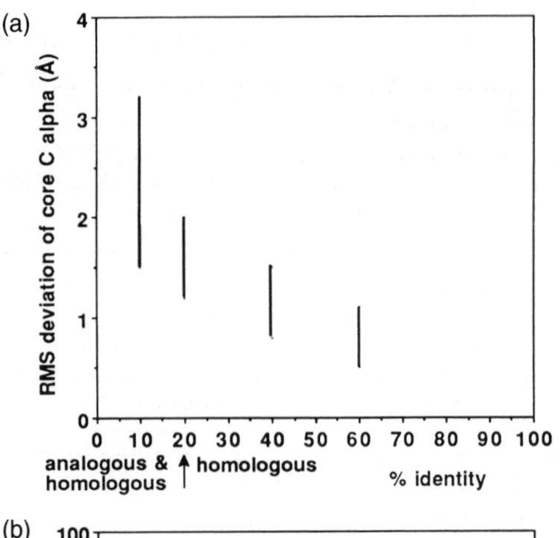

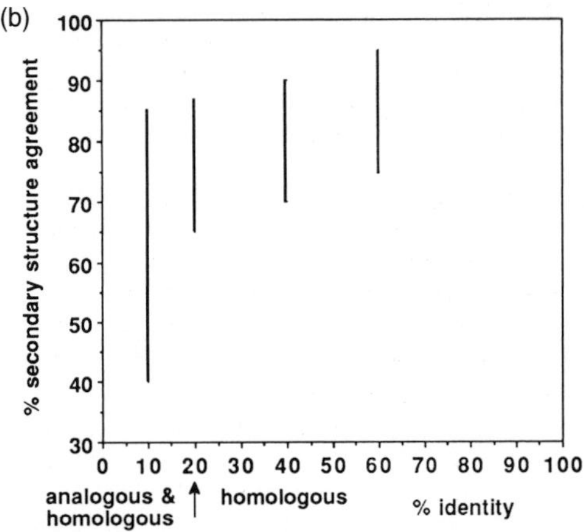

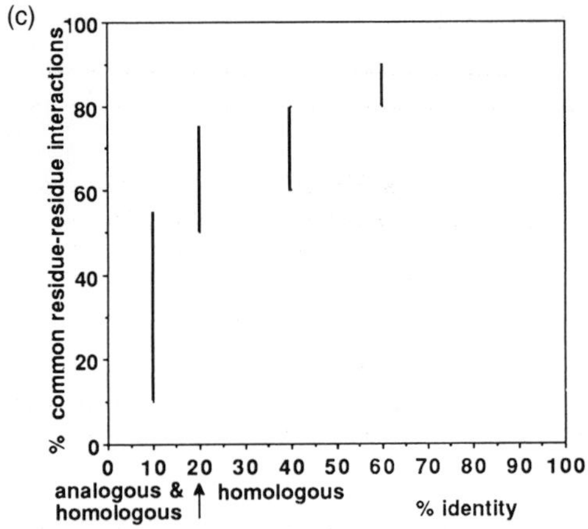

Figure 6. Variation of structural conservation in homologous and analogous folds with sequence identity. Plotted against per cent residue identity of the two chains are: (a) the root mean square (rms) deviation of C^{α} atoms equivalenced in the core; (b) the per cent of residues with the same class of secondary structure (α, β, or coil); (c) the per cent of residues with equivalent side chain/side chain contacts. Adapted from ref. 84.

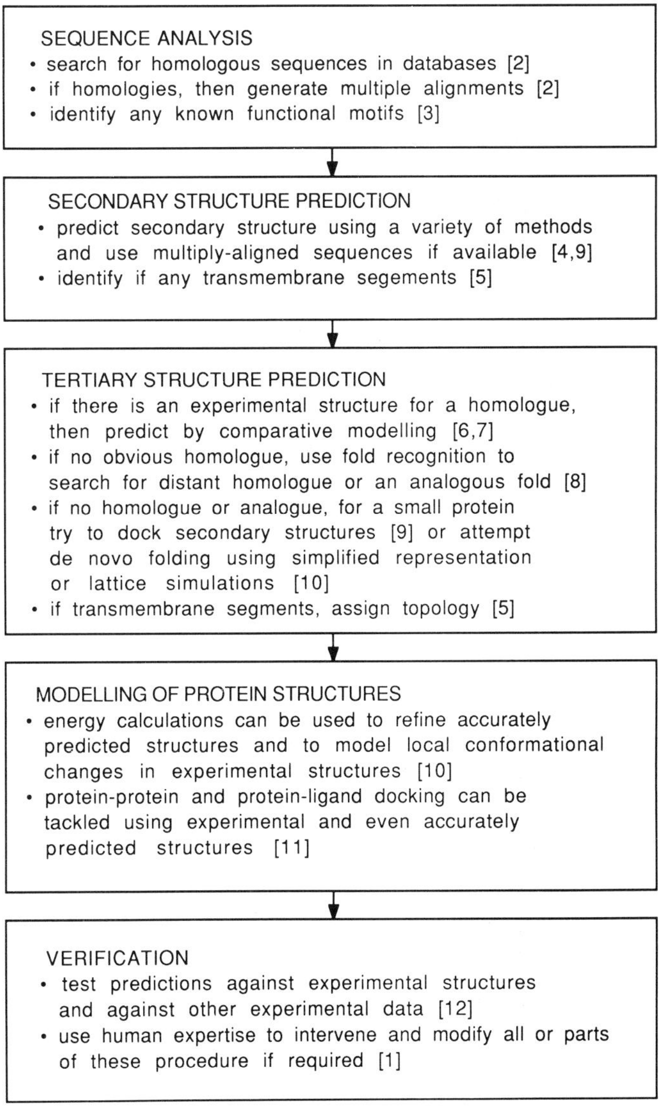

SEQUENCE ANALYSIS
- search for homologous sequences in databases [2]
- if homologies, then generate multiple alignments [2]
- identify any known functional motifs [3]

↓

SECONDARY STRUCTURE PREDICTION
- predict secondary structure using a variety of methods and use multiply-aligned sequences if available [4,9]
- identify if any transmembrane segements [5]

↓

TERTIARY STRUCTURE PREDICTION
- if there is an experimental structure for a homologue, then predict by comparative modelling [6,7]
- if no obvious homologue, use fold recognition to search for distant homologue or an analogous fold [8]
- if no homologue or analogue, for a small protein try to dock secondary structures [9] or attempt de novo folding using simplified representation or lattice simulations [10]
- if transmembrane segments, assign topology [5]

↓

MODELLING OF PROTEIN STRUCTURES
- energy calculations can be used to refine accurately predicted structures and to model local conformational changes in experimental structures [10]
- protein-protein and protein-ligand docking can be tackled using experimental and even accurately predicted structures [11]

↓

VERIFICATION
- test predictions against experimental structures and against other experimental data [12]
- use human expertise to intervene and modify all or parts of these procedure if required [1]

Figure 7. The approach for protein structure prediction. Numbers in square brackets refer to chapters in the book. Arrows denote the main direction of processing but there is feedback between later steps and earlier stages to refine the predictions.

wise sequence match that by examination of the sequence alone one cannot (at present) distinguish between a chance match and a common ancestor.

The occurrence of this twilight zone stems, in part, from the degeneracy of structural information in a local sequence. Identical pentapeptides and hexapeptides can adopt quite different secondary structures (e.g. the same sequence can be found in an α-helix in one known structure and in a β-sheet in another) (85). In addition there can be runs of some 60 residues with 25% identity and a further 25% conservative substitutions that adopt different conformations in unrelated proteins (see ref. 86 and Chapter 2). Sander and Schnieder (87) provided a simple approach to identify the percentage identity in a local sequence match of a given length that can be expected to result only from a true homology with its implication of a similar three-dimensional structure (see also Chapter 2).

4.2 Multiple sequence alignment

If homologies are found in the sequence database, then the next step is the generation of a multiple sequence alignment with homologues of the target sequence. This is helpful in identification of conserved motifs that are functionally and/or structurally important. Multiple alignments provide increased accuracy for secondary structure prediction and subsequent tertiary modelling. Various widely-used approaches are described in Chapter 2, Section 5.2. A recent development is the use of hidden Markov models to obtain multiple alignments and to identify sequence motifs (88).

Despite their widespread use, there have been few tests of the accuracies of the different methods. Accordingly McClure *et al.* (89) applied 12 algorithms to four families and evaluated the approaches in terms of their ability correctly to identify an order set of sequence motifs. The conclusion was that for sequences of > 50% identity most methods are successful. However < 30%, the presence of several insertions/deletions (i.e. indels) can lead to incorrect multiple alignments. In general global methods that consider the entire sequence performed better than algorithms that used local consensus words.

4.3 Sequence motifs

Even if an homologous sequence cannot be found, clues to the probable function of a target protein can be obtained by the identification of previously characterized local sequence motif(s) (see Chapter 3). This has proved a powerful tool with the motifs having a remarkable degree of discrimination power—few false positives and few misses. The recent use of hidden Markov models is leading to further improvements in the discrimination power of motif search (88). However there is a danger in extrapolating from a common function to a common structure. The same sequence motif might adopt the same local structure (e.g. the loop between a β-strand and an α-helix) but this local structure can be found in different domain folds, e.g. the ATP binding loop (90).

4.4 Sequence analysis of the yeast III chromosome

The power of sequence analysis to provide information about the possible structure and/or function of a new sequence was highlighted by the study of

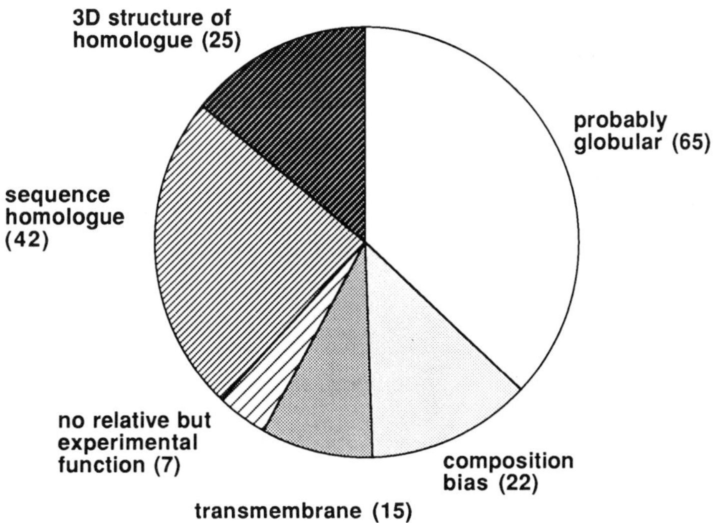

Figure 8. Classification of the 176 ORFs of yeast chromosome III. Redrawn from ref. 91.

Bork *et al.* (91) who analysed the 176 confirmed open reading frames (ORFs) of yeast chromosome III. Using a variety of sequence search software, they identified homologies in the database for 67 (i.e. 38%) of the ORFs (see *Figure 8*). In their study 44 ORFs were classified as clearly significant with a BLAST probability of $< 1 \times 10^{-10}$. The remaining 23 matches were initially observed in the 'twilight zone' with a lower BLAST probability than this cut-off but further investigation assigned these matches as probably biologically meaningful. To try to distinguish true matches from the noise in this zone, a variety of other approaches and scoring schemes, such as from FASTA, were used. If the new sequence matched against a family of other sequences, a multiple sequence alignment could be generated and the presence of conserved residues spanning the family provides additional support for a presumed homology. Additionally searching for classified functional local sequence motifs, such as PROSITE database (see Chapter 3) helped to verify presumed homologies. Nine of the 67 sequence homologies found by Bork *et al.* also had an exact match with a PROSITE motif.

This study highlights how the integration of the results sequence analysis requires human intervention and judgement. Many matches fall in the twilight zone. In addition, the significance of common patterns of residues in multiple alignments with gaps inserted cannot yet be reliably estimated. Given these problems of obtaining statistics that reliably model the evolutionary process, it remains crucial to assess the biological sense of presumed matches.

5. Secondary structure prediction

5.1 Globular proteins

In the mid 1970s as the structures of several proteins were determined crystallographically, different groups developed algorithms to predict the location of α-helices, β-sheet strands, and sometimes β-turns from sequence. The algorithms could be divided into empirically-based, such as those of Chou and Fasman (40), and of Garnier–Osguthorpe–Robson (GOR) (92), and those that identified patterns of hydrophobic residues, particularly the scheme of Lim (93). Some of these algorithms, with minor updates, are frequently provided as sequence analysis tools in molecular biology software packages. These schemes achieved about 58% per residue accuracy (94) for a three-state prediction of α-, β-, or coil (i.e. Q_3 reported in Chapter 4). Since then improvements in this accuracy was limited despite developments that included the use of neural networks, use of structural class-specific prediction, and more extensive pattern recognition (see Chapters 4 and 9).

The major change over the last few years was the recognition of the value of the additional information provided by aligned homologous sequences. Early work (95) suggested an improvement of 4% from the additional statistical information using several sequences and a further 5% identifying that conserved region without gaps are likely to be the α- and β-structures. Today, approaches for prediction from an alignment are of two types. The first are computer-based algorithms such as extension of GOR (96), the neural network method PHD of Rost and Sander (97), and the nearest neighbour approach of Salamov and Solovyev (98). Q_3 accuracies for three-state of just over 70% can be obtained. These algorithms are available via World Wide Web sites.

The second approach is the use expert examination first to obtain the multiple alignment and then to parse the information into a prediction. The key concept for the prediction is the conservation of hydrophobic residues on the faces of the secondary structures that form the core of the protein. The concepts identified by Lim (93) for a single sequence are extended to conserved hydrophobic residues: i, $i + 3$, and $i + 4$ for an α-helix; i, $i + 2$, $i + 4$ for a strand with one exposed and one buried face typical of β/β proteins; and i, $i + 1$, $i + 2$.. for a buried β-strand typical of the core of α/β proteins. Several such predictions have been performed by Benner and co-workers (99,100).

In the evaluation of the different methods, certain aspects have recently been highlighted:

(a) There is a wide variation in accuracy by any approach on different proteins, e.g. Rost and Sander (97) obtain an accuracy of 71.0% with a standard deviation of 9.3%. Thus the success of one method over another on one protein may simply reflect statistical variation.

(b) The evaluation of the number of residues correctly predicted (Q_3) fails to quantify the relative seriousness of different errors. It is far less serious for each N-terminal residue of five α-helices to be mispredicted as a coil than for a five residue strand to be identified as a coil, yet there is the same effect on Q_3. Although many approaches are suggested to consider this type of problem (101), there is no agreement on which approach to use and so the readily understandable Q_3 still dominates in evaluating predictions.

(c) The secondary structures of homologous proteins show considerable variation, particularly at the termini of the elements. Thus when the prediction uses homologous sequences, there is a limit to the maximum expected accuracy. Generally for a diverse homologous family (around 20% identity), there is between 65% and 85% agreement between the three secondary states (see *Figure 6b*). Thus 85% accuracy is likely to be the practical upper limit although in theory 100% could be obtained (99–101).

(d) Blind trials with a predecided measures of success made prior to knowledge of the structure remain the best method to evaluate the different methods.

5.2 Blind trials

In this book two comparisons of secondary structure predictions are presented. Chapter 4 reports the use of several fully or largely automatic algorithms on the adaC O6 Methyl G.DNA Methyltransferase from *E. coli*. Here, *Figure 9* shows two blind manual predictions (102,103) on the pleckstrin homology domain (PH) compared to subsequent fully automatic predictions (104). Both the manual prediction of Jenny and Benner (102) and of Musacchio *et al.* (103) accurately predict the location and type of the core secondary structures and respectively achieve a Q_3 of 78–94% and 75–90% depending against which determined protein the prediction is compared. Importantly despite being manual methods, the two groups obtained very similar predictions suggesting (at least for this protein) a consensus amongst the experts in interpreting sequences. In contrast two automatic methods, the PHD and GOR predictions are poorer and failed correctly to identify some secondary structures. The benefit of the manual approaches is mainly from the careful construction of the sequence alignment.

The conclusion from several trials is that recent manual predictions are achieving a similar accuracy to those of the best automatic approaches and one must wait for the results of further blind trials before drawing a conclusion about which approach to follow. Today I would modify the results of the algorithm to encapsulate the complex, and as yet not fully quantified, principles of protein architecture.

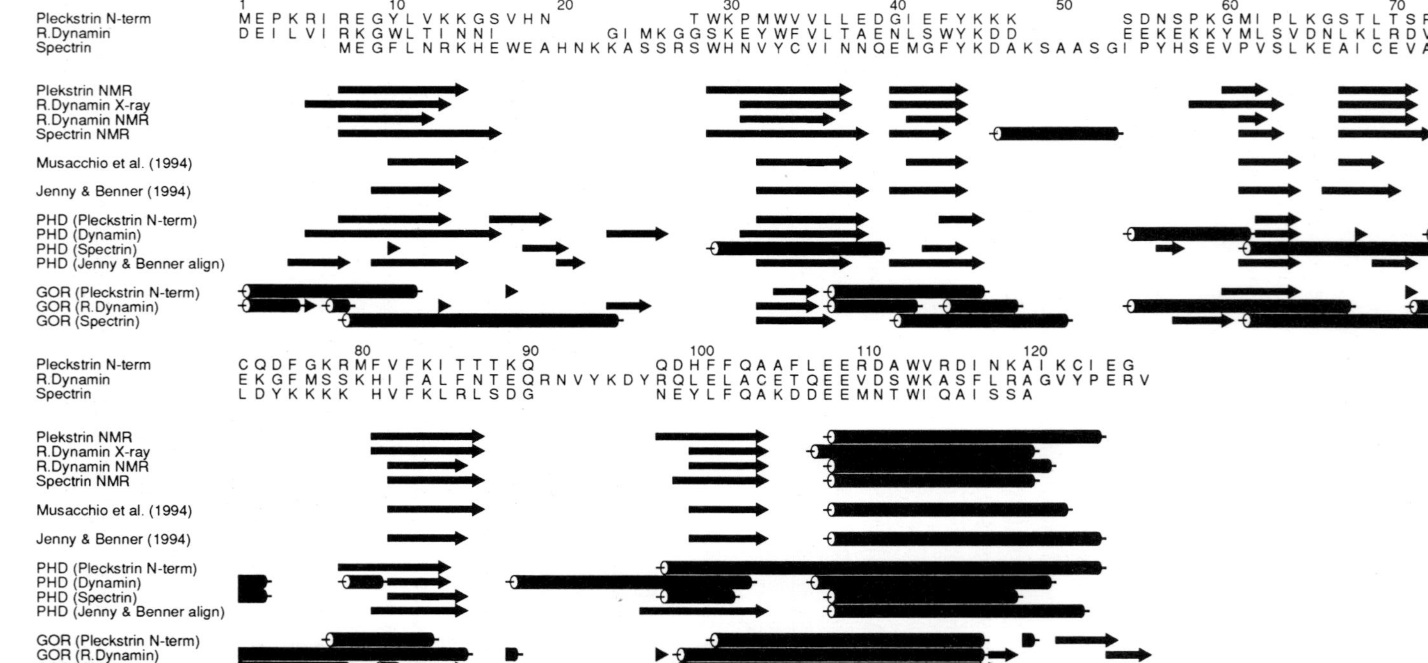

Figure 9. Comparison of the known with the predicted structures of the PH domain (the pleckstrin sequence is from the N-terminal PH domain). The figure shows: aligned sequences of three PH domains, their experimentally-determined secondary structures, and the predictions. Arrows denote β-strands, cylinders denote α-helices. Reproduced from ref. 104.

5.3 Transmembrane regions

Out of the 176 ORFs in yeast chromosome III, 15 were ultimately assigned as transmembrane highlighting the importance of predictions of this protein class. However, in contrast to globular proteins with a substantial database of solved structures, there is limited three-dimensional information for transmembrane proteins. Some structures, e.g. bacteriorhodopsin and the photosynthetic reaction centre exhibit one type of structure—a bundle of α-helices typically of 20 residues that span the membrane. Another three-dimensional architecture is exhibited by porin—a 16 stranded β-sheet barrel (see Chapter 5). Both arrangements ensure that there are no unsatisfied potential hydrogen bonding groups buried in the hydrophobic membrane.

Early work on transmembrane prediction was based on identification of hydrophobic runs or faces of residues. Recently there have been improvements in transmembrane prediction. One approach recognizes that positive charges dominate on the cytoplasmic face generating an asymmetry (see Chapter 5 and ref. 105). The other uses information from multiply aligned homologous sequences (106,107). These predictions are achieving a per residue accuracy of around 95% with around 95% of all segments being predicted. These predictions could be the start of modelling three-dimensional structures (see Section 6.4).

6. Tertiary structure prediction

6.1 Comparative modelling

Today the most powerful approach for tertiary structure prediction remains comparative modelling from the experimental structure of an homologue, i.e. homology modelling. The applicability of this approach is illustrated by the results of the Bork *et al.* search—25 of the 176 ORFs had at least one domain with an homology to a known structure and thus could have their tertiary structure modelled. Currently over 50 publications a year report a new comparative modelling study (P. Harrison, personal communication).

Many algorithms for comparative modelling use a fragment-based assembly of components from known structures (see Chapter 6, Section 2.1). This approach is increasingly being automated into packages such as COMPOSER (108) and the e-mail homology building server at Geneva SWISS-MODEL (109). The key steps are:

(1) Establish one or a set of suitable experimental parent structures that are homologous to the sequence of the unknown structure.
(2) Establish a sequence alignment between the unknown and the parent(s).
(3) Identify main chain segments that are expected to be structurally conserved between the parent(s) and the unknown.

(4) Model the structurally variable regions—these will tend to be the loops connecting the conserved secondary structures.
(5) Model the side chains of the unknown on to the modelled main chain.
(6) Refine the structure using energy calculations.

An alternative approach is to use three-dimensional distance constraints as in MODELER (110). Steps 1, 2, and 6 remain the same as in fragment modelling but restraints from homologous parents and from experimental data are used for steps 3–5.

The results of comparative modelling presented at the Asilomar prediction test (see Chapter 12) were rather disappointing and some general conclusions are:

(a) Automatic approaches that were followed often failed at the alignment step and consequently the entire subsequent modelling was incorrect.
(b) Modelling the structurally variable loop regions still presented problems.
(c) The energy minimization moved the predicted structure further from the correct conformation than the pre-minimized coordinates.

Suggestions for comparative modelling are:

(a) Try to generate several alternative alignments and subsequent models and select between them at the end of the procedure and in the light of experimental evidence.
(b) Use human intervention to provide feedback between the various stages of the algorithms. For example the sequence alignment could be modified as a result of attempting to model loops.
(c) Augment the general algorithms by knowledge of the principles of protein architecture that apply to the particular parent/unknown structures being modelled (see Section 3).
(d) Consider the expected errors in the model given the sequence conservation (see *Figure 6*). These errors will vary for the same model between the region that tends to be well predicted due to evolutionary conservation, particularly the active site, and the most variable regions such as loops.

6.2 Modelling antibody combining sites

A specialist area of homology modelling is the prediction of antibody combining sites. The structural classification by Chothia and Lesk (111) into canonical forms for most classes of complementary determining regions (CDRs) has provided an approach of considerable reliability to translate loop sequences into C^{α} coordinates (see Chapter 7). There still remains difficulties in modelling CDR heavy 3, the angle between a loop and the β-sheet, and the packing variable light and variable heavy chains. Nevertheless the models tend to be of sufficient accuracy to direct the identification of residues to alter in antibody engineering (e.g. humanization).

6.3 Fold recognition

A major development in structure prediction is fold recognition also known as threading or the inverse folding problem (see Chapter 8). This approach stems from the recognition that proteins with a level of sequence similarity that could not be detected by conventional homology searches sometimes adopt an analogous three-dimensional fold such as the $(\beta\alpha)_8$-barrel.

The concept behind threading is that by aligning of a residue in the unknown structure (the probe residue) on to a residue in a known fold (the target residue), the tertiary location of the target can specify not only the amino acid type (as in conventional sequence searches) but also the secondary structure, the exposure, and the spatial interaction with other residues. These spatial interactions are often modelled by empirically-derived values quantifying the favourability of a residue of type i being a given distance from a residue of type j (112). The results of threading is a list of possible hits with folds with a measure of significance similar to the lists from conventional sequence searches.

The results from Asilomar were that the threading algorithm were often able correctly to identify an analogous fold. The key problem was that in general the residue alignments between the probes and the targets were very poor. Entire secondary structures could be wrongly equivalenced even if the correct fold was identified. Clearly a three-dimensional model built from such an incorrect alignment would be of little use even to map epitopes or an active site from sequentially distant chain segments.

A recent blind test of applying fold recognition by users rather than developers is the work of Edwards and Perkins (113) on the von Willebrand factor type A domain. A combination of secondary structure prediction followed by threading suggested a close structural similarity with the GTP binding domain of ras-p21. Human intervention was then used to establish the alignment of the unknown structure with ras-p21. The subsequent X-ray structure (114) confirmed this prediction.

The extent of structural conservation between truly analogous fold is limited (see *Figures 5* and *6*). For some pairs of proteins only 40% of the residues will adopt the same (three-state) secondary structure and only 10% of interresidue contacts will be conserved. There can be substantial shifts in packing geometry even for the core secondary structures and there can be large insertions and deletions. The challenge in threading is to identify conserved features amongst these variations. Over the next few years, the development of threading algorithms is likely to be an area of progress extending comparative modelling from homology into and beyond the twilight zone.

6.4 *De novo* folding

In the absence of a template structure for prediction, a variety of approaches are being developed. Chapter 9 describes the combinatorial approach in which first secondary structures are predicted, then a series of alternative docking generated, and finally the tertiary models filtered. This strategy is best suited to modelling small, mainly α-helical proteins and can yield structures to guide experiments. Transmembrane helical proteins are also amenable to a similar combinatorial generation and selection (115). The inside/outside topology can often be assigned from consideration of charges (see ref. 105 and Chapter 5). However, given that only a few architectures are known for membrane protein, there remains a danger of forcing models to fit into one of these few templates.

Other approaches (see Chapter 10, Section 3.4) employ a simplified model for the protein such as a single sphere for the side chain together with appropriate terms for residue–residue interactions similar to the model introduced by Levitt and Warshel in the mid 1970s (116). With the subsequent leap in computing power, this approach is being pursued again. For example, Scheraga and co-workers (117) have obtained a good prediction (C^{α} rms deviation of 3.8 Å, for the 36 residues of avian pancreatic polypeptide). They started with a simplified model, progressively converted to an all-atom representation, and then used conventional minimization. Srinivasan and Rose (118) used a Monte Carlo simulation to sample populated region of main chain ϕ,ψ space and estimated the stability of the fold with simplified side chains. Segments of 50 residues that resemble the native were obtained. Other groups use lattice simulations to obtain starting compact structures for subsequent refinement . For example, Kolinski and Skolnick (119) simulated small all-α and α + β proteins and obtained a limited set of compact folds some of which showed a C^{α} rms agreement in the range 2.25–3.65 Å to the native.

These approaches promise a general solution to structure prediction, rather than one based on a known fold. Their development is reaching an exciting phase due to the anticipation of computing power that will soon be available. This should promote their use from mainly in the laboratories of the developers to the wider community.

7. Simulations with protein structures

Once the structure of a protein has been reliably determined by NMR or crystallography, one wishes to simulate conformational changes or to dock the protein to a ligand. These topics are now considered.

7.1 Molecular mechanics and dynamics

Molecular mechanics and dynamics provide powerful tools to complement the knowledge-based approach that is the main theme of this book. The theory and several applications to model perturbation from a determined structure, e.g. site-directed mutations, are reviewed in Chapter 10. An important area is the use of the potential functions to refine a predicted structure. The Asilomar tests showed that simply applying these techniques to a model-built structure is likely to lead to a deterioration in the quality of the model. An important question is whether the errors in a predicted structure are likely to be greater than the expected radius of convergence of molecular mechanics and dynamics simulations. Many of the standard packages have facilities to restrain parts of the structure and allow greater freedom for other regions. Clearly, more systematic studies of the application of molecular mechanics and dynamics to refine model-built structures are required.

7.2 Docking ligands to proteins

Modelling the conformation of the interaction of a protein with either small molecule ligands or with other proteins is central to understanding and regulating its function. The problem is complex—even a rigid body search has six degrees of freedom. Additionally, conformational changes occur on docking that eventually must be included. The developments of computing power have lead to several algorithms to tackle the docking problem including starting with molecules in their unbound conformations (see Chapter 11).

For protein–protein interactions, the current status is that many algorithms can generate a limited set of docked complexes one of which is close to the native arrangement (see Chapter 11, *Table 2* and ref. 120). The approach generally is first to search for complementarity of surface, then to model the burial of hydrophobic surface and electrostatic complementarity. The problem is to distinguish the correct solution from several false positives. Recently we (121) have obtained encouraging results for screening by evaluating:

- the hydrophobic contribution via its proportionality to molecular surface area (see Section 2.2)
- both electrostatic desolvation and interactions using the finite difference algorithm of DELPHI (122)

Recently there was a blind trial of seven docking algorithms. The dockers were given the coordinates of the unbound β-lactamase and its inhibitor challenged to predict the complex (120). All groups successfully identified a solution that was close (between 1.1 Å and 2.5 Å main chain rms) to the crystallographic complex. However, most groups submitted a few entries some of which were far from the correct solution (typically up to 15 Å).

Of particular importance to this book is the question whether predicted protein structures can be used in docking simulations. Predicted antibody structures are of sufficient accuracy to justify docking them against antigens (123). Cohen and co-workers (124) have applied homology building to predict structures for two major protease families. Computationally they then identified possible non-peptide inhibitors, some of which exhibited micromolar inhibition. Their success relies on the greater conservation of the active site. Today if a protein structure is expected to be accurately predicted, it could be used (with caution) in subsequent docking studies.

8. Conclusion

It is becoming easier for the non-experts to implement the approaches described in this book. First, commercial companies are producing documented and robust suites for sequence analysis and modelling. Secondly, there is the explosion in the use of e-mail servers and in particular the World Wide Web. Sequences can be sent to a host of sites for structure prediction. The problem for the non-specialist may soon be which algorithm to use rather than how to obtain the program. This decision should be guided by whether the algorithm has been consistently applied to several systems not included in its development and whether the algorithm has been subjected to blind, comparative trials.

Faster computers, the increased sizes of the sequence and structural databases, our deeper understanding of the principles of protein structures, and more accurate models for atomic interactions have contributed to the development of prediction algorithms. These trends will continue leading to the wider applicability and to improved accuracies for prediction. However, the prediction problem remains difficult and challenging. The key question is: how many sequences will we be able to translate into a three-dimensional model?

Acknowledgements

I thank Suhail Islam, Paul Bates, Ross King, Richard Jackson, Robert Russell, Paul Harrison, Matthew Betts, and Hedi Hegyi (Biomolecular Modelling Laboratory) for stimulating discussions and for comments on this chapter. I also thank Dr Christine Orengo (University College, London) for supplying the data in *Figure 1b*.

References

1. Anfinsen, C. B. (1973). *Science*, **181**, 223.
2. Seckler, R. and Jaenicke, R. (1992). *FASEB J.*, **6**, 2545.
3. Bairoch, A. and Boeckmann, B. (1991). *Nucleic Acids Res.*, **19**, 2247.
4. Orengo, C. A., Jones, D. T., and Thornton, J. M. (1994). *Nature*, **372**, 631.
5. Bernstein, F. C., Koetzle, T. F., Williams, G., Meyer, D. J., Brice, M. D., Rodgers, J. R., *et al.* (1977). *J. Mol. Biol.*, **112**, 535.

6. Dill, K. A. (1990). *Biochemistry*, **29**, 7133.
7. Honig, B. and Yang, A.-S. (1995). *Adv. Protein Chem.*, **46**, 27.
8. Kauzmann, W. (1959). *Adv. Protein Chem.*, **14**, 1.
9. Fauchère, J.-L. and Pliska, V. (1983). *Eur. J. Med. Chem. Chim. Ther.*, **18**, 369.
10. Cornette, J. L., Cease, K. B., Margalit, H., Spouge, J. L., Berzofsky, J. A., and DeLisi, C. (1987). *J. Mol. Biol.*, **195**, 659.
11. Richards, F. M. (1977). *Annu. Rev. Biophys. Bioeng.*, **6**, 151.
12. Chothia, C. (1974). *Nature (London)*, **248**, 338.
13. Miller, S., Janin, J., Lesk, A. M., and Chothia, C. (1987). *J. Mol. Biol.*, **196**, 641.
14. Eisenberg, D. and McLachlan, A. D. (1986). *Nature*, **319**, 199.
15. Serrano, L., Kellis, J. T., Cann, P., Matouschek, A., and Fersht, A. R. (1992). *J. Mol. Biol.*, **224**, 783.
16. Sharp, K. A., Nicholls, A., Friedman, R., and Honig, B. (1991). *Biochemistry*, **30**, 9686.
17. Jackson, R. M. and Sternberg, M. J. E. (1994). *Protein Eng.*, **7**, 371.
18. Nicholls, A., Sharp, K. A., and Honig, B. (1991). *Proteins*, **11**, 281.
19. Schellman, J. A. (1955). *Trav. Lab. Carsberg, Ser. Chim.*, **29**, 230.
20. Doig, A. J. and Sternberg, M. J. E. (1995). *Protein Sci.*, **4**, 2247.
21. Creamer, T. P. and Rose, G. D. (1992). *Proc. Natl. Acad. Sci. USA*, **89**, 5937.
22. Pickett, S. D. and Sternberg, M. J. E. (1993). *J. Mol. Biol.*, **231**, 825.
23. Sternberg, M. J. E. and Chickos, J. S. (1994). *Protein Eng.*, **7**, 149.
24. Barlow, D. J. and Thornton, J. M. (1983). *J. Mol. Biol.*, **168**, 867.
25. Baker, E. N. and Hubbard, R. E. (1984). *Prog. Biophys. Mol. Biol.*, **44**, 97.
26. Harrison, P. and Sternberg, M. J. E. (1994). *J. Mol. Biol.*, **244**, 448.
27. Lesk, A. (1991). *Protein architecture: a practical approach.* IRL Press at Oxford University Press, Oxford.
28. Branden, C. and Tooze, J. (1991). *Introduction to protein structure* (1st edn) Garland Publishing Inc, New York.
29. Richardson, J. S. (1981). *Adv. Protein Chem.*, **34**, 167.
30. Chothia, C. and Finkelstein, A. V. (1990). *Annu. Rev. Biochem.*, **59**, 1007.
31. Ramachandran, G. N. and Sasisekharen, V. (1968). *Adv. Protein Chem.*, **23**, 283.
32. Efimov, A. V. (1993). *Curr. Opin. Struct. Biol.*, **3**, 379.
33. Rooman, M. J., Kocher, J.-P. A., and Wodak, S. J. (1991). *J. Mol. Biol.*, **221**, 961.
34. MacArthur, M. W. and Thornton, J. M. (1991). *J. Mol. Biol.*, **218**, 397.
35. McGregor, M. J., Islam, S. A., and Sternberg, M. J. E. (1987). *J. Mol. Biol.*, **198**, 295.
36. Dunbrack, R. L. and Karplus, M. (1993). *J. Mol. Biol.*, **230**, 543.
37. Eisenhaber, F. and Argos, P. (1993). *J. Mol. Biol.*, **230**, 592.
38. Barlow, D. J. and Thornton, J. M. (1988). *J. Mol. Biol.*, **201**, 601.
39. Keefe, L. J., Sondek, J., Shortle, D., and Lattman, E. E. (1993). *Proc. Natl. Acad. Sci. USA*, **90**, 3275.
40. Chou, P. Y. and Fasman, G. D. (1974). *Biochemistry*, **13**, 222.
41. Richardson, J. S. and Richardson, D. C. (1988). *Science*, **240**, 1648.
42. Presta, L. G. and Rose, G. D. (1988). *Science*, **240**, 1632.
43. Harper, E. T. and Rose, G. D. (1993). *Biochemistry*, **32**, 7605.
44. Chothia, C. (1973). *J. Mol. Biol.*, **75**, 295.
45. Richardson, J. S., Getzoff, E. D., and Richardson, D. C. (1978). *Proc. Natl. Acad. Sci. USA*, **75**, 2574.

46. Chan, A. W. E., Hutchinson, E. G., Harris, D., and Thornton, J. M. (1993). *Protein Sci.*, **2**, 1574.
47. Karpen, M. E., de Haseth, P. L., and Neet, K. E. (1992). *Protein Sci.*, **1**, 1333.
48. Adzhubei, A. A. and Sternberg, M. J. E. (1993). *J. Mol. Biol.*, **229**, 472.
49. Venkatachalam, C. M. (1968). *Biopolymers*, **6**, 1425.
50. Wilmot, C. M. and Thornton, J. M. (1990). *Protein Eng.*, **3**, 479.
51. Milner-White, E. J. (1990). *J. Mol. Biol.*, **216**, 385.
52. Sibanda, B. L., Blundell, T. L., and Thornton, J. M. (1989). *J. Mol. Biol.*, **206**, 759.
53. Thornton, J. M., Sibanda, B. L., Edwards, M. S., and Barlow, D. J. (1988). *Bioessays*, **8**, 63.
54. Edwards, M. S., Sternberg, M. J. E., and Thornton, J. M. (1987). *Protein Eng.*, **1**, 173.
55. Seale, J. W., Srinivasan, R., and Rose, G. D. (1994). *Protein Sci.*, **3**, 1741.
56. Colloc'h, N. and Cohen, F. E. (1991). *J. Mol. Biol.*, **221**, 603.
57. Leszczynski, J. F. and Rose, G. D. (1986). *Science*, **234**, 849.
58. Ring, C. S., Kneller, D. G., Langridge, R., and Cohen, F. E. (1992). *J. Mol. Biol.*, **224**, 685.
59. Chothia, C. (1976). *J. Mol. Biol.*, **105**, 1.
60. McDonald, I. K. and Thornton, J. M. (1994). *J. Mol. Biol.*, **238**, 777.
61. Singh, J. and Thornton, J. M. (1990). *J. Mol. Biol.*, **211**, 595.
62. Chothia, C., Levitt, M., and Richardson, D. (1981). *J. Mol. Biol.*, **145**, 215.
63. Richmond, T. and Richards, F. M. (1978). *J. Mol. Biol.*, **119**, 537.
64. Zhou, N. E., Kay, C. M., and Hodges, R. S. (1994). *Protein Eng.*, **7**, 1365.
65. Cohen, F. E., Sternberg, M. J. E., and Taylor, W. R. (1980). *Nature (London)*, **285**, 378.
66. Chothia, C. and Janin, J. (1982). *Biochemistry*, **21**, 3955.
67. Janin, J. and Chothia, C. (1980). *J. Mol. Biol.*, **143**, 95.
68. Cohen, F. E., Sternberg, M. J. E., and Taylor, W. R. (1982). *J. Mol. Biol.*, **156**, 821.
69. Murzin, A. G., Lesk, A. M., and Chothia, C. (1994). *J. Mol. Biol.*, **236**, 1382.
70. Richardson, J. S. (1976). *Proc. Natl. Acad. Sci. USA*, **73**, 2619.
71. Sternberg, M. J. E. and Thornton, J. M. (1977). *J. Mol. Biol.*, **110**, 269.
72. Sternberg, M. J. E. and Thornton, J. M. (1976). *J. Mol. Biol.*, **105**, 367.
73. Orengo, C. A. and Thornton, J. M. (1993). *Structure*, **1**, 105.
74. Wetlaufer, D. B. (1973). *Proc. Natl. Acad. Sci. USA*, **70**, 697.
75. Islam, S. A., Luo, J., and Sternberg, M. J. E. (1995). *Protein Eng.*, **8**, 513.
76. Levitt, M. and Chothia, C. (1976). *Nature (London)*, **261**, 552.
77. Harris, N. L., Presnell, S. R., and Cohen, F. E. (1994). *J. Mol. Biol.*, **236**, 1356.
78. Harpaz, Y. and Chothia, C. (1994). *J. Mol. Biol.*, **238**, 528.
79. Lasters, I., Wodak, S. J., Alard, P., and Cutsem, E. (1988). *Proc. Natl. Acad. Sci. USA*, **85**, 3338.
80. Orengo, C. A., Flores, T. P., Taylor, W. R., and Thornton, J. M. (1993). *Protein Eng.*, **6**, 485.
81. Murzin, A. G., Brenner, S. E., Hubbard, T., and Chothia, C. (1995). *J. Mol. Biol.*, **247**, 536.
82. Murzin, A. G. (1994). *Curr. Opin. Struct. Biol.*, **4**, 441.
83. Chothia, C. and Murzin, A. G. (1993). *Structure*, **1**, 217.
84. Russell, R. B. and Barton, G. J. (1994). *J. Mol. Biol.*, **244**, 332.

85. Kabsch, W. and Sander, C. (1984). *Proc. Natl. Acad. Sci. USA*, **81**, 1075
86. Sternberg, M. J. E. and Islam, S. A. (1990). *Protein Eng.*, **4**, 125.
87. Sander, C. and Schneider, R. (1991). *Proteins*, **9**, 56.
88. Krogh, A., Brown, M., Mian, I.S., Sjölander, K., and Haussler, D. (1994). *J. Mol. Biol.*, **235**, 1501.
89. McClure, M. A., Vasi, T. K., and Fitch, W. M. (1994). *Mol. Biol. Evol.*, **11**, 571.
90. Saraste, M., Sibbald, P. R., and Wittinghofer, A. (1990). *Trends Biochem. Sci.*, **15**, 430.
91. Bork, P., Ouzounis, C., Sander, C., Scharf, M., Schneider, R., and Sonnhammer, E. (1992). *Protein Sci.*, **1**, 1677.
92. Garnier, J., Osguthorpe, D. J., and Robson, B. (1978). *J. Mol. Biol.*, **120**, 97.
93. Lim, V. I. (1974). *J. Mol. Biol.*, **88**, 873.
94. Kabsch, W. and Sander, C. (1983). *FEBS Lett.*, **155**, 179.
95. Zvelebil, M. J. J. M., Barton, G. J., Taylor, W. R., and Sternberg, M. J. E. (1987). *J. Mol. Biol.*, **195**, 957.
96. Levin, J. M., Pascarella, S., Argos, P., and Garnier, J. (1993). *Protein Eng.*, **6**, 849.
97. Rost, B. and Sander, C. (1993). *J. Mol. Biol.*, **232**, 584.
98. Salamov, A. A. and Solovyev, V. V. (1995). *J. Mol. Biol.*, **247**, 11.
99. Russell, R. B. and Barton, G. J. (1993). *J. Mol. Biol.*, **234**, 951.
100. Jenny, T. F. and Benner, S. A. (1994). *Biochem. Biophys. Res. Commun.*, **200**, 149.
101. Rost, B., Sander, C., and Schneider, R. (1994). *J. Mol. Biol.*, **235**, 13.
102. Jenny, T. F. and Benner, S. A. (1994). *Proteins*, **20**, 1.
103. Musacchio, A., Gibson, T., Rice, P., Thompson, J., and Saraste, M. (1993). *Trends Biochem. Sci.*, **18**, 343.
104. Russell, R. B. and Sternberg, M. J. E. (1995). *Curr. Biol.*, **5**, 488.
105. Jones, D. T., Taylor, W. R., and Thornton, J. M. (1994). *Biochemistry*, **33**, 3038.
106. Rost, B., Casadio, R., Fariselli, P., and Sander, C. (1995). *Protein Sci.*, **4**, 521.
107. Persson, B. and Argos, P. (1994). *J. Mol. Biol.*, **237**, 182.
108. Blundell, T., Carney, D., Gardner, S., Hayes, F., Howlin, B., Hubbard, T., *et al.* (1988). *Eur. J. Biochem.*, **172**, 513.
109. Peitsch, M. C. (1995). *Biotechnology*, **13**, 658.
110. Sali, A. and Blundell, T. L. (1993). *J. Mol. Biol.*, **234**, 779.
111. Chothia, C., Lesk, A. M., Tramontano, A., Levitt, M., Smith, G. S., Air, G., *et al.* (1989). *Nature (London)*, **342**, 877.
112. Sippl, M. J. (1995). *Curr. Opin. Struct. Biol.*, **5**, 229.
113. Edwards, Y. J. K. and Perkins, S. J. (1995). *FEBS Lett.*, **358**, 283.
114. Lee, J.-O., Rieu, P., Arnaout, M. A., and Liddington, R. (1995). *Cell*, **80**, 631.
115. Taylor, W. R., Jones, D. T., and Green, N. M. (1994). *Proteins*, **18**, 281.
116. Levitt, M. and Warshel, A. (1975). *Nature*, **253**, 694.
117. Liwo, A., Pincus, M. R., Wawak, R. J., Rackovsky, S., and Scheraga, H.A. (1993). *Protein Sci.*, **2**, 1715.
118. Srinivasan, R. and Rose, G. D. (1995). *Proteins*, **22**, 81.
119. Kolinski, A. and Skolnick, J. (1994). *Proteins*, **18**, 353.
120. Strynadka, N. C. J., Eisenstein, M., Katchalski-Katzir, E., Shoichet, B., Kuntz, I., Abagyan, R., *et al.* (1996). *Nature Struct. Biol.*, **3**, 233.
121. Jackson, R. M. and Sternberg, M. J. E. (1995). *J. Mol. Biol.*, **250**, 258.

122. Gilson, M. K. and Honig, B. (1988). *Proteins*, **4**, 7.
123. Walls, P. H. and Sternberg, M. J. E. (1992). *J. Mol. Biol.*, **228**, 277.
124. Ring, C. S., Sun, E., McKerrow, J. H., Lee, G. K., Rosenthal, P. J., Kuntz, I. D., *et al.* (1993). *Proc. Natl. Acad. Sci. USA*, **90**, 3583.
125. Varghese, J. N. and Colman, P. M. (1991). *J. Mol. Biol.*, **221**, 473.
126. Yoder, M. D., Lietzke, S. E., and Jurnak, F. (1993). *Curr. Biol.*, **1**, 241.
127. Kobe, B. and Deisenhofer, J. (1993). *Nature*, **366**, 751.
128. Steigemann, W. and Weber, E. (1979). *J. Mol. Biol.*, **127**, 309.
129. Fermi, G. and Perutz, M. F. (1984). *J. Mol. Biol.*, **175**, 159.
130. Duerring, M., Schmidt, G. B., and Huber, R. (1991). *J. Mol. Biol.*, **217**, 577.

2

Protein sequence alignment and database scanning

GEOFFREY J. BARTON

1. Introduction

In the context of protein structure prediction, there are two principle reasons for comparing and aligning protein sequences:

(a) To obtain an accurate alignment. This may be for protein modelling by comparison to proteins of known three-dimensional structure.

(b) To scan a database with a newly determined protein sequence and identify possible functions for the protein by analogy with well-characterized proteins.

In this chapter I review the underlying principles and techniques for sequence comparison as applied to proteins and used to satisfy these two aims.

2. Amino acid scoring schemes

All algorithms to compare protein sequences rely on some scheme to score the equivalencing of each of the 210 possible pairs of amino acids, (i.e. 190 pairs of different amino acids plus 20 pairs of identical amino acids). Most scoring schemes represent the 210 pairs of scores as a 20 $\times$ 20 matrix of similarities where identical amino acids and those of similar character (e.g. I, L) give higher scores compared to those of different character (e.g. I, D). Since the first protein sequences were obtained, many different types of scoring scheme have been devised. The most commonly used are those based on observed substitution and of these, the 1976 Dayhoff matrix for 250 PAMS (1) has until recently dominated. This and other schemes are discussed in the following sections.

2.1 Identity scoring

This is the simplest scoring scheme: amino acid pairs are classified into two types; identical and non-identical. Non-identical pairs are scored zero and identical pairs given a positive score (usually one). The scoring scheme is generally considered less effective than schemes that weight non-identical pairs, particularly for the detection of weak similarities (2,3). The normalized sum of identity scores for an alignment is popularly quoted as 'percentage identity', but although this value can be useful to indicate the overall similarity between two sequences, there are pitfalls associated with the measure. These are discussed in Section 4.1.1.

2.2 Genetic code scoring

Whereas the identity scoring scheme considers all amino acid transitions with equal weight, genetic code scoring as introduced by Fitch (4) considers the minimum number of DNA/RNA base changes (0, 1, 2, or 3) that would be required to interconvert the codons for the two amino acids. The scheme has been used both in the construction of phylogenetic trees and in the determination of homology between protein sequences having similar three-dimensional structures (5). However, today it is rarely the first choice for scoring alignments of protein sequences.

2.3 Chemical similarity scoring

The aim with chemical similarity scoring schemes is to give greater weight to the alignment of amino acids with similar physico-chemical properties. This is desirable since major changes in amino acid type could reduce the ability of the protein to perform its biological role and hence the protein would be selected against during the course of evolution. The intuitive scheme developed by McLachlan (6) classified amino acids on the basis of polar or non-polar character, size, shape, and charge, and gives a score of six to interconversions between identical rare amino acids (e.g. F, F) reducing to zero for substitutions between amino acids of quite different character (e.g. F, E). Feng *et al.* (3) encode features similar to McLachlan by combining information from the structural features of the amino acids and the redundancy of the genetic code.

2.4 Observed substitutions

Scoring schemes based on observed substitutions are derived by analysing the substitution frequencies seen in alignments of sequences. This is something of a chicken and egg problem, since in order to generate the alignments, one really needs a scoring scheme but in order to derive the scoring scheme one needs the alignments! Early schemes based on observed substitutions worked from closely related sequences that could easily be aligned by eye. More recent schemes have had the benefit of the earlier substitution matrices to generate alignments on which to build. Long experience with scoring schemes based on observed substitutions suggests that they are superior to simple identity, genetic code, or intuitive physico-chemical property schemes.

2.4.1 The Dayhoff mutation data matrix

Possibly the most widely used scheme for scoring amino acid pairs is that developed by Dayhoff and co-workers (1). The system arose out of a general model for the evolution of proteins. Dayhoff and co-workers examined alignments of closely similar sequences where the likelihood of a particular mutation (e.g. A–D) being the result of a set of successive mutations (e.g. A–x–y–D) was low. Since relatively few families were considered, the resulting matrix of accepted point mutations included a large number of entries equal to zero or one. A complete picture of the mutation process including those amino acids which did not change was determined by calculating the average ratio of the number of changes a particular amino acid type underwent to the total number of amino acids of that type present in the database. This was combined with the point mutation data to give the mutation probability matrix (M) where each element $M_{i,j}$ gives the probability of the amino acid in column j mutating to the amino acid in row i after a particular evolutionary time, for example after two PAM (percentage of acceptable point mutations per 10^8 years).

The mutation probability matrix is specific for a particular evolutionary distance, but may be used to generate matrices for greater evolutionary distances by multiplying it repeatedly by itself. At the level of 2000 PAM Schwartz and Dayhoff suggest that all the information present in the matrix has degenerated except that the matrix element for Cys–Cys is 10% higher than would be expected by chance. At the evolutionary distance of 256 PAMs one amino acid in five remains unchanged but the amino acids vary in their mutability; 48% of the tryptophans, 41% of the cysteines, and 20% of the histidines would be unchanged, but only 7% of serines would remain.

When used for the comparison of protein sequences, the mutation probability matrix is usually normalized by dividing each element $M_{i,j}$ by the relative frequency of exposure to mutation of the amino acid i. This operation results in the symmetrical 'relatedness odds matrix' with each element giving the probability of amino acid replacement per occurrence of i per occurrence of j. The logarithm of each element is taken to allow probabilities to be summed over a series of amino acids rather than requiring multiplication. The resulting matrix is the 'log odds matrix' which is frequently referred to as 'Dayhoff's matrix' and often used at a distance of close to 256 PAM since this lies near to the limit of detection of distant relationships where approximately 80% of the amino acid positions are observed to have changed (2).

2.4.2 PET91—an updated Dayhoff matrix

The 1978 family of Dayhoff matrices was derived from a comparatively small set of sequences. Many of the 190 possible substitutions were not observed at all and so suitable weights were determined indirectly. Recently, Jones *et al.* (7) have derived an updated substitution matrix by examining 2621 families of sequences in the SWISS-PROT database release 15.0. The principal differences between the Jones *et al.* matrix (PET91) and the Dayhoff matrix are for substitutions that were poorly represented in the 1978 study. However, the overall character of the matrices is similar. Both reflect substitutions that conserve size and hydrophobicity, which are the principle properties of the amino acids (8).

2.4.3 BLOSUM—matrix from ungapped alignments

Dayhoff-like matrices derive their initial substitution frequencies from global alignments of very similar sequences. An alternative approach has been developed by Henikoff and Henikoff using local multiple alignments of more distantly related sequences (9). First a database of multiple alignments without gaps for short regions of related sequences was derived. Within each alignment in the database, the sequences were clustered into groups where the sequences are similar at some threshold value of percentage identity. Substitution frequencies for all pairs of amino acids were then calculated between the groups and this used to calculate a log odds BLOSUM (blocks substitution matrix) matrix. Different matrices are obtained by varying the clustering threshold. For example, the BLOSUM 80 matrix was derived using a threshold of 80% identity.

2.4.4 Matrices derived from tertiary structure alignments

The most reliable protein sequence alignments may be obtained when all the proteins have had their tertiary structures experimentally determined. Comparison of three-dimensional structures also allows much more distantly related proteins to be aligned accurately. Analysis of such alignments should therefore give the best substitution matrices. Accordingly, Risler *et al.* (10) derived substitution frequencies from 32 proteins structurally aligned in 11 groups. On similar lines, Overington *et al.* (11) aligned seven families for which three or more proteins of known three-dimensional structure were known and derived a series of substitution matrices. Overington *et al.* also subdivided the substitution data by the secondary structure and environment of each amino acid, however this led to rather sparse matrices due to the lack of examples. Bowie *et al.* (12) have also derived substitution tables specific for different amino acid environments and secondary structures.

2.5 Which matrix should I use?

The general consensus is that matrices derived from observed substitution data (e.g. the Dayhoff or BLOSUM matrices) are superior to identity, genetic code, or physical property matrices (for example see ref. 3). However, there are Dayhoff matrices of different PAM values and BLOSUM matrices of different percentage identity and which of these should be used?

Schwartz and Dayhoff (2) recommended a mutation data matrix for the distance of 250 PAMs as a result of a study using a dynamic programming procedure (13) to compare a variety of proteins known to be distantly related. The 250 PAM matrix was selected since in Monte Carlo studies (see Section 4.1) matrices reflecting this evolutionary distance gave a consistently higher significance score than other matrices in the range 0–750 PAM. The matrix also gave better scores when compared to McLachlan's substitution matrix (6), the genetic code matrix, and identity scoring. Recently, Altschul (14) has examined Dayhoff-style mutation data matrices from an information theoretical perspective. For alignments that do not include gaps he concluded, in broad agreement with Schwarz and Dayhoff, that a matrix of 200 PAMS was most appropriate when the sequences to be compared were thought to be related. However, when comparing sequences that were not known in advance to be related, for example when database scanning, a 120 PAM matrix was the best compromise. When using a local alignment method (Section 6.7) Altschul suggests that three matrices should ideally be used: PAM40, PAM120, and PAM250, the lower PAM matrices will tend to find short alignments of highly similar sequences, while higher PAM matrices will find longer, weaker local alignments. Similar conclusions were reached by Collins and Coulson (15) who advocate using a compromise PAM100 matrix, but also suggest the use of multiple PAM matrices to allow detection of local similarities of all types.

Henikoff and Henikoff (16) have compared the BLOSUM matrices to PAM, PET, Overington, Gonnet (17), and multiple PAM matrices by evaluating how effectively the matrices can detect known members of a protein family from a database when searching with the ungapped local alignment program BLAST (18). They conclude that overall the BLOSUM 62 matrix is the most effective. However, all the substitution matrices investigated perform better than BLOSUM 62 for a proportion of the families. This suggests that no single matrix is the complete answer for all sequence comparisons. It is probably best to complement the BLOSUM 62 matrix with comparisons using PET91 at 250 PAMS, and Overington structurally-derived matrices. It seems likely that as more protein three-dimensional structures are determined, substitution tables derived from structure comparison will give the most reliable data.

3. Comparison of two sequences

Given a scoring scheme, the next problem is how to compare the sequences, decide how similar they are, and generate an alignment. This problem may be subdivided into alignment methods for two sequences, multiple alignment methods, and methods that incorporate additional non-sequence information, for example from the tertiary structure of the protein.

The simplest two sequence comparison methods do not explicitly consider insertions and deletions (gaps). More sophisticated methods make use of dynamic programming to determine the best alignment including gaps (see Section 3.2).

3.1 Sequence comparison without gaps—fixed length segments

Given two sequences A and B of length m and n, all possible overlapping segments having a particular length (sometimes called a 'window length') from A are compared to all segments of B. This requires of the order of $m \times n$ comparisons to be made. For each pair of segments the amino acid pair scores are accumulated over the length of the segment. For example, consider the comparison of two seven residue segments, ALGAWDE and ALATWDE, using identity scoring. The total score for this pair would be $1 + 1 + 0 + 0 + 1 + 1 + 1 = 5$.

In early studies of protein sequences, statistical analysis of segment comparison scores was used to infer homology between sequences. For example, Fitch (4) applied the genetic code scoring scheme to the comparison of α- and β-haemoglobin and showed the score distribution to be non-random. Today, segment comparison methods are most commonly used in association with a 'dot plot' or 'diagram' (19) and can be a more effective method of finding repeats than using dynamic programming.

The scores obtained by comparing all pairs of segments from A and B may be represented as a comparison matrix R where each element $R_{i,j}$ represents the score for matching an odd length segment centred on residue A_i with one centred on residue B_j. This matrix can provide a graphic representation of the segment comparison data particularly if the scores are contoured at a series of probability levels to illustrate the most significantly similar regions. Collins and Coulson (20) have summarized the features of the 'dot plot'. The runs of similarity can be enhanced visually by placing a dot at all the contributing match points in a window rather than just at the centre.

McLachlan (6) introduced two further refinements into segment comparison methods. The first was the inclusion of weights in the comparison of two segments in order to improve the definition of the ends of regions of similarity. For example, the scores obtained at each position in a five residue segment comparison might be multiplied by 1, 2, 3, 2, 1 respectively before being summed. The second refinement was the development of probability distributions which agreed well with experimental comparisons on random and unrelated sequences and which could be used to estimate the significance of an observed comparison.

3.1.1 Correlation methods

Several experimental, and semi-empirical properties have been derived associated with amino acid types, for example hydrophobicity (21), and propensity to form an α-helix (22). Correlation methods for the comparison of protein sequences exploit the large number of amino acid properties as an alternative to comparing the sequences on the basis of pair scoring schemes.

Kubota *et al.* (23) gathered 32 property scales from the literature and through application of factor analysis selected six properties which for carp parvalbumin gave good correlation for the comparison of the structurally similar CE- and EF-hand region Ca^{2+} binding sites and poor correlation in other regions. They expressed their sequence comparisons in the form of a comparison matrix similar to that of McLachlan (6) and demonstrated that their method could identify an alignment of α-lytic protease and *Stryptomyces griseus* protease A which agrees with that determined from comparison of the available crystal structures.

Argos (24) determined the most discriminating properties from a set of 55 by calculating correlation coefficients for all pairs of sequences within 30 families of proteins that had been aligned on the basis of their three-dimensional structures. The correlation coefficients for each property were then averaged over all the families to leave five representative properties. Unlike Kubota *et al.* (23), Argos applied the correlation coefficients from the five properties in addition to a more conventional segment comparison method using the Dayhoff matrix scoring scheme. He also combined the result of using more than one segment length on a single diagram such that the most significant scores for a particular length always prevail.

3.1.2 Variable length segments

The best local ungapped alignments of variable length may be found either by dynamic programming with a high gap penalty, or using heuristic methods. Since the heuristic methods are primarily used for database searching they are described in Section 6.

3.2 Sequence comparison with gaps

The segment-based techniques described in Section 3.1 do not consider explicitly insertions and deletions. Deletions are often referred to as 'gaps', while insertions and deletions are collectively referred to as 'indels'. Insertions and deletions are usually needed to align accurately even quite closely related sequences such as the α- and β-globins. The naive approach to finding the best alignment of two sequences including gaps is to generate all possible alignments, add up the scores for equivalencing each amino acid pair in each alignment, then select the highest scoring alignment. However, for two sequences of 100 residues there are more than 10^{75} alternative alignments, so such an approach would be time-consuming and not feasible for longer sequences. Fortunately, there is a group of algorithms that can calculate the best score and alignment in the order of *mn* steps. These *dynamic programming* algorithms were first developed for protein sequence comparison by Needleman and Wunsch (13), though similar methods were independently devised during the late 1960s and early 1970s for use in the fields of speech processing and computer science (25).

3.2.1 Finding the best alignment with dynamic programming

Dynamic programming algorithms may be divided into those that find a global alignment of the sequences and those that find local alignments. The difference between global and local alignment is illustrated in *Figure 1*. Global alignment is appropriate for sequences that are known to share similarity over their whole length. Local alignment is appropriate when the sequences may show isolated regions of similarity, for example multiple domains or repeats. Local alignment is best applied when scanning a database to find similarities or when there is no a priori knowledge that the protein sequences are similar.

There are many variations on the theme of dynamic programming applied to protein comparisons. Here I give a brief account of a basic method for finding the *global* best score for aligning two sequences. For a clear and detailed explanation of dynamic programming see Sankoff and Kruskal (26).

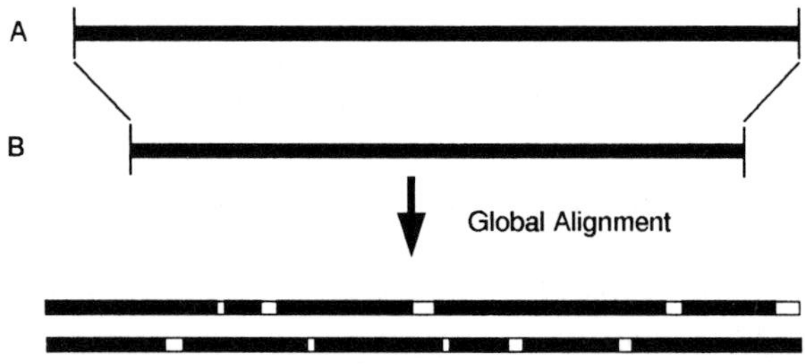

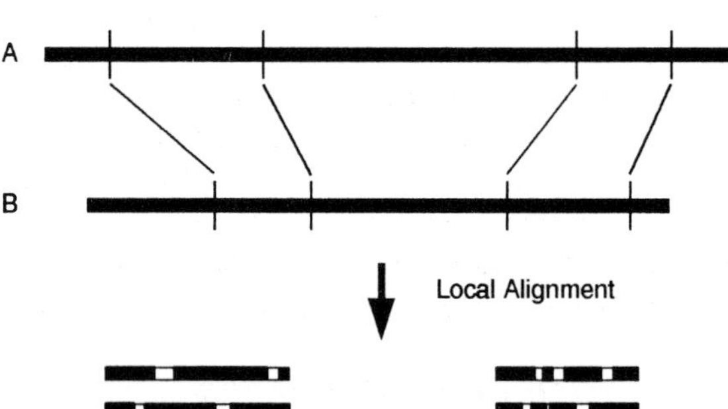

Figure 1. Comparison of global and local alignment. Global alignment optimizes the alignment over the full-length of the sequences A and B. Local alignment locates the best alignment between subregions of A and B. There may be a large number of distinct local alignments.

Let the two sequences of length m and n be $A = (A_1, A_2, \ldots A_m)$, $B = (B_1, B_2, \ldots B_n)$, and the symbol for a single gap be Δ. At each aligned position there are three possible events.

$w(A_i, B_j)$ substitution of A_i by B_j.
$w(A_i, \Delta)$ deletion of A_i.
$w(\Delta, B_j)$ deletion of B_j.

The substitution weight $w(A_i, B_j)$ is derived from the chosen scoring scheme—perhaps Dayhoff's matrix. Gaps Δ are normally given a negative weight often referred to as the 'gap penalty' since insertions and deletions are usually less common than substitutions.

The maximum score M for the alignment of A with B may be represented as $s(A_{1\cdots m}, B_{1\cdots n})$. This may be found by working forward along each sequence successively finding the best score for aligning $A_{1\ldots i}$ with $B_{1\ldots j}$ for all i, j where $1 \leqslant i \leqslant m$ and $1 \leqslant j \leqslant n$. The values of $s(A_{1\cdots i}, B_{1\cdots j})$ are stored in a matrix H where each element of H is calculated as follows:

$$H_{i,j} = \max \begin{Bmatrix} H_{i-1,j-1} + w_{Ai,\, Bj} \\ H_{i,j-1} + w_{Ai,\, \Delta} \\ H_{i-1,j} + w_{\Delta,\, Bj} \end{Bmatrix}$$

The element $H_{m,n}$ contains the best score for the alignment of the complete sequences.

If the alignment is required as well as the best score, then the alignment path may be determined by tracing back through the H matrix. Alternatively a matrix of pointers is recorded to indicate which of the three possibilities was the maximum at each value $H_{i,j}$.

3.2.2 Alternative weighting for gaps

The above scheme showed a simple length-dependent weighting for gaps. Thus two isolated gaps give the same score as two consecutive gaps. It is possible to generalize the algorithm to allow gaps of length greater than one to carry weights other than the simple sum of single gap weights (27). Such gap weighting can give a more biologically meaningful model of transitions from one sequence to another since insertions and deletions of more than one residue are not uncommon events between homologous protein sequences. Most computer programs that implement dynamic programming allow gaps to be weighted with the form $v + uk$ where k is the gap length and v and u are constants $\geqslant 0$, since this can be computed efficiently (28).

3.3 Identification of local similarities

Although segment-based comparison methods (see Section 3.1) rely on local comparisons, if insertions and deletions have occurred, the match may be disrupted for a region of the order of the length of the segment. In order to circumvent these difficulties algorithms which are modifications of the basic global alignment methods have been developed to locate common subsequences including a consideration of gaps (29–31). For protein sequences, the most commonly used local alignment algorithm that allows gaps is that described by Smith and Waterman 930). This is essentially the same as the global alignment algorithm described in Section 3.2.1, except that a zero is added to the recurrence equation.

$$H_{i,j} = \max \begin{Bmatrix} H_{i-1,j-1} + w_{Ai,Bj} \\ H_{i,j-1} + w_{Ai,\Delta} \\ H_{i-1,j} + w_{\Delta,Bj} \\ 0 \end{Bmatrix}$$

Thus all $H_{i,j}$ must have a value $\geqslant 0$. The score for the best local alignment is simply the largest value of H and the corresponding alignment is obtained by tracing back from this cell.

3.3.1 Finding second and subsequent best local alignments

The Smith–Waterman algorithm returns the single best local alignment, but two proteins may share more than one common region. Waterman and Eggert (32) have shown how all local alignments may be obtained for a pair of sequences with minimal recalculation. Recently, Barton (33) has described how for a simple length-dependent gap penalty, nearly all locally optimal alignments may be determined in the order of *mn* steps without recalculation.

4. Evaluation of alignment accuracy

What is a good alignment? The amino acid sequence codes for the protein three-dimensional structure. Accordingly, when an alignment of two or more sequences is made, the implication is that the equivalenced residues are performing similar structural roles in the native folded protein. The best judge of alignment accuracy is thus obtained by comparing alignments resulting from sequence comparison with those derived from protein three-dimensional structures. There are now many families of proteins for which two or more members have been determined to atomic resolution by X-ray crystallography or NMR. Accurate alignment of these proteins by consideration of their tertiary structures (34–36) provides a set of test alignments against which to compare sequence-only alignment methods. Care must be taken when performing the comparison since within protein families, some regions show greater similarity than others. For example, the core β-strands and α-helices are normally well conserved, but surface loops vary in structure and alignments in these regions may be ambiguous, or if the three-dimensional structures are very different in a region, alignment may be meaningless. Accordingly, evaluation of alignment accuracy is best concentrated on the core secondary structures of the protein and other conserved features (37);

such regions may automatically be identified by the algorithm of Russell and Barton (36).

4.1 Predicting overall alignment accuracy

It is important to know in advance what the likely accuracy of an alignment will be. A common method for assessing the significance of a global alignment score is to compare the score to the distribution of scores for alignment of random sequences of the same length and composition. The result (the SD score) is normally expressed in standard deviation units above the mean of the distribution.

Comparison of the SD score for alignment to alignment accuracy obtained by comparison of the core secondary structures, suggests that for proteins of 100–200 amino acids in length, a score above 15 SD indicates a near ideal alignment, scores above 5.0 SD a 'good' alignment where $\geqslant$ 70% of the residues in core secondary structures will be correctly equivalenced, while alignments with scores below 5.09 SD should be treated with caution (37,38).

Figure 2 shows the distribution of SD scores for 100 000 optimal alignments of length $\geqslant$ 20 between proteins of unrelated three-dimensional structure.

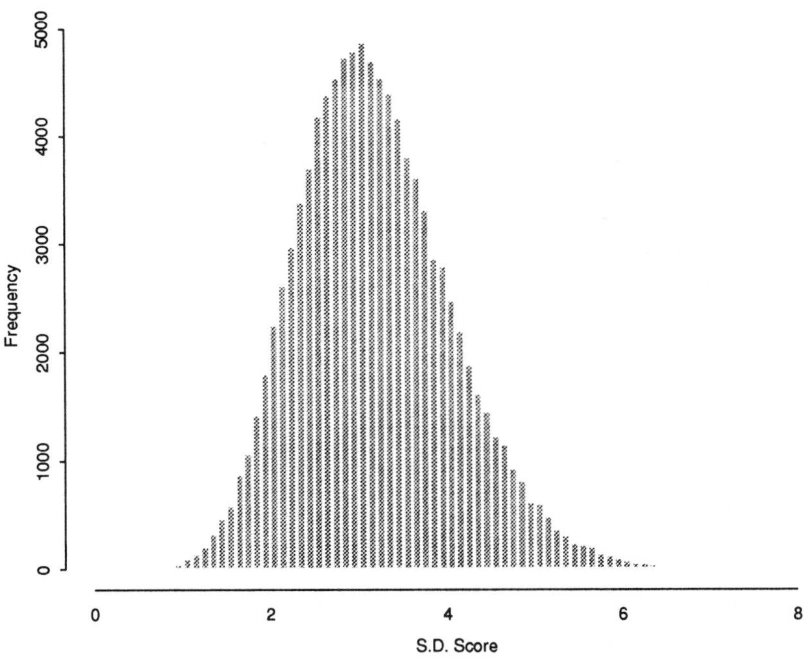

Figure 2. Distribution of SD scores obtained for 100 000 alignments of length > 20 between unrelated proteins. The SD scores were calculated from 100 randomizations using a global alignment method (13), PAM250 matrix with eight added to each element, and length-independent gap penalty of eight.

Figure 3. A local alignment found between citrate synthase (Brookhaven code: 2cts) and transthyritin (2paba). The SD score for this alignment is 7.55, its length is 54 residues, and the identity is 25.9%. Despite this apparently high similarity, the sequences are of completely different secondary structure.

From *Figure 2*, the mean SD score expected for the comparison of unrelated protein sequences is 3.2 SD with a SD of 0.9. However, the distribution is skewed with a tail of high SD scores. In any large collection of alignments it is possible to have a rare, high scoring alignment that actually shares no structural similarity. For example, *Figure 3* illustrates an optimal local alignment between regions of citrate synthase and transthyritin which gives 7.55 SD though the secondary structure of these two protein segments are completely different.

4.1.1 Predicting quality using percentage identity

Percentage identity is a frequently quoted statistic for an alignment of two sequences. However, the expected value of percentage identity is strongly dependent upon the length of alignment (39) and this is frequently overlooked. *Figure 4* shows the percentage identities found for a large number of locally optimal alignments of differing length between proteins known to be of unrelated three-dimensional structure. Clearly, an alignment of length 200 showing 30% identity is more significant than an alignment of length 50 with the same identity. Applying this to the alignment shown in *Figure 3* shows that although the alignment scores over 7.0 SD it has a percentage identity that one would often see by chance between unrelated proteins.

4.2 Predicting the reliable regions of an alignment

Although the overall accuracy of an alignment may be estimated from the SD score (see Section 4.1) this value does not indicate which regions of the alignment are correct. Experience suggests that the reliable regions of an alignment are those that do not change when small changes are made to the gap penalty and matrix parameters. An alternative strategy is to examine the suboptimal alignments of the sequences to find the regions that are shared by suboptimal alignments within a scoring interval of the best alignment. For any two sequences, there are usually many alternative alignments with scores similar to the best. These alignments share common regions and it is these regions that are deemed to be the most reliable. For example, the simple alignment of ALLIM with ALLM scoring 2 for identities, 1 for mismatch, and −1 for a gap gives:

```
A  L  L  I  M
A  L  L     M
```

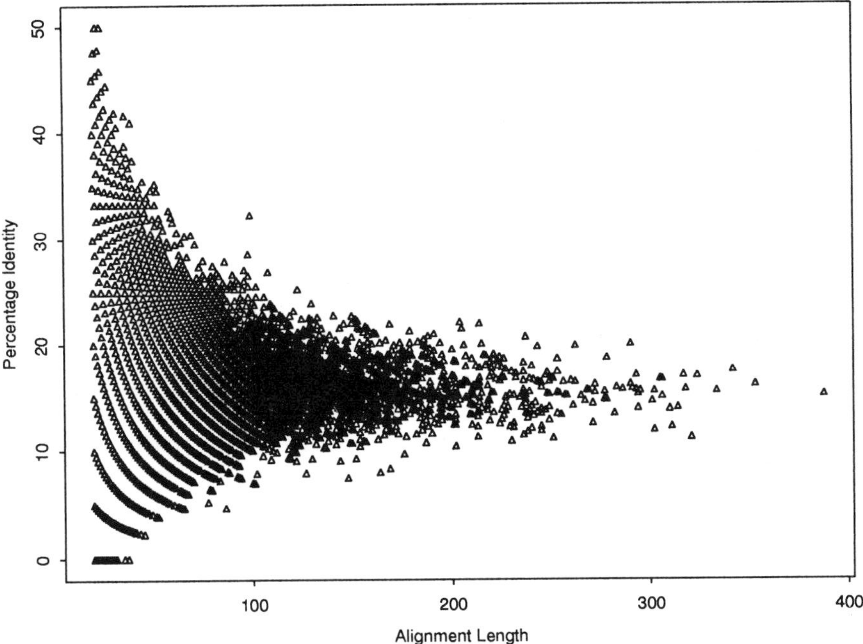

Figure 4. Plot of percentage identity versus alignment length for the 100 000 alignments from *Figure 2*.

with a score of 2 + 2 + 2 − 1 + 2 = 7. The suboptimal alignment:

```
A  L  L  I  M
A  L     L  M
```

gives a score of 2 + 2 − 1 + 1 + 2 = 6 but shares the alignment of AL and M with the optimal alignment. Rather than calculate all suboptimal alignments, Vingron and Argos (40) use an elegant and simple method to identify the reliable regions in an alignment by calculating the comparison matrix H both forwards and backwards and summing the two matrices. The cells in $H_{i,j}$ that are equal to the best score for the alignment delineate the optimal alignment path. Cells within a selected value of the best score are flagged and reliable regions defined as those for which there is no other cell $H_{i,k}$ or $H_{l,j}$ with $k \neq j$ and $l \neq i$. The results of the analysis are displayed in the form of a dot plot with larger dots identifying the reliable regions.

Although the details of his calculation differ from Vingron and Argos, Zuker (41) produces a dot plot that highlights the regions where there are few alternative local alignments. He also caters for optimal local alignments with gaps. Zuker shows that the alignment of distantly related sequences such as *Streptomyces griseus* proteinase A and porcine elastase may be clearly seen to be unstable with many suboptimal alignments close to the optimal.

Rather than use the dot plot representation, Saqi and Sternberg (42) directly determine alternative suboptimal alignments. They first calculate the H matrix and best path, then identify the cells that contributed to the best path, and reduce these by a preset value (usually 10% of the typical scoring matrix value). A new H matrix is calculated and a new best path and alignment. This process is repeated iteratively to generate a series of global suboptimal alignments.

Investigating suboptimal alignments by one or more of these methods allows:

(a) The most reliable regions of an alignment to be identified and by inference the overall quality of the alignment.

(b) Alternative alignments close to the optimum to be generated. These can be useful when building three-dimensional models of proteins by homology.

4.3 Incorporating non-sequence information into alignment

If the three-dimensional structure of one of the proteins to be aligned is known, then this information may be encoded in the form of a modified gap penalty (37,38,43). The penalty reduces the likelihood of insertions/deletions occurring in known secondary structure regions, or conversely increases the likelihood of placing gaps in known loop regions. This approach increases the usual accuracy of alignment and has the additional bonus of reducing the sensitivity of the alignments to changes in gap penalty (37).

A stricter constraint on the alignment is possible if specific residues are known to be equivalent in the two proteins. The weight for aligning these specific residues may be increased to force them to align. However, if this type of treatment is really necessary, then it is likely that the alignment will have a low significance score and must be treated with caution.

5. Multiple sequence alignment

So far I have only considered methods to align two sequences. However, when the sequence data is available, a multiple alignment is always preferable to pairwise alignment.

Techniques for the alignment of three or more sequences may be divided into four categories:

- extensions of pairwise dynamic programming algorithms
- hierarchical extensions of pairwise methods
- segment methods
- consensus or 'regions' methods

Of these, the second is by far the most practical and widely-used method. Consensus methods are not greatly used for protein sequence alignment and so are not discussed further.

5.1 Extension of dynamic programming to more than two sequences

Needleman and Wunsch (13) suggested that their dynamic programming algorithm could be extended to the comparison of many sequences. Waterman *et al.* (27) also described how dynamic programming could be used to align more than two sequences. In practice, the need to store an N-dimensional array (where N is the number of sequences) limits these extensions to three-sequence applications. In addition, the time required to perform the comparison of three sequences is proportional to N^5. Murata *et al.* (44) described a modification of the Needleman and Wunsch procedure for three sequences which ran in time proportional to N^3; unfortunately this approach required an additional three-dimensional array thus further limiting its application to short sequences. One of the earliest practical applications of dynamic programming to multiple alignment was the work of Sankoff *et al.* (45) who aligned nine 5S RNA sequences that were linked by an evolutionary tree. Their algorithm which also constructed the protosequences at the interior nodes of the tree was made computationally feasible by decomposing the nine-sequence problem into seven three-sequence alignments. The alignments were repeatedly performed working in from the periphery of the tree until no further change occurred to the protosequences.

5.2 Tree or hierarchical methods using dynamic programming

Practical methods for multiple sequence alignment based on a tree have been developed in several laboratories (38–49). The principle is that since the alignment of two sequences can be achieved very easily, multiple alignments should be built by the successive application of pairwise methods.

The steps are summarized here and illustrated in *Figure 5*:

(a) Compare all sequences pairwise. For N sequences there are $N \times (N-1)/2$ pairs.

(b) Perform cluster analysis on the pairwise data to generate a hierarchy for alignment. This may be in the form of a binary tree, or a simple ordering.

(c) Build the multiple alignment by first aligning the most similar pair, and so on. Once an alignment of two sequences has been made, then this is fixed. Thus for a set of sequences A, B, C, D having aligned A with C and B with D the alignment of A, B, C, D is obtained by comparing the alignments of A and C with that of B and D using averaged scores at each aligned position.

This family of methods gives good usable alignments with gaps, it can be applied to large numbers of sequences, and with the exception of the initial pairwise comparison step is very fast.

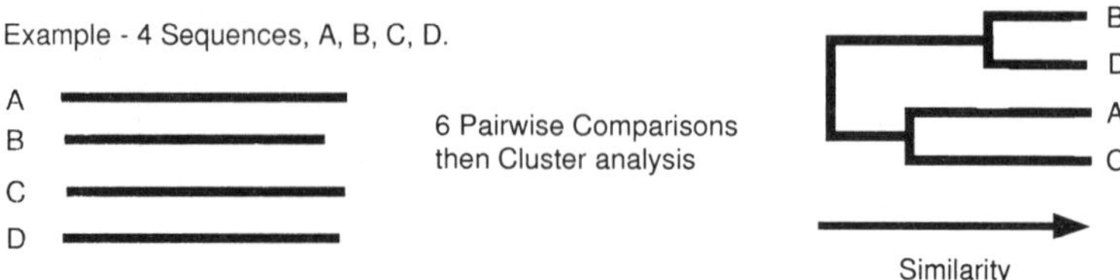

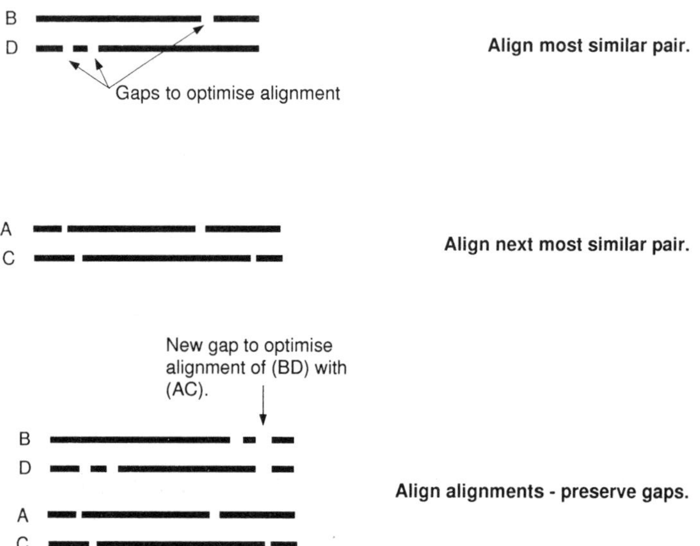

Figure 5. The stages in generating a multiple sequence alignment using a hierarchical method (see text).

Although based on the successive application of pairwise methods, multiple alignment will often yield better alignments than any pair of sequences taken in isolation. This effect was illustrated by Barton and Sternberg (46) for the alignment of immunoglobulin and globin domains. *Figure 6* shows that for some alignment pairs there is a marked improvement in accuracy over optimal pairwise alignment (e.g. variable versus constant domains).

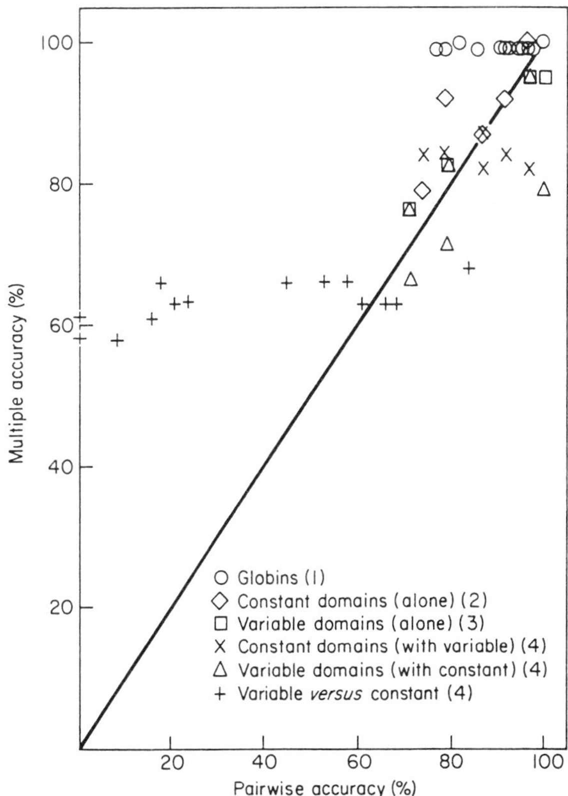

Figure 6. Comparison of alignment accuracy (as judged by comparison of alignments to those generated by tertiary structure comparison) for optimal pairwise and hierarchical multiple alignment by the method of Barton and Sternberg (46). Figure adapted from ref. 46.

5.3 Extension of segment methods to multiple alignment

A naive extension of the segment comparison methods described in Section 3.1 to N sequences would require a number of comparisons in the order of the product of the sequence lengths. Clearly, as with dynamic programming methods, such an approach is not practical. Bacon and Anderson (50) reduced the magnitude of this problem by considering the alignment in one specific order. First sequence one is compared to sequence two and the top M scoring pairs of segments are stored. The next sequence is then compared to these top scoring segments, and the top scoring segments from he three sequences are kept. This process is continued and leads to a list of M alignments of top scoring segments from N sequences. Bacon and Anderson also extended the statistical models of McLachlan (6) to N sequences, and used this model as well as one based on random sequences to assess the significance of the highest scoring segment alignment found. They suggested that these techniques allow sequences to be objectively grouped, even when most of the pairwise interrelationships are weak, and cite examples of applications to five ribonucleases, three FAD binding enzymes, and five -cro like DNA binding proteins. The Bacon and Anderson (1986) algorithm shows considerable promise for the location of significant short sequence similarities. However, the method does not provide an overall alignment of the sequences and does not explicitly consider gaps. Johnson and Doolittle (51) reduce the number of segment comparisons that must be performed by progressively evaluating selected segments from each sequence within a specified 'window'. Their method generates a complete alignment of the sequences with a consideration of gaps. Unfortunately, time constraints limit its application to four-way alignments whilst five-way alignments become unreasonably expensive for sequence lengths above 50 residues.

A variation on segment methods is employed by the alignment tool Macaw (52). Macaw applies the BLAST algorithm (see Section 6.7) to locate the most significant ungapped similarities irrespective of length. This facility is coupled with a flexible alignment display tool under Microsoft Windows. The program works well for small numbers of sequences, but lacks the convenience of the hierarchical dynamic programming methods (see Section 5.2).

5.4 Representation and analysis of multiple alignments

How do we extract the maximum information from a multiple protein sequence alignment?

When making a multiple sequence alignment a crude tree is normally generated. The tree shows the gross relationships between the sequences. It may show that sequences A, D, and C are more similar to each other than they are to B and E. However, it does not show which individual residues have changed in order to make A, D, and C different from B and E. These residues may be the most important ones to investigate by site-directed mutagenesis. Livingstone and Barton (53) have described a set-based strategy to identify such differences by comparing pairs of groups of aligned residues. Their method automatically provides a text summary of the similarities and a boxed and shaded or coloured alignment. An example of the graphical output of this analysis is illustrated in *Figure 7* for the SH2 domain family.

Providing the alignment is accurate then the following may be inferred about the secondary structure of the protein family:

(a) The position of insertions and deletions suggests regions where surface loops exist in the protein.

(b) Conserved glycine or proline suggests a β-turn.

(c) Residues with hydrophobic properties conserved at i, $i + 2$, $i + 4$ separated by unconserved or hydrophilic residues suggest a surface β-strand.

(d) A short run of hydrophobic amino acids (four residues) suggests a buried β-strand.

Pairs of conserved hydrophobic amino acids separated by pairs of unconserved, or hydrophilic residues suggests an α-helix with one face packing in the protein core. Likewise, an i, $i + 3$, $i + 4$, $i + 7$ pattern of conserved hydrophobic residues.

These patterns are not always easy to see in a single sequence, but given a multiple alignment, they often stand out and allow secondary structure to be assigned with a degree of confidence. For example, patterns were used to aid the accurate prediction of the secondary structure and position of buried residues for the annexins and SH2 domains prior to knowledge of their tertiary structures (54–56).

6. Database scanning

The techniques described in the previous sections all assume that we already have two or more sequences to align. However, if we have just determined a new sequence, then our first task is to find out whether it shares similarities with other proteins that have already been sequenced. To do this we must compare our sequence to the sequence database(s) using some computer algorithm. Any of the methods described in the previous sections may be used, but database scanning presents special problems that have led to the development of specialist algorithms. In this section I will review the options and goals of these methods.

6.1 Basic principles of database searching

When scanning a database we take a query sequence, and use an algorithm to compare the query to each sequence in the database. Every pair comparison yields a score where larger scores usually indicate a higher degree of similarity. Thus, a scan of a database containing 60 000 sequences will typically provide 60 000 scores for analysis. If a local alignment method is used, then the total number of scores may be much larger since more than one 'hit' may occur with each sequence. *Figure 8* illustrates three score distributions from such a scan. The dark shaded bars show scores with sequences known to be structurally related to the query sequence whereas the light shaded bars show scores with proteins that are thought not to be related to the query. A perfect database scanning method would completely separate these two distributions as shown in *Figure 8c*. Normally, there is some overlap between the genuinely related and unrelated sequence distributions as shown in *Figures 8c* and *8b*. There are a number of methods for ranking and rescaling the scores to improve separation and remove artefacts due to different sequence lengths and compositions. In their most highly developed form, these methods provide an estimate of the probability of seeing a score by change given a

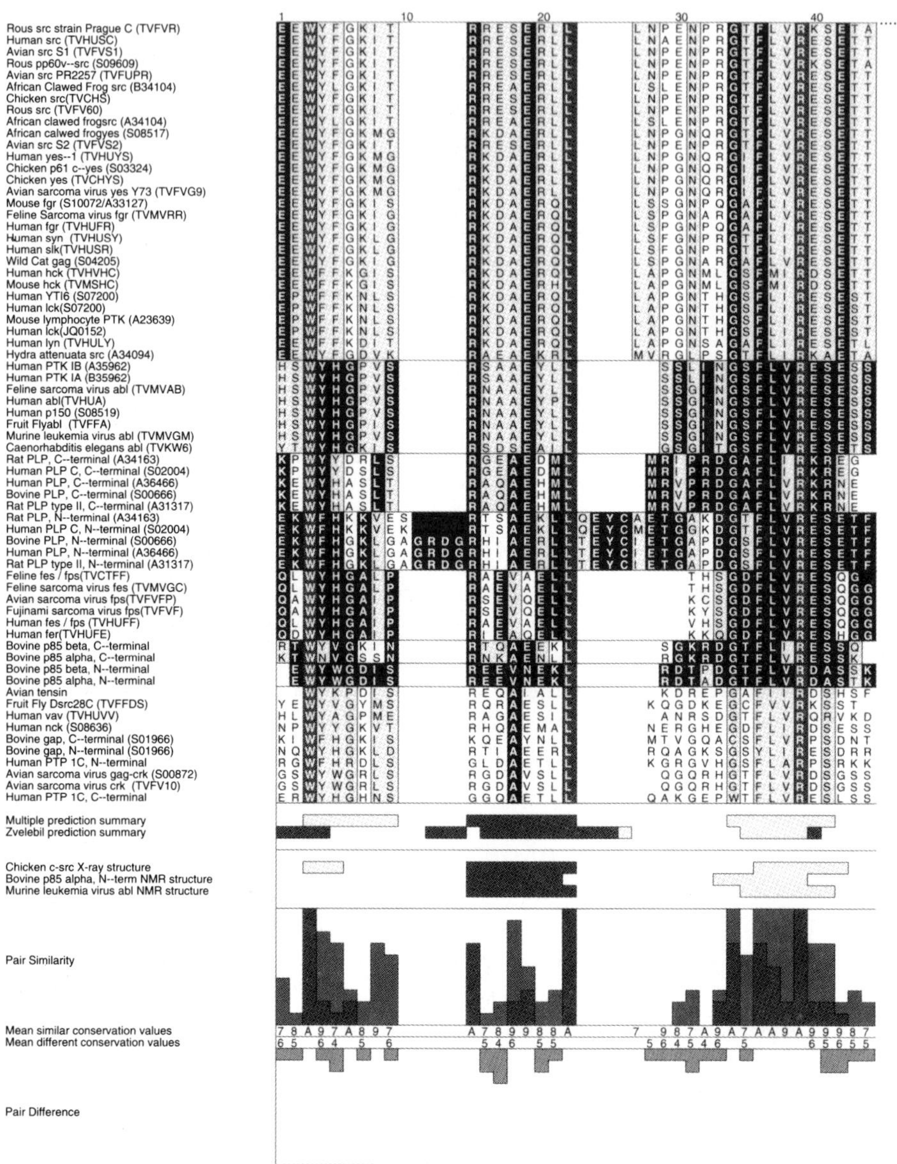

Figure 7. SH2 domain analysis performed using the program AMAS (53). The aim of this analysis is to locate patterns of amino acid conservation within each subgroup of related sequences and across all sequences in the set. The sequences are clustered by their overall similarity, then a set-based method is used to find the positions that have conserved physico-chemical properties within each group and between pairs of groups. The

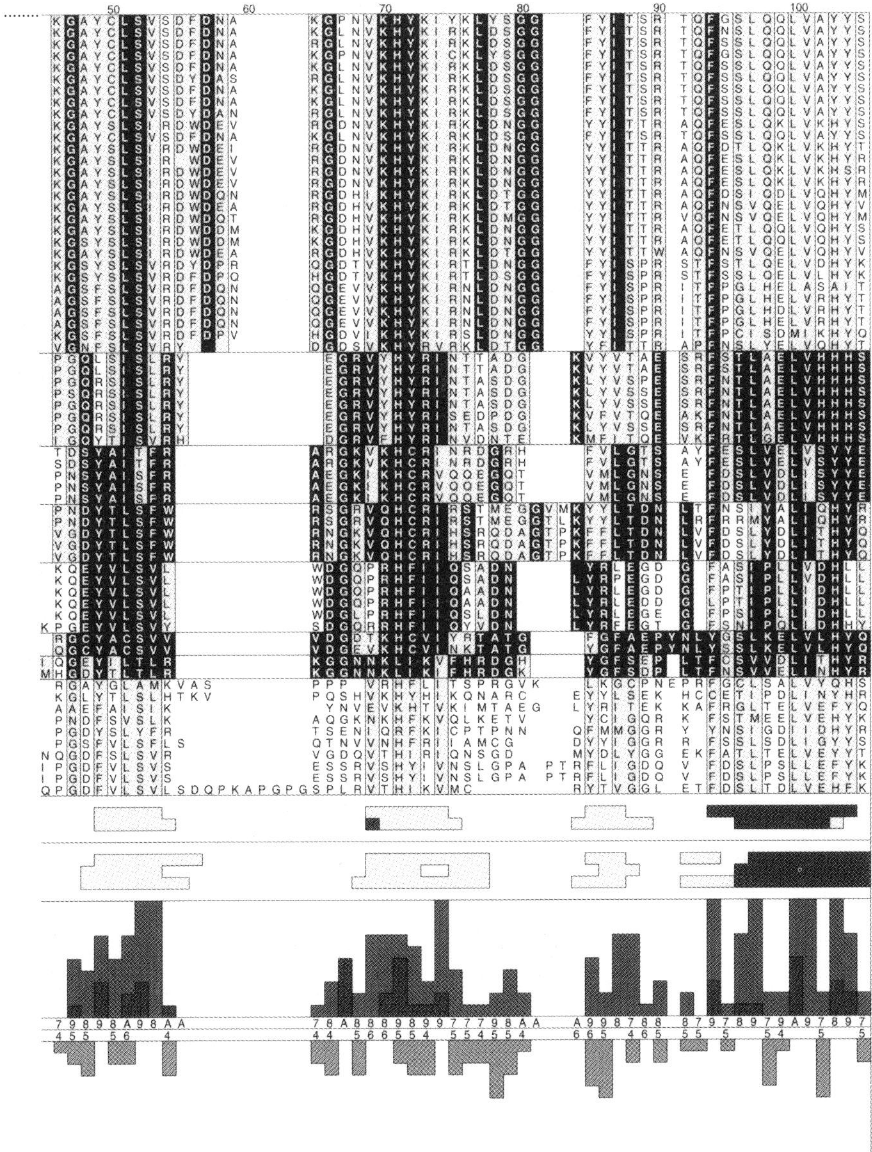

conservation is summarized by colour coding the alignment (shown here as grey shading). Pair conservation is summarized as a histogram. The histogram helps to locate conservation patterns characteristic of α-helix and β-strand. For full details see ref. 53. For details of how to obtain AMAS and other programs from the author's group please download the file README from `geoff.biop.ox.ac.uk`.

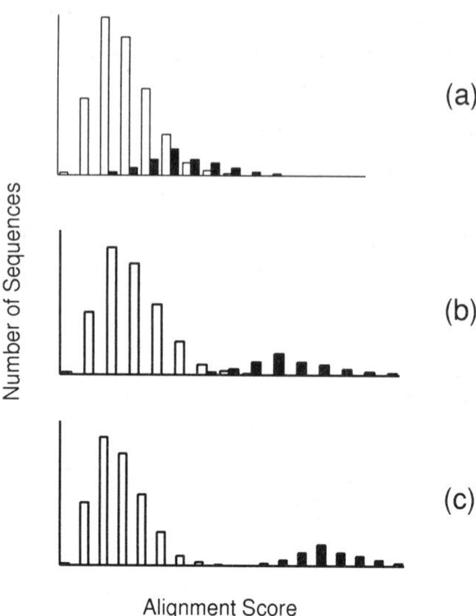

Figure 8. Schematic representation of typical alignment score distributions resulting from a database scan. The black bars represent proteins that are known to be similar to the query sequence, the white bars are not related to the query. (a) Shows a scan that does not discriminate well, (c) shows perfect discrimination, while (b) is the more usual intermediary result of database searching.

database of the size used and the query length. However, regardless of the method of ranking, there are nearly always some proteins giving scores in the overlap region that in fact are structurally related to the query. In practice, since no method succeeds for all protein queries, the aim is to minimize the overlap and ensure that potentially interesting similarities are scored high enough that they will be noticed by the user. Of course, what constitutes an 'interesting' match is dependent upon the subjective biological context of the query.

6.2 Time considerations

In the early days of database scanning, the computer time required to execute the scan was a major consideration. Today, the ready availability of cheap, high performance computers means that computer resources are rarely a limiting factor. In the early 1980s computers with sufficient memory and processor speed to compare a query to a database using dynamic programming were expensive shared resources. Over the last ten years, the speed of typical institutional computers has increased by a factor of 70 while the sequence database has only grown by a factor of nine. This disparity coupled with the dramatic fall in the cost of computing means that it is currently feasible to perform protein database scans in a few hours on a personal computer using dynamic programming algorithms (57).

For occasional use, high scanning speed is not essential. After all, if it has taken months to obtain the sequence data, what is a few hours to check for similarities? However, much greater speed is helpful when providing a national or regional database scanning service and when carrying out analyses that require very large numbers of sequences to be compared. For example, the comparison of a 25 000 sequence database to itself would require 4.5 months using dynamic programming on a typical workstation (57). The algorithms discussed in Sections 6.6 and 6.7 that make approximations, or implementations on specialist hardware may reduce this time by a factor of 10–100.

6.3 Which database should I search? Local or network?

The answer to this question depends a little on why you are performing the search. If you have just determined a new sequence then it is essential that you search the most recent and up to date databases available to test if your new protein is unique. Having the most up to date database is less important if the aim is to gather a well-known family of proteins together for multiple alignment as an aid to modelling.

The nucleic acid and protein sequence databases are collated by EMBL in Europe, the NCBI in the USA, and the DDBJ in Japan. In addition, the NBRF in the USA also provide a database of nucleic acid and protein sequences. The databases are distributed in CD-ROM by EMBL, NCBI, and NBRF organizations and if you require a local database to scan, this is the preferred method of obtaining it. Some of the database distributions include software for searching the databases (e.g. NBRF-ATLAS program). The disks are normally updated every three months but since over 1000 new protein sequences are deposited per month, even the current disk is out of date as soon as it arrives! To overcome this problem, the database providers also maintain daily or weekly updates to the databases since the last CD release. If searching with a newly-determined sequence one should ideally scan a database that includes all available sequences up to today and if nothing is found, periodically rescan the updated database. Maintaining the regular updates of the sequence databases is usually beyond the scope of an individual investigator, however major data centres do maintain such updated databases and software for searching them. Indeed, providing you have e-mail access to the Internet and are prepared to accept the scanning tools provided by the database centre, then there is no compelling reason for maintaining the databases locally. However, while network access to a database may provide the most up to date version of the data, it does not necessarily give the most effective scanning method for your sequence.

6.4 Searching with dynamic programming

Dynamic programming requires a matrix of pair scores and a gap penalty and will return the best score for aligning the two sequences (see Section 3.2).

Both local and global alignment methods may be applied to database scanning, but local alignment methods are more useful since they do not make the assumption that the query protein and database sequence are of similar length.

Although it is feasible to use dynamic programming to search databases on a desktop computer (Section 6.2) the technique has not generally been adopted for database searching. This is mainly because fast implementations of the Smith–Waterman and similar algorithms (33) have not been widely available until recently.

6.4.1 Scanning with parallel computers Prosrch, MPsrch, and others

Collins *et al.* (15,58) are responsible for much of the early work on scanning sequence databases with dynamic programming. They implemented a variant of the Smith and Waterman (30) algorithm on the parallel AMT-DAP computer. This provided them with sufficient processing speed to not only record the top scoring local alignment between the query and each sequence, but also to record alternative local alignments. As such, the DAP implementation of local alignment with gaps is currently the only program to provide this service on the Internet (send the text HELP to `dapmail@biocomp.ed.ac.uk`). Collins *et al.* (15,59) also provide a method to estimate the statistical significance of their alignments. They fit a straight line to log(N) where N is the number of alignments with a given score versus score to the lower 97% of the top 16 384 alignments, then express the score for alignment as a probability derived from the distance from this line. This scoring method was used since there is currently no formal statistical method of estimating the expected score for a local alignment with gaps. The Collins *et al.* approach provides a convenient way of correcting for changes in the score distribution for unrelated alignments due to differences in composition and length of the query sequence database.

A development of Collins' work is a parallel Smith–Waterman implementation for the MasPar range of massively parallel computers (60). Scans can be made using this program using a service at EMBL Heidelberg (send the text HELP to `BLITZ@embl-heidelberg.de`). Unfortunately, the BLITZ service currently only returns a single top scoring alignment, but like the DAP program it gives an estimate of the alignment significance. A further Smith–Waterman implementation (BLAZE) has been developed by Intelligenetics and is commercially available for the MasPar. The GenQuest system at Oak Ridge National Laboratory, USA, also supports database searching with the Smith–Waterman algorithm using a specialized parallel computing environment (send the text HELP to `grail@ornl.gov` for instructions).

6.5 Index methods

6.5.1 Simple index for identical matching

Indexing has long been used for identifying identical ungapped regions in sequences. For example the SCAN facility in the PSQ and ATLAS programs distributed with the NBRF-PIR data bank allows the rapid identification of short identical strings (61). This is achieved by pre-processing the entire data bank once to identify the locations of all unique tripeptides. These data are stored in a direct access file together with pointers to the sequence identifier codes. The query peptide is also divided into a series of tripeptides and identification of the sequence in the data bank then becomes a simple matter of looking up the starting positions of each peptide in the list held on file. There is a tradeoff with indexing methods between the time and space taken to build and store the index and the number of queries expected. Search times are usually very fast and involve a few disk accesses, the drawback with simple indexes is that they are restricted to exact matching without gaps.

6.5.2 Indexing with gaps—the FLASH algorithm

Recently, Rigoutsos and Califano of the IBM T. J. Watson Research Center have extended the idea of indexing to allow for gaps and mismatches (62). The indexes, or look up tables are highly redundant and based on a probabilistic model. As a consequence the index files are very large and the problem is less one of absolute CPU speed, and more a question of fast disk access. For example, the index for SWISS-PROT release 25 requires 2.8 GBytes of disk space (I. Rigoutsos personal communication). However, the Rigoutsos and Califano FLASH algorithm permits very rapid scans to be performed on protein databases with claimed sensitivity and accuracy close to dynamic programming. The algorithm has been implemented on a network of seven non-dedicated RISC workstations which provides sufficient speed to service a searching facility *via* e-mail (send the text SEND HELP to `dflash@watson,ibm.com` for information). As databases grow in size with a large amount unchanging, and the cost of disk storage falls, it seems likely that indexing techniques will become increasingly important methods of searching.

6.6 Approximations: the FASTP and FASTA algorithm

The early personal computers had insufficient memory and were too slow to carry out a database scan using dynamic programming. Accordingly, Wilbur and Lipman (63) developed a fast procedure for DNA scans that in concept searches for the most significant diagonals in a dot plot. The initial step in the algorithm is to identify all exact matches of length k (k-tuples) or greater between the two sequences. Speed is achieved by employing a look up procedure. For example, for proteins, if $k = 3$ then there are 8000 (20^3) possible k-tuples and each element of an array C of length 8000 is set to represent one of these k-tuples. Sequence A is scanned once and the location of each k-tuple in A is recorded in the corresponding element of C. Sequence B is then scanned and by reference to C the location of all k-tuple matches common to A and B may be identified. If two k-tuples are present on the same diagonal then the difference between their starting position (offset) is also the same, thus the diagonals with the most significant number of matches may be identified. Since runs of identity are relatively rare even between related proteins, Lipman and Pearson (64) first identified the five diagonals of highest similarity with *k* set to 1, or 2. They then applied Dayhoff's scoring scheme (Section 2.4.1) to the amino acid pairs over these regions. The region giving the highest score for the protein comparison was used to rank order the sequences located in the data bank for further study by more rigorous procedures. Pearson and Lipman (65) have refined these ideas in the program FASTA. FASTA saves the ten highest regions of identity which are then rescored with the PAM250 matrix (see Section 2.4.1). If there are several initial regions above a preset cut-off score then those that could form a longer alignment are joined, allowing for gaps and a score *initn* is calculated by subtracting a penalty for each gap. *initn* is used to rank the database sequences by similarity. Finally, dynamic programming is used over a narrow region of the high scoring diagonal to produce an alignment with score *opt*. These steps are illustrated in *Figure 9*.

FASTA only shows the top scoring region, it does not locate all high scoring alignments between two sequences. As a consequence FASTA may not identify directly repeats or multiple domains that are shared between two proteins. The FASTA software can be obtained by anonymous ftp from `virginia.edu` and a number of sites offer searching facilities with FASTA (e.g. EMBL).

6.7 Approximations: BLAST basic local alignment search tool

BLAST (18) is a heuristic method to find the highest scoring locally optimal alignments between a query sequence and a database. The important simplification that BLAST makes is not to allow gaps, but the algorithm does allow multiple hits to the same sequence. The BLAST algorithm and family of programs rely on work on the statistics of ungapped sequence alignments by Karlin and Altschul (66). The statistics allow the probability of obtaining an ungapped alignment (MSP—maximal segment pair) with a particular score to be estimated. The BLAST algorithm permits nearly all MSPs above a cut-off to be located efficiently in a database.

The algorithm operates in three steps:

1. For a given word length w (usually three for proteins) and score matrix (see Section 2) a list of all words (w-mers) that can score $> T$ when compared to w-mers from the query is created.
2. The database is searched using the list of w-mers to find the corresponding w-mers in the database (hits).
3. Each hit is extended to determine if an MSP that includes the w-mer scores $> S$, the preset threshold score for an MSP. Since pair score matrices

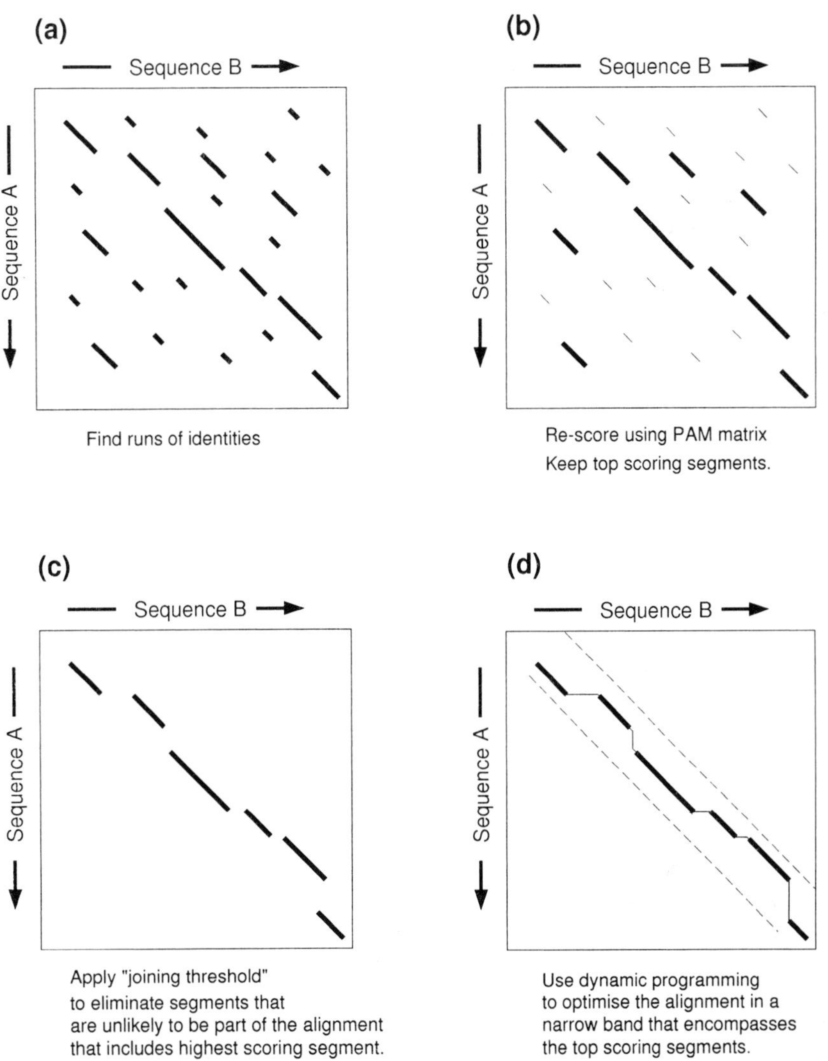

Figure 9. Summary of the steps in the FASTA sequence comparison program.

typically include negative values, extension of the initial *w*-mer hit may increase or decrease the score. Accordingly, a parameter *X* defines how great an extension will be tried in an attempt to raise the score above *S*.

The steps involved in the BLAST algorithm are illustrated in *Figure 10*.

A low value for *T* reduces the possibility of missing MSPs with the required *S* score, however lower *T* values also increase the size of the hit list generated in step 2 and hence the execution time and memory required. In practice, the BLAST program used for protein searches sets compromise

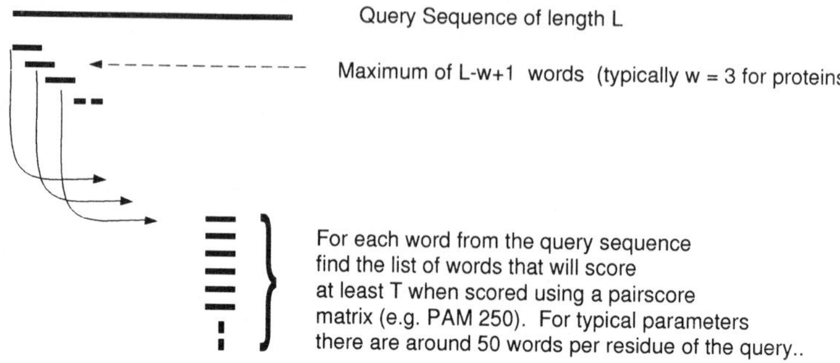

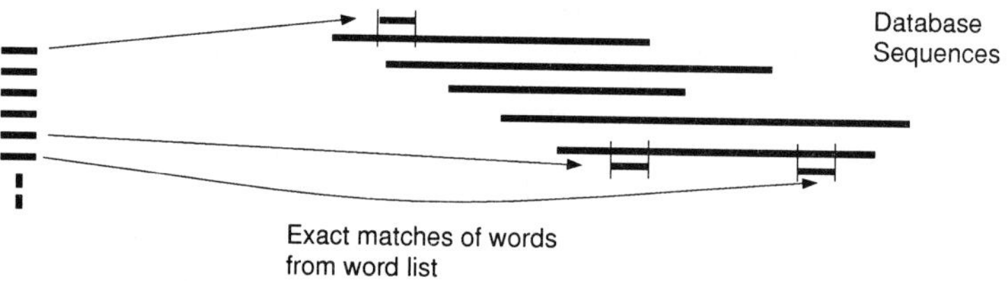

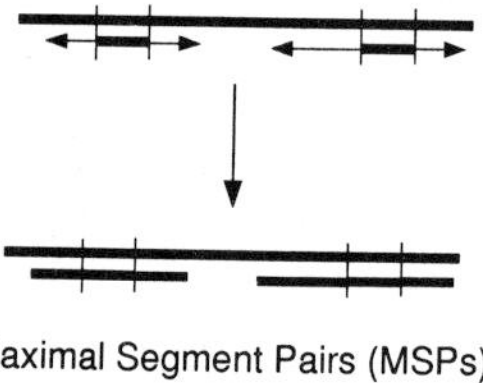

Figure 10. BLAST algorithm.

values of T and X to balance the processor requirements and sensitivity.

BLAST is unlikely to be as sensitive for all protein searches as a full dynamic programming algorithm. However, the underlying statistics provide a direct estimate of the significance of any match found. The program was developed at the NCBI and benefits from strong technical support and continuing refinement. For example, filters have recently been developed to exclude automatically regions of the query sequence that have low compositional complexity, or short periodicity internal repeats. The presence of such sequences can yield extremely large numbers of statistically significant but biologically uninteresting MSPs. For example, searching with a sequence that contains a long section of hydrophobic residues will find many proteins with transmembrane helices.

BLAST runs on virtually all types of computer (the program may be obtained by anonymous ftp from `ncbi.nlm.nih.gov`). Parallel processing is also supported on multiprocessor computers such as the 4 or 8-processor Silicon Graphics POWER series. Searches using BLAST may be conducted by e-mail at NCBI (send the text HELP to `blast@ncbi.nlm.nih.gov` for instructions). Recently, specialist hardware has been developed to implement the BLAST algorithm at even higher speed. A network service based on these developments is available from the University of North Carolina at Chapel Hill (send the text HELP on the subject line to `bioscan@cs.unc.edu`).

6.8 Guidelines for database scanning

Which is the best method for database scanning? Sadly, there is not a straightforward answer to this question. Attempts have been made to make comparisons but the process is complicated by the difficulty of designing suitable test cases and the number of adjustable parameters. The most effective method of assessing the success of a scanning technique is to test its ability to find all the members of a known protein family from the database of all known sequences (67,68). The principle is simple.

- record the identifier codes of all proteins known to be in the family
- select a member to scan with (the query)
- perform the scan using the method of choice
- count how many of the known members are found with higher scores than known non-members

A less strict criterion is to count the number of members that score as high as the top 0.5% of the non-members in the data bank (68). The best scanning method will give the most members before non-members, i.e. will have the fewest false positives. of course, evaluation is not as simple as this appears. First one must choose well-characterized protein families with which to test. Do we really know all the members? A high scoring non-member may in fact be a previously undiscovered family member. Further difficulties arise for scans where there are many false negatives. If two methods both miss 30 known members, are they missing the same 30? Ideally, evaluation should also explore alternative parameter combinations, but this greatly increases the number of tests that need to be done and complicates the data analysis. For example, if we consider scanning with dynamic programming, then there is a choice of pair-score matrix and gap penalty, local or global alignment.

The best gap penalty depends on the matrix in use. If both length-dependent and independent penalties are used, then the number of alternative combinations increases dramatically. The best combination of matrix and penalty may not be appropriate for other algorithms. BLAST does not consider gaps, so the situation is a little easier and this feature was exploited by Henikoff and Henikoff to evaluate different substitution matrices (16), however we still have the choice of other parameters special to the BLAST algorithm.

When given a newly-determined sequence, a search with BLAST or FASTA will quickly tell you if a close homologue exists. Although a scan with full dynamic programming takes longer on a local workstation, the turn-round time from e-mail servers such as BLITZ are similar to BLAST searches at NCBI. Accordingly, it is worth scanning one of these services as well. If no similar sequences are found then alternative PAM matrices should be tried. Start with PAM120, then try PAM250, and in each case vary the gap penalty around the minimum value of the matrix. For PAM250 this is eight, values of seven to ten are worth trying. Care should always be taken to consider the likely significance of an apparent match. The methods for predicting the accuracy of alignment are discussed in Section 4.1.

7. Summary

In the early years of sequence searching, only a few specialized centres had access to the necessary computing facilities and programming expertise to perform the scans. In the early–mid 1980s, the availability of personal computers and software that could perform useful analyses on them (e.g. FASTP) meant that it was normally most efficient for searches to be performed locally. Today, the optimum choice is again swinging towards databases maintained at a few centres, but now fast networks and windowing workstations allow the user to use software locally and be unaware that the search is being carried out on a computer in another country. Perhaps the best example of this to date is the Entrez software (69) available from the U.S. NCBI (ask the information from `info@ncbi.nlm.nih.gov`). Entrez provides a windowing interface to a database that integrates the nucleotide and protein sequence databases with associated references and abstracts. Entrez will either use the database on CD-ROM or alternatively, with suitable network connection can interrogate the master database at the NCBI in Washington. While Entrez does not provide searching facilities for a new sequence it stores pre-computed similarities between pairs of sequences in the database. Thus, one can quickly navigate between a protein name, the sequence, its close homologues, the corresponding DNA sequence, and all relevant publications. Network Entrez was heavily used when compiling this chapter!

The advantages of centralized databases for the user are:

(a) That he need only have a comparatively low powered computer and network connection.

(b) The database centres can keep the database up to date far more effectively than the individual investigator.

The centres can provide access to a range of software to interrogate the data, either as a simple text search (e.g. find all entries with the word 'kinase') or using sequence comparison algorithms. They can update the software as new algorithms become available.

The drawback with a centralized service is that one has to accept the service providers view of the best way to perform the search. However, with more database centres giving public access to search facilities every year there is an increasing choice of algorithms available.

References

1. Dayhoff, M. O., Schwartz, R. M., and Orcutt, B. C. (1978). In *Atlas of protein sequence and structure* (ed. M. O. Dayhoff), Vol. 5, pp. 345–58. National Biomedical Research Foundation, Washington DC.
2. Schwartz, R. M. and Dayhoff, M. O. (1978). In *Atlas of protein sequence and structure* (ed. M. O. Dayhoff), Vol. 5, pp. 353–62. National Biomedical Research Foundation, Washington DC.
3. Feng, D. F., Johnson, M. S., and Doolittle, R. F. (1985). *J. Mol. Evol.*, **21**, 112.
4. Fitch, W. M. (1966). *J. Mol. Biol.*, **16**, 9.
5. Cohen, F. E., Novotny, J., Sternberg, M. J. E., Campbell, D. G., and Williams, A. F. (1981). *Biochem. J.*, **195**, 31.
6. McLachlan, A. D. (1972). *J. Mol. Biol.*, **64**, 417.
7. Jones, D. T., Taylor, W. R., and Thornton, J. M. (1992). *Comput. Appl. Biosci.*, **8**, 275.
8. Taylor, W. R. (1986). *J. Theor. Biol.*, **119**, 205.
9. Henikoff, S. and Henikoff, J. G. (1992). *Proc. Natl. Acad. Sci. USA*, **89**, 10915.
10. Risler, J. L., Delorme, M. O., Delacroix, H., and Henaut, A. (1988). *J. Mol. Biol.*, **204**, 1019.
11. Overington, J., Johnson, M. S., Sali, A., and Blundell, T. L. (1990). *Proc. R. Soc. Lond. Ser. B*, **241**, 132.
12. Bowie, J. U., Luthy, R., and Eisenberg, D. (1991). *Science*, **253**, 164.
13. Needleman, S. B. and Wunsch, C. D. (1970). *J. Mol. Biol.*, **48**, 443.
14. Altschul, S. (1991). *J. Mol. Biol.*, **219**, 555.
15. Collins, J. F., Coulson, A. F. W., and Lyall, A. (1988). *Comput. Appl. Biosci.*, **4**, 67.
16. Henikoff, S. and Henikoff, J. G. (1993). *Proteins: Struct. Funct. Genet.*, **17**, 49.
17. Gonnet, G. H., Cohen, M. A., and Benner, S. A. (1992). *Science*, **256**, 1443.
18. Altschul, S. F., Gish, W., Miller, W., Myers, E. W., and Lipman, D. J. (1990). *J. Mol. Biol.*, **215**, 403.
19. Gibbs, A. J. and McIntyre, G. A. (1970). *Eur. J. Biochem.*, **16**, 1.
20. Collins, J. F. and Coulson, A. F. W. (1987). In *Nucleic acid and protein sequence analysis: a practical approach* (ed. M. J. Bishop and C. J. Rawlings), pp. 323–58. IRL Press, Oxford.

21. Jones, D. D. (1975). *J. Theor. Biol.*, **50**, 167.
22. Chou, P. Y. and Fasman, G. D. (1978). *Adv. Enzymol.*, **47**, 45.
23. Kubota, Y., Nishikawa, K., Takahashi, S., and Ooi, T. (1982). *Biochim. Biophys. Acta*, **701**, 242.
24. Argos, P. (1987). *J. Mol. Biol.*, **193**, 385.
25. Kruskal, J. B. (1983). In *Time warps, string edits and macromolecules: the theory and practice of sequence comparison* (ed. D. Sankoff and J. B. Kruskal), pp. 1–44. Addison Wesley.
26. Sankoff, D., and Kruskal, J. B. (ed.) (1983). *Time warps, string edits and macromolecules: the theory and practice of sequence comparison.* Addison Wesley.
27. Waterman, M. S., Smith, T. F., and Beyer, W. A. (1976). *Adv. Math.*, **20**, 367.
28. Gotoh, O. (1982). *J. Mol. Biol.*, **162**, 705.
29. Sellers, P. H. (1984). *Bull. Math. Biol.*, **46**, 501.
30. Smith, T. F. and Waterman, M. S. (1981). *J. Mol. Biol.*, **147**, 195.
31. Boswell, D. R. and McLachlan, A. D. (1984). *Nucleic Acids Res.*, **12**, 457.
32. Waterman, M. S. and Eggert, M. (1987). *J. Mol. Biol.*, **197**, 723.
33. Barton, G. J. (1993). *Comput. Appl. Biosci.*, **9**, 729.
34. Taylor, W. and Orengo, C. (1989). *J. Mol. Biol.*, **208**, 1.
35. Sali, A. and Blundell, T. L. (1990). *J. Mol. Biol.*, **212**, 403.
36. Russell, R. B. and Barton, G. J. (1992). *Proteins: Struct. Funct. Genet.*, **14**, 309.
37. Barton, G. J. and Sternberg, M. J. E. (1987). *Protein Eng.*, **1**, 89.
38. Barton, G. J. (1990). In *Methods in enzymology* (ed. R. Doolittle), Vol. 183, pp. 403–28. Academic Press.
39. Sander, C. and Schneider, R. (1991). *Proteins: Struct. Funct. Genet.*, **9**, 56.
40. Vingron, M. and Argos, P. (1990). *Protein Eng.*, **3**, 565.
41. Zuker, M. (1991). *J. Mol. Biol.*, **221**, 403.
42. Saqi, M. A. and Sternberg, M. J. (1991). *J. Mol. Biol.*, **219**, 727.
43. Lesk, A. M., Levitt, M., and Chothia, C. (1986). *Protein Eng.*, **1**, 77.
44. Murata, M., Richardson, J. S., and Sussman, J. L. (1985). *Proc. Natl. Acad. Sci. USA*, **82**, 3073.
45. Sankoff, D., Cedergren, R. J., and Lapalme, G. (1976). *J. Mol. Evol.*, **7**, 133.
46. Barton, G. J. and Sternberg, M. J. E. (1987). *J. Mol. Biol.*, **198**, 327.
47. Feng, D. F. and Doolittle, R. F. (1987). *J. Mol. Evol.*, **25**, 351.
48. Taylor, W. R. (1990). In *Methods in enzymology* (ed. R. Doolittle), Vol. 183, pp. 456–74. Academic Press.
49. Higgins, D. G. and Sharp, P. M. (1989). *Comput. Appl. Biosci.*, **5**, 151.
50. Bacon, D. J. and Anderson, W. F. (1986). *J. Mol. Biol.*, **191**, 153.
51. Johnson, M. S. and Doolittle, R. F. (1986). *J. Mol. Evol.*, **23**, 267.
52. Schuler, G. D., Altschul, S. F., and Lipman, D. J. (1991). *Proteins: Struct. Funct. Genet.*, **9**, 180.
53. Livingstone, C. D. and Barton, G. J. (1993). *Comput. Appl. Biosci.*, **9**, 745.
54. Barton, G. J., Newman, R. H., Freemont, P. F., and Crumpton, M. J. (1991). *Eur. J. Biochem.*, **198**, 749.
55. Russell, R. B., Breed, J., and Barton, G. J. (1992). *FEBS Lett.*, **304**, 15.
56. Barton, G. J. and Russell, R. B. (1993). *Nature (London)*, **361**, 505.
57. Barton, G. J. (1992). *Science*, **257**, 1609.
58. Coulson, A. F. W., Collins, J. F., and Lyall, A. (1987). *Comput. J.*, **30**, 420.

59. Collins, J. F. and Coulson, A. W. F. (1990). In *Methods in enzymology* (ed. R. Doolittle), Vol. 183, pp. 474–87. Academic Press.
60. Sturrock, S. S. and Collins, J. F. (1993). MPsrch version 1.3.
61. Orcutt, B. C. and Barker, W. C. (1984). *Bull. Math. Biol.*, **46**, 545.
62. Rigoutsos, I. and Califano, A. (1993). In Proceedings of the First International Conference on Intelligent Systems for Molecular Biology.
63. Wilbur, W. J. and Lipman, D. J. (1983). *Proc. Natl. Acad. Sci. USA*, **80**, 726.
64. Lipman, D. J. and Pearson, W. R. (1985). *Science*, **227**, 1435.
65. Pearson, W. R. and Lipman, D. J. (1988). *Proc. Natl. Acad. Sci. USA*, **85**, 2444.
66. Karlin, S. and Altschul, S. F. (1990). *Proc. Natl. Acad. Sci. USA*, **87**, 2264.
67. Barton, G. J. and Sternberg, M. J. E. (1990). *J. Mol. Biol.*, **212**, 389.
68. Pearson, W. (1991). *Genomics*, **11**, 635.
69. Entrez: Sequences Bldg. 38A, NIH, Bethesda, MD 20894, USA. (1992).

3

Identification of protein sequence motifs

MANSOOR A. S. SAQI

1. Introduction

One of the main aims of computational biology is to address the question: "Given a newly-determined sequence what can be said about the structure and function of the corresponding protein". If a similarity search (1–3) against the sequence database reveals a global similarity with another sequence of known function, much can immediately be inferred (see Chapter 2). If the similarity is with a sequence of known three-dimensional structure homology modelling studies can be initiated (4) (see Chapter 6).

Standard homology searches will fail however, if the relationships between proteins is characterized by the common occurrence of only a few critical residues. Such residues taken together specify a pattern or signature and groups of functionally related sequences will share similar patterns. The residues that make up the pattern do not have to be contiguous in sequence as a functional site may be composed of side chains in close proximity in the three-dimensional structure but far apart in sequence. Structural constraints mean that each position in the pattern may be specified by a single residue or one of a number of similar residues. As the patterns are not fixed substrings some positions will accommodate any residue. Methods of finding such patterns among groups of functionally related proteins are important as the motifs extracted provide insight into the role of a protein and its relationship with other sequences and serve as useful diagnostic tools.

An analysis of yeast chromosome III (5,6) showed that only 14% of the proteins had similarity to proteins of known three-dimensional structure (see Chapter 1). The use of sensitive sequence analysis methods and database searches suggested that a known or probable function could be ascribed to a further 28% of the proteins. It is interesting to speculate that, with the development of algorithms to detect more subtle sequence signals, the function of a significant proportion of the remaining 58% might be revealed.

This chapter reviews methods for the automatic identification of a given sequence motif in a sequence or set of sequences. Such methods require a library of motifs or patterns and computer programs which efficiently carry out the search.

How are motifs that characterize a family of protein sequences constructed? This is a complex problem and is often carried out by inspection. Attempts have been made to automate this procedure and automatic methods for pattern extraction will also be reviewed. The term pattern extraction is used for finding patterns of conserved residues among groups of functionally related proteins. The detection of these subtle sequence signals is a major challenge for molecular biology.

2. Identification of sequence motifs

2.1 The Prosite database

The Prosite database (7) is a compilation of sites and patterns found in protein sequences. There are two files associated with Prosite, namely `prosite.doc` and `prosite.dat`. The patterns in Prosite are extremely well documented (`prosite.doc`) and the pattern file itself (`prosite.dat`) is a file containing the patterns and associated information which facilitates the use of progams that scan sequences or sequence databases with patterns. The syntax adopted in Prosite is as follows:

- each position in the motif is separated by a hyphen
- one character denotes a specified residue is allowed at that position
- {...} denotes a set of disallowed residues
- [...] denotes a set of allowed residues
- (*n*) denotes a repeat of *n*
- (*n,m*) denotes a repeat of between *n* and *m* inclusive

For example, the well-known ATP/GTP binding site motif, (P-loop) (8) is described by [AG]-X(4)-G-K-[ST]. Only half the positions are fixed. Release 10.1 of PROSITE (April 1993) contained 635 documentation entries describing 803 different patterns.

2.2 Matching motifs against sequences

Several computer programs exist which look for the occurrence of a given motif in a sequence or set of sequences.

Cockwell and Giles (9) describe two programs that scan sequences for matches to user-defined motifs and patterns respectively. Their definition of a motif is a string of specified residues, or a group of fixed residues together with sets of allowed or disallowed residues. Their syntax is somewhat similar to the current Prosite syntax with square brackets [...] enclosing a set of allowed residues and a caret ^ preceding a character or set of characters meaning that character or set is disallowed. A match to any residue is denoted by 'x', 'X', or '–'. In addition a dot (.) as the first or last character restricts the search to the N- or C-terminus of the sequence. A pattern is defined as a collection of separate motifs separated by variable gaps. The pattern finding program allows a maximum of five motifs in a pattern.

The algorithm is based on representing a motif by a set membership matrix where rows correspond to the type of residue and the columns to positions in the motif. Acceptable residues at a given motif position are given a score of 1 in the matrix and 0 otherwise. A sliding window from the query sequence of length equal to that of the motif is matched against the set membership matrix.

The algorithm of Cockwell and Giles in employed in MacPattern (10). MacPattern searches protein sequences for instances of Prosite (7) motifs. Any database which conforms to the Prosite pattern syntax is also accepted. MacPattern uses the graphical user interface of the Apple Macintosh computers. Although the input sequence files to be scanned for patterns can contain many sequences the program is not intended to search entire protein sequence data banks.

When searching sequence databases for an instance of a motif or pattern a given number of hits will be obtained by chance and the number of such hits will depend on the size and amino acid composition of the database. The Prosite database gives the number of incorrect matches to a given motif and these false positives have been used to establish criteria for assessing the significance of a motif. A classification of Prosite motifs according to the calculated expected number of random matches (E) based on the SWISS-PROT14 database, and the observed number of matches, has been carried out by Sternberg (11). 210 motifs in Prosite had a value of E less than 0.5. Out of these 201 corresponded to cases with no false positives. This suggests that there is a 95% (201/210) chance that a motif is not a false positive if the calculated expected number of random matches is less than 0.5. In such a case it is likely that the motif detects a biologically meaningful pattern.

Motif statistics are reported in the program PROMOT (12). PROMOT matches a sequence against the Prosite pattern database or against a set of user-defined patterns. It will also scan a Prosite or user-defined pattern against a protein sequence database. The probability of a motif matching by chance is computed using the method of Staden (13). The program allows some flexibility in defining the search. Selected motifs from Prosite can be specified by their accession number and the program can be run interactively or in batch which is useful for large database scans.

Other motif identification programs can be found as part of larger packages. Scrutineer (14) is a Pascal program that can search for motifs and patterns. Other constraints can also be specified such as restricting the search to particular regions within each sequence and including predicted secondary structure and chemical information (e.g. hydrophobicity) in the pattern specification. The GCG package (15) includes a routine (*motifs*) that searches for Prosite motifs in a protein sequence, allowing mismatches if desired.

Searching for instances of a particular pattern may miss functionally related sequences if the pattern does not adequately describe all sequences with that function, for example if the pattern is too specific. However, if a pattern is too general, or a large number of mismatches are tolerated, many false positives will be found. The use of weight matrix scoring (16) allows for conservative substitutions and can increase sensitivity.

For a sequence of known structure it is sometimes instructive to visualize a motifs position in the protein structure. The motifs instance can then be classified as functional, structural, or coincidental (a false positive). PdbMotif (17) is a program that automatically identifies Prosite or user-defined motifs in a protein data bank file and produces a script file for input to the molecular rendering program RasMol (18). An optional parameter (*-strict* threshold) filters out motifs that occur with a probability greater than a user-defined threshold.

3. Pattern extraction

The patterns in Prosite have been compiled in a non-automatic fashion generally by inspection. Given the rate at which new sequences are being determined there is a need for automatic methods to extract patterns from primary sequence information.

The problem of extracting a diagnostic pattern is complex. X-ray structural studies of proteins with similar function can indicate the binding site and hence lead to a knowledge of the amino acid elements which are essential for function. Due to the limited number of proteins with known structure, the primary sequence alone is generally used to extract patterns. Conserved regions of sequence are often pointers to functionally important residues. A particular family of proteins which carry out a given general function (e.g. kinase) may have associated with it a variety of motifs. Such proteins do not have any obvious homology but may only have in common short regions of sequence around the motif. These short regions themselves are not exact repeats but will have a given number of positions conserved. It is necessary to be able to extract these motifs automatically and several previous approaches to this problem will be reviewed (see *Table 1*).

3.1 Global sequence alignment methods

The traditional approach to pattern extraction is to use multiple sequence alignment methods (19–21). From a multiple sequence alignment features can be recognized as regions with greater than average homology and consensus patterns can be constructed. A portion of the alignment can be used to construct a profile or position-dependent scoring table (22,23). This can be

Table 1. Summary of motif extraction methods

Methods	**Comments**
Multiple sequence alignment	Good first step in locating motifs but will fail to locate more subtle patterns
Pattern induced multi sequence alignment (PIMA) (24,25)	Finds patterns of conserved residues far apart in sequence
SAPS (28)	Unusual sequence features easily recognized
Information theory (34,35)	Fast but results may depend on the order in which the sequences are presented
Expectation maximization (39–41)	A variety of motif models can be investigated
Unique peptide words (42–46)	A dictionary of peptide words characteristic of protein families can provide a fast method for prediction of function
Smith algorithm (48)	Fast automated method for motif extraction
Saqi and Sternberg algorithm (53)	Identifies and ranks significant motifs but slow run time
Gibbs sampler (54)	Stochastic EM method; fast; prevents solutions becoming trapped in local minima

used along with dynamic programming algorithms to search for similar patterns in sequence databases. The profile approach is more applicable to the detection of common overall folds rather than searching for common local regions. Multiple sequence methods are best suited to limited sets of related proteins. They can also be sensitive to choice of gap penalty parameters and similarity scoring matrix.

As an example consider a multiple sequence alignment of a set of DNA ligase sequences. *Figure 1* shows part of an alignment performed with the *pileup* program within the GCG package (15). A consensus sequence is then obtained using the GCG program *pretty* and requiring at least 10 out of the 15 characters at each position to be identical. The known (Prosite) motifs for this group of sequences are:

- [EDQH]-x-K-x-[DN]-G-x-R-[GACV]
- E-G-[LIVMA]-[LIVM](2)-[KR]-x(5,8)-[YW]-[QEK]-x(2,6)-[KRH]-x(3,5)-K-[LIVMFY]-K

In this case *pileup* performs well in aligning the known motifs correctly but other regions of conservation also exist and we are challenged to identify the significant patterns.

Smith and Smith (24,25) have developed a method to extract diagnostic patterns from sets of related sequences using locally optimal pairwise alignments. Their approach involves first carrying out all pairwise comparisons between members of a set of related sequences. The similarity scores are then clustered. Based on the resulting dendrogram, a series of progressive

```
Plurality: 10.00  Threshold: 1.00  AveWeight 1.00  AveMatch 0.54  AvMisMatch -0.40

                  541                                                             600
L.msf{Dnlj_Ecoli} lgfdidgvvi kvnslaqqeq lgFvaraprw aVafkFpaqe qmtFvrDveF qvgRtgaitp
L.msf{Dnlj_Theth} lpfeadgvvv kldelalwre lgYtaraprf aIaykFpaee ketrllDvvF qvgRtgrvtp
L.msf{Dnli_Zymmo} ldfdidgvvy kldqldwqqr fgFsaraprf alahkFpaek aqttllDiei qvgRtgvltp
 L.msf{Dnli_Bpt4} cgasvsiank vwPglipeqp qMLAssydek gInkni...k fpaFa.qlKa DGaRcfaeVr
 L.msf{Dnli_Bpt6} cgasvsiank vwPglipeqp qMLAssydek gInkni...k fpaFa.qlKa DGaRcfaeVr
L.msf{Dnli_Asfm2} rharqkrgah tnrgmi...p pMLvkyfni. .IpktFfeee tdpiv.hgKr nGvRavacqq
 L.msf{Dnli_Bpt3} .....mnifn tnP....... .fkAvsfves aVkkaLet.. sgyLiaDcKY DGvR..gnIv
 L.msf{Dnli_Bpt7} .....mmnik tnP....... .fkAvsfves aIkkaLdn.. agyLiaEiKY DGvR..gnIc
L.msf{Dnli_Vaccc} ...sisvmtp inP....... .MLAesc..d sVnkaFkkfp sgmFa.EvKY DGeR..vqVh
L.msf{Dnli_Vaccv} ...sisvmtp inP....... .MLAesc..d sVnkaFkkfp sgmFa.EvKY DGeR..vqVh
 L.msf{Dnli_Varv} ...svsvmtp inP....... .MLAesc..d sVnkaFkkfp sgmFa.EvKY DGeR..vqVh
L.msf{Dnli_Schpo} etckltpgip tkP....... .MLAkpt..k qIsevLntfd qaaFtcEyKY DGeR..aqVh
L.msf{Dnli_Yeast} kyctlrpgip lkP....... .MLAkpt..k aInevLdrfq getFtsEyKY DGeR..aqVh
L.msf{Dnli_Human} ehcklspgip lkP....... .MLAhpt..r gIsevLkrfe eaaFtcEyKY DGqR..aqIh
L.msf{Dnli_Desam} knikpqpgip irP....... .MLAerl..s dpaemLskvg niaLv.DyKY DGeR..gqIh
        Consensus ---------- --P------- -MLA------ -I---F---- ---F--E-KY DG-R----V-
```

Figure 1. Part of a multiple sequence alignment of a set of DNA ligase sequences. The alignment was obtained using the GCG programs *pileup* and *pretty* (15). Examination of the consensus sequence can be a first step in locating motifs.

pairwise alignments are performed, starting with the most closely related pair of sequences. At each stage the pairwise alignment is replaced by a common pattern. In this way a root covering pattern is obtained. The common pattern at each step is obtained through the use of amino acid classes which are based on residue chemical similarity. Together with a novel gap penalty function the alogrithm is able to reduce the dendrogram down to a single root covering pattern. The method is powerful in that it finds patterns of conserved residues far apart in sequence. In addition although the root pattern may not be optimal for the complete set of proteins, it is not very sensitive to the initial alignment order.

The problem of motif recognition has also been addressed by comparing dot plots (26). The procedure is less dependent on gap penalty values than conventional multiple alignment methods and has the advantage that suboptimal scoring schemes can be used to calculate the dot plots. A weakness of the method as pointed out by the authors is the failure to find common motifs among distantly related sequences that display weak conservation signals.

3.2 Statistical analyses of sequences

Another approach to the problem is to use statistical techniques to characterize biologically meaningful patterns and relationships (27,28). These methods distinguish the occurrence of particular sequence relationships observed with that expected by chance. Statistical approaches have been used to find long repeats of contiguous pieces of sequence (words) within and between sequences (29,30), the occurrence of long runs of letters of a given type (31), and to analyse clusters of charged residues (32,33).

Patterns of particular biological importance can be characterized by words that occur with a frequency greater than expected by chance among a set of sequences. The distribution of high frequency oligonucleotides within sequences in human, mouse, and rabbit Ig κ DNA is given in Karlin and Ghandour (29). For multiple sequences high frequency words are located using a measure based on cumulative occurrences of the word (of length k) across the sequences. An analysis of high frequency shared words for $k = 6$–9 for Ig κ sequences of human, mouse, and rabbit shows that all high frequency shared words for $k \geq 7$ base pairs relate to a consensus nonamer GGTTTTTGT (29). Statistical methods allow many possible alphabets to be employed in addition to the standard four letter code for nucleotides and 20 letter code for amino acids.

The computer program SAPS (statistical analysis of protein sequences) (28) reports statistically significant features of a protein sequence.

3.3 Information theory approaches

For most motifs however not all positions in the motif are conserved. For example in the ATP/GTP binding site motif (P-loop) (8) [AG]-X-X-X-X-G-K-[ST] motif only half the positions are fixed. This prompts the use of methods that do not rely entirely on the identification of fixed common substrings. An information theory approach to detect patterns in DNA sequences has been developed by Stormo and co-workers (34,35) which addresses this problem. It has been applied to the extraction of protein binding sites from a set of N = 18 unaligned DNA sequences (34). The procedure involves associating an information matrix with an alignment of k-words. The matrix is of size $n*w$, where n is 4 for DNA sequences and w is the size of the k-word. Initially with only one sequence of length L there are $L - k + 1$ matrices. Sequences are added one at a time. Each of the k-words of the initial sequence is compared to each of the k-words of the new sequence and the best match is retained and the matrix updated. When all sequences have been added the total information content of each of the matrices is evaluated. A consensus binding pattern is extracted from the matrix with the highest information content. Because the method does not compare all possible alignments it is computionally fast ($O(L^2N)$ in time). Generally however the order in which the sequences are added may affect the result and in applying the method to a new problem the order of the sequences should be randomized and the analysis repeated to assess the variation in the extracted patterns (34). Other methods for DNA pattern recognition are given by Mengeritsky and Smith (36) and Gallas *et al.* (37) but these require some approximate alignment of the sequences.

The approach of Stormo and co-workers (34,35) has been used to find a new region of sequence similarity in the integrase family of recombination proteins (38). In this study several steps were taken to test that any patterns found were meaningful, namely:

(a) Recognizing that the order in which the sequences are presented may affect the results, three different orderings were used.

(b) Comparisons were performed with randomized sequences with the same amino acid frequency distribution.

(c) Three different *k*-word sizes, namely 10, 15, and 20 residues were used.

A new region containing a conserved arginine was located and mutagenesis experiments confirmed the importance of this residue.

Significant correlations can occur among residues in a sequence (e.g. palindromic patterns) and a statistical technique which takes such effects into account is expectation maximization (39). Expectation maximization (EM) is a two-step iterative process: starting from an initial guess of the location of the pattern (for example that the pattern is equally likely to occur at any position) an information matrix is constructed. Application of this matrix to the sequence data can then lead to a revised estimate of the probabilities for the pattern starting at each position. The use of EM allows a variety of motif models to be explored and the model which is most consistent with the data can be used as a probe for discovering new sequences with similar function (40). For sequences with several motifs the spacing between the patterns is not always conserved. This is the case with the consensus patterns in *E. coli* promoter sites which occur at approximately 35 and 10 bases before the transcription initiation site. An extension to the EM method which handles variable length spacings is described by Cardon and Stormo (41).

3.4 Identification of unique peptide words

Peptide words common to members of a protein family or superfamily are useful for prediction of function. A set of 2581 peptide words, compiled manually, has been compared against the NBRF-PIR protein sequence database (42) and a library of words that are characteristic of particular superfamilies has been compiled (43).

An automatic procedure to extract sequence motifs based on the identification of unique peptide words has been proposed by Ogiwara *et al.* (44). The method involves searching for short oligopeptide patterns that are conserved and found exclusively in one protein group. The order in which these peptide words occur in each sequence is then examined and a peptide sentence constructed. A multiple alignment of sentences leads to a consensus pattern.

One way of extracting a peptide vocabulary for functional classification is to measure the deviation of the frequency of peptide words in a set of protein sequences from that expected in a random set of sequences. Solovyev and Makarova (45) use two measures, namely the Malahonobis distance (46) and the Poisson approximation to do this, and obtain a set of peptide words (a classification set) and a corresponding classification vector containing the frequencies of these words. The comparison of classification vectors can distinguish among various protein families (45).

3.5 Local alignment methods

Any motif or pattern to some extent represents a degree of local homology among a set of sequences. Sometimes a family of sequences may be described by a number of motifs separated by variable regions. The motifs themselves will vary in significance the more significant having a greater number of conserved positions. In an analysis of methyltransferase sequences (47) the location of the most significant pattern among the sequences was used to divide the sequences into two segments. This process was repeated recursively, lowering the stringency criteria for a pattern. The analysis revealed five highly conserved motifs and five less well-conserved patterns.

An alignment of residues in the conserved regions from the sequences making up the family is generally referred to as a block (47). A block does not contain any gaps. A well-conserved motif will have many conserved positions. For the methyltransferase sequences (47) each of the five highly conserved motifs had at least three conserved residues. A stringent motif is a pattern that occurs in all the sequences in the block.

Smith *et al.* (48) have developed an automatic method for locating patterns common to a majority of sequences in a group of proteins and then aligning segments of each sequence with respect to the common motifs. The algorithm defines a motif as three amino acids separated by two distances, of the form a1.d1.a2.d2.a3 where a1, a2, and a3 are the residues and d1 and d2 the distances. It requires three parameters namely the maximum distance value, the number of matches selected as significant, and the maximum number of allowed internal repeats. The algorithm is fast since an array is created that stores all the a1.d1.a2.d2.a3 possible patterns and only one pass through the sequence is needed to update the patterns. Patterns that occur frequently are candidates for motifs. Aligned motifs (blocks) are assigned a score to reflect the degree of conservation. A consensus pattern describing the block is obtained.

Henikoff and Henikoff (49) using an automated application of the Smith algorithm to protein families in the Prosite database have constructed a database of blocks or short regions of ungapped alignment. A collection of best blocks is obtained for each Prosite entry. Searching a database of blocks can provide a sensitive method for characterizing the function of a newly-determined sequence. Analysis of residue frequencies in the BLOCKS database has led to the construction of a new amino acid substitution matrix, BLOSUM (50) which can be used with standard homology searching algorithms (see Chapter 2).

Sheridan and Venkataraghavan (51) have carried out a systematic search for protein signature sequences using the BLAST3 algorithm (52) to find local similarities between all sequences in a non-redundant database of about 6300 sequences. The large number of matches found by BLAST3 were reduced down to an 'interesting' set which were then clustered. The 30 largest clusters are reported. Apart from a few miscellaneous clusters, most contain sequence segments that correspond to known function.

What defines a match between two imperfect substrings? Two words of length k will match if they share r characters, ($r \leq k$). For a given family of sequences, a high frequency word has more matches than expected by chance. This expectation value can be given by the probability of a match multiplied the number of k-words in the database of sequences. The probablity of a match is dependent on the amino acid frequencies in the set of sequences under consideration and the values of k and r. More generally, probability generating functions can be employed (13):

$$F(x) = \Pi_j \Sigma_i f_i .x^{w_{ig}} \qquad i = 1,M; j = 1,J;$$

where the coefficient of x^N is the probability of getting a score of N. M is the sequence character set size (20 for proteins) and J is the length of the motif. f_i are the residue frequencies, j marks the position in the motif ($1 \leq j \leq J$), and w^{ig} are the score matrix weights, where g labels the chatacter index for position j ($1 \leq g \leq M$).

A method to extract patterns based on the extraction and clustering of high frequency k-words is given by Saqi and Sternberg (53). A flow diagram illustrating the basic steps in the method is given in *Figure 2*. *Figure 3* shows the application of the algorithm to the set of DNA ligase sequences in *Figure 1* for values of $(k,r) = (12,5)$ with identity scoring. A number of motifs are recovered the most significant being xx[DE]xKYDGxRx[NQ]. Note the existence of phase shifts of the correct solution.

An important consideration in both the approach of Sheridan and Venkataraghavan (51) and also Saqi and Sternberg (53) is the elimination of proteins with substantial sequence similarity resulting from homology as this could swamp the search for motifs.

A stochastic version of the expectation maximization algorithm known as the Gibbs sampler has been developed by Lawrence *et al.* (54). It prevents solutions being trapped in local minima and has been shown to be a fast and sensitive way to extract patterns from a family of sequences. The sampler is included in the MACAW program (55), a sequence alignment workbench written for relatively inexpensive PC platforms and distributed by the NCBI. MACAW has a useful interface allowing visualization of motifs at the sequence level and their location on a schematic of the sequences showing their relative positions in the sequences.

4. Summary

There are a number of programs available that will search for a Prosite or user-defined motif in a given sequence or a sequence database. Fast algorithms will greatly facilitate searches of many patterns against a sequence database. The ability to include mismatches and/or weight matrix scoring is

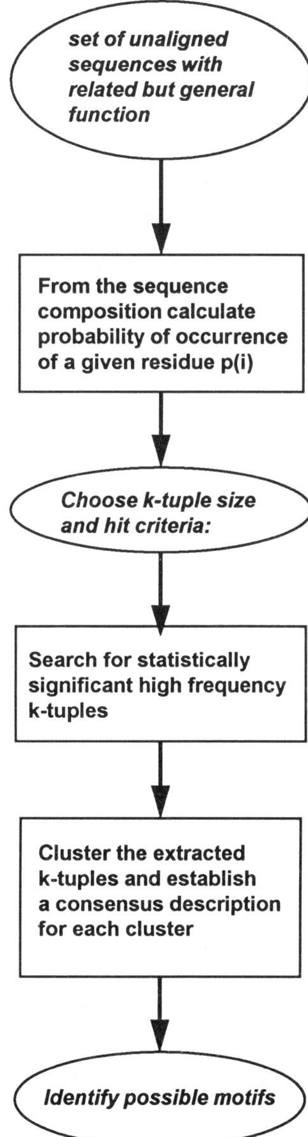

Figure 2. Schematic diagram illustrating the main steps in the motif extraction algorithm of Saqi and Sternberg (53).

also important to allow greater flexibility. But probably most important of all is some measure of statistical significance to assess whether or not a given hit is likely to be biologically meaningful.

The pattern extraction process, that is obtaining a pattern that is chatacteristic of a group of functionally related sequences, is difficult. There may exist

Rank	Pattern
1	--a-KYDG-R-c
2	-KYDG-Rgce--
3	YDG-R-QeH---
4	--g---E-KYDG
5	--g-eb--Gabe
6	----a-KYDG-R
7	EG-IVb-----Y
8	RFPRFIR-R-DK
9	LFVFDCLYFDGF

Amino Acid Covering Classes :
a=DE, b=KRH, c=NQ, d=ST, e=ILV, f=FWY, g=AG

Figure 3. Application of the algorithm of Saqi and Sternberg (53) to a set of unaligned DNA ligase sequences. The most significant being xx[DE]xKYDGxRx[NQ]. The DNA_LIGASE_1 Prosite motif is [EDQH]xKx[DN]GxR[GACV]. Further down the list the motif EGxIV[KRH]x(5)Y is also recovered which corresponds closely to part of the second Prosite motif for the DNA ligases, namely EG[LIVMA][LIVM](2)[KR]x(5,8)[YW][QEK]x(2,6)[KRH]x(3,5)K[LIVMFY]K.

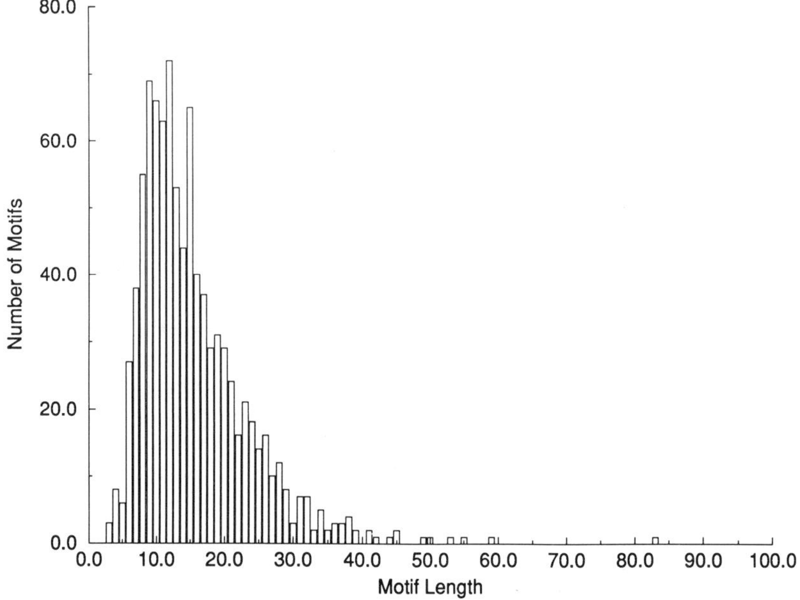

Figure 4. Motif length distribution in the Prosite11 database. For variable repeats such as x(2,3) the first choice was always taken.

several patterns which charcaterize subsets of the sequence family and traditional multiple sequence alignment methods, whilst being a good starting point, will generally fail to correctly identify such patterns.

Attempting to unify a set of sequences by detecting a common motif can be misleading particularly if the alignments are forced or obtained by visual inspection. A comparison of a proposed motif block from a set of aminoacyl tRNA sequences suggested to reveal a distant relationship among the sequences, with blocks from a randomized set of the same sequences has been carried out by Henikoff (56). One of the blocks derived from the random set had more conserved positions and the discriminatory power of the blocks in identifying the randomized sequences from a database was comparable with that of the proposed motif in detecting the synthetases.

The solutions offered by many automatic methods are somewhat parameter-dependent and the best strategy is to use use a variety of algorithms with different parameter settings. Some methods require an assumption to be made about the pattern width. A plot of motif lengths in the Prosite11 database (*Figure 4*) shows a wide range of values although the distribution peaks for lengths of between 10 to 12. When using the Gibbs sampler (54), for example, one practical step to detect convergence which is suggested by the authors is to look for the occurrence of the same pattern with different starting seeds. Finally, care must be taken in ascribing functional significance to a proposed motif and mutagenesis experiments where possible may establish the importance of particular residues.

References

1. Needleman, S. and Wunsch, C. (1970). *J. Mol. Biol.*, **48**, 443.
2. Smith, T. F. and Waterman, M. (1981). *J. Mol. Biol.*, **147**, 195.
3. Altschul, S. F., Gish, W., Miller, W., Myers, E. W., and Lipman, D. J. (1990). *J. Mol. Biol.*, **215**, 403.
4. Blundell, T. L., Sibanda, B. L., Sternberg, M. J. E., and Thornton, J. M. (1987). *Nature*, **326**, 347.
5. Bork, P., Ouzounis, C., Sander, C., Scharf, M., Schneider, R., and Sonhammer, E. (1992). *Nature*, **358**, 287.
6. Bork, P., Ouzounis, C., Sander, C., Scharf, M., Schneider, R., and Sonhammer, E. (1992). *Protein Sci.*, **1**, 1677.
7. Bairoch, A. (1991). *Nucleic Acids Res.*, **19**, 2241.
8. Saraste, M., Sibbald, P. R., and Wittinghofer, A. (1990). *Trends Biochem. Sci.*, **15**, 430.
9. Cockwell, K. Y. and Giles, I. G. (1989). *CABIOS*, **5**, 227.
10. Fuchs, R. (1990). *CABIOS*, **7**, 105.
11. Sternberg, M. J. E. (1991). *Nature*, **349**, 111.
12. Sternberg, M. J. E. (1991). *CABIOS*, **7**, 257.
13. Staden, R. (1989). *CABIOS*, **5**, 89.
14. Sibbald, P. R. and Argos, P. (1990). *CABIOS*, **6**, 279.
15. Devereux, J., Haeberli, P., and Smithies, O. (1984). *Nucleic Acids Res.*, **12**, 387.
16. Venezia, D. and O'Hara, P. J. (1993). *CABIOS*, **9**, 65.
17. Saqi, M. A. S. and Sayle, R. (1994). *CABIOS*, **10**, 545.
18. Sayle, R. and Blisset, A. (1992). *Proceedings of the 10th Eurographics UK92 Conference.*
19. Lipmann, D. J. and Pearson, W. R. (1985). *Science*, **227**, 1435.
20. Barton, G. J. and Sternberg, M. J. E. (1990). *J. Mol. Biol.*, **212**, 389.

21. Depiereux, E. and Fetymans, E. (1992). *CABIOS,* **7**, 195.
22. Gribskov, M., Homyak, M., Edenfield, J., and Eisenberg, D. (1988). *CABIOS,* **4**, 61.
23. Gribskov, M., McLachlan, A. D., and Eisenberg, D. (1987). *Proc. Natl. Acad. Sci. USA,* **84**, 4355.
24. Smith, R. F. and Smith, T. F. (1990). *Proc. Natl. Acad. Sci. USA,* **87**, 118.
25. Smith, R. F. and Smith, T. F. (1992). *Protein Eng.*, **5**, 35.
26. Vingron, M. and Argos, P. (1991). *J. Mol. Biol.,* **218**, 33.
27. Karlin, S. and Brendel, V. (1992). *Science,* **257**, 39.
28. Brendel, V., Bucher, P., Nourbakhsh, I. R., Blaisdell, B. E., and Karlin, S. (1992). *Proc. Natl. Acad. Sci. USA,* **89**, 2002.
29. Karlin, S. and Ghandour, G. (1985). *Proc. Natl. Acad. Sci. USA,* **82**, 6186.
30. Karlin, S. and Ghandour, G. (1985). *Proc. Natl. Acad. Sci. USA,* **82**, 5800.
31. Karlin, S., Ost, F., and Blaisdell, B. E. (1989). In *Mathematical methods for DNA sequences* (ed. M. S. Waterman). CRC Press, Florida.
32. Karlin, S. (1990). In *Structure and methods* (ed. R. H. Sarma and M. H. Sarma). Academic Press.
33. Karlin, S. and Brendel, V. (1990). *Oncogene,* **5**, 85.
34. Stormo, G. D. and Hartzell, G. W. (1989). *Proc. Natl. Acad. Sci. USA,* **86**, 1183.
35. Hertz, G. Z., Hartzell, G. W., and Stormo, G. D. (1990). *CABIOS,* **6**, 81.
36. Mengeritsky, G. and Smith, T. F. (1987). *CABIOS,* **3**, 223.
37. Gallas, D., Eggert, M., and Waterman, M. (1985). *J. Mol. Biol.,* **186**, 117.
38. Abremski, K. E. and Hoess, R. H. (1992). *Protein Eng.,* **5**, 87.
39. Little, R. J. A. and Rubin, D. B. (1987). *Statistical analysis with missing data*, pp. 1–278. John Wiley and Son, New York.
40. Lawrence, C. F. and Reilly, A. A. (1990). *Proteins,* **7**, 41.
41. Cardon, L. R. and Stormo, G. D. (1992). *J. Mol. Biol.,* **223**, 159.
42. Dayhoff, M. O. (1978). *Atlas of protein sequence and structure*, Vol. 5, Suppl. 3. National Biomedical Research Foundation, Washington DC.
43. Seto, Y., Ikeuchi, Y., and Kanehisa, M. (1990). *Proteins*, **8**, 341.
44. Ogiwara, A., Uchiyama, I., Seto, Y., and Kanehisa, M. (1992). *Protein Eng.,* **5**, 479.
45. Solovyev, V. V. and Makarova, K. S. (1993). *CABIOS,* **9**, 17.
46. Bolch, B. W. and Huang, C. J. (1979). *Multivariate statistical methods for business and economics*. Statistics, Moskow.
47. Posfai, J., Bhagwat, A. S., Posfai, G., and Roberts, R. J. (1989). *Nucleic Acids Res.*, **17**, 2421.
48. Smith, H. O., Annau, T. M., and Chandrasegaran, S. (1990). *Proc. Natl. Acad. Sci. USA,* **87**, 826.
49. Henikoff, S. and Henikoff, J. G. (1991). *Nucleic Acids Res.,* **19**, 6565.
50. Henikoff, S. and Henikoff, J. G. (1992). *Proc. Natl. Acad. Sci. USA,* **89**, 10915.
51. Sheridan, R. P. and Venkataraghavan, R. (1992). *Proteins,* **14**, 16.
52. Altschul, S. F. and Lipman, D. J. (1990). *Proc. Natl. Acad. Sci. USA,* **87**, 5509.
53. Saqi, M. A. S. and Sternberg, M. J. E. (1994). *Protein Eng*., **7**, 165.
54. Lawrence, C. F., Altschul, S. F., Boguski, M. S., Liu, J. S., Neuwald, A. F., and Wooton, J. C. (1993). *Science,* **262**, 208.
55. Schuler, G. D., Altschul, S. F., and Lipman, D. J. (1991). *Proteins,* **9**, 180.
56. Henikoff, S. (1991). *New Biol.,* **3**, 1148.

4

Prediction of secondary structure

ROSS D. KING

1. Introduction: the need for secondary structure prediction

The prediction of protein secondary structure is a major part of the general protein folding problem and is the most general method of obtaining some structural information from any newly-determined sequence. Secondary structure prediction is useful in a wide variety of problems concerned with proteins.

(a) In the design of novel proteins, the rules governing α-helix and β-sheet structures provide guidelines for selecting specific mutants.
(b) The assignment of secondary structure can help to confirm a structural and functional relationship between proteins when there is a weak sequence relationship, e.g. the prediction of the retroviral protease (1).
(c) Secondary structure prediction is important in establishing alignments during model building by homology (2) (see Chapter 6), and in checking the validity of crystallographic models (3).
(d) Secondary structure prediction is also the first step in attempting to generate tertiary models by docking α-helices and β-sheets, e.g. predictions of ATPases (4) and interleukin 2(5) (see Chapter 9).
(e) Prediction of local secondary structure can be also be helpful in structure determination by 2D NMR.

2. Methods of secondary structure prediction

There exist a bewildering variety of methods for predicting the secondary structure of proteins from primary structure, with almost every conceivable data analysis method having been applied to the problem. The literature is vast. A quick survey of my library produced over 100 specific papers on different prediction methods; and I would estimate that this is between a half and a third of the literature. Secondary structure prediction methods can be roughly divided into statistical, knowledge-based, and hybrid systems.

Statistical methods are based on studies of the database of proteins of known primary and secondary structure. These studies are aimed at finding empirical relationships between the two types of structure. Statistical methods have the advantages of consistently and explicitly using the large database of known protein structures: they have the disadvantages of not fully taking into account physico-chemical knowledge about proteins, and in having poor explanatory power. The amount of available data is very important for statistical prediction methods. The earliest proposed methods tended to suffer from an acute shortage of data, as the conformations of very few proteins were then available; there is now more data available (~ 20 000 residues), but the shortage of data is still a fundamental limit to the accuracy of statistical methods. In this paper we examine four broad secondary structure prediction methods based on statistics: Chou–Fasman (6), GOR (7), nearest neighbour (8), and neural networks (9).

Physico-chemical methods are based on knowledge about the physical and chemical basis of protein structure. Several important structural features of proteins have been recognized to be useful in predicting protein secondary structure and are commonly included in protein sequence analysis program suites, e.g. helical wheels and hydrophobicity profiles. In this paper we examine two secondary structure prediction methods based on the physical and chemical analysis of protein structure, those of Lim (10) and that of Cohen *et al.* (11).

One set of methods that have the potential for combining the best features of statistical and physico-chemical methods come from the field of symbolic machine learning. These methods are statistical, in that they learn rules by generalizing from examples in a database (through induction). But they also incorporate physico-chemical knowledge about protein structure, through the use of the background knowledge used in the generalization; with the added advantage that the rules learnt are more easily understood than those generated by statistical means.

2.1 The Chou–Fasman method

The Chou–Fasman method of predicting secondary structure was, and may still be, the most popular and commonly used of all prediction methods (6). It owes its popularity to being simple to use, intuitively easy to understand, and reasonably successful in practice. The Chou–Fasman method is a statistical prediction method based on the calculation of the statistical propensities of each residue forming either an α-helix or a β-strand. These propensities are used to classify the residues, into six classes depending on their likelihood of forming an α-helix and six classes depending on their likelihood of forming a β-strand. The class designations are then used to find areas of probable α-helix and β-strand in the protein sequence to be predicted. Probable areas of α-helix and β-strand are then modified by a series of rules to produce the final prediction.

Advantages:

- the method is quite easy to understand

Disadvantages:

- the statistics used are naive
- the prediction rules are somewhat arbitrary
- the reasons for designation of secondary structure are not directly related to chemical or physical theory

2.2 The GOR method

The GOR (Garnier–Osguthorpe–Robson) method of Robson and co-workers is, perhaps, the most impressive of all the statistical prediction methods (7). Its theory is based on sound statistical principles, is well defined, and has few artificially introduced variables. It has also been successful in practice and computer implementations are readily available. The GOR method is based on the idea of treating the primary sequence and the sequence of secondary structure as two messages related by a translation process, this translation process is then examined using information theory. The biological process of translation, the folding of the protein, is regarded as a 'black box' defined only by the observed relation between the input symbols and output symbols (*Figure 1*). Structure prediction depends on measuring the amount of information residues carry about their secondary structure and other residues secondary structure; the theory behind this is complex, but it results in a simple method in practice.

To avoid a need for enormous amounts of data, the information function can be expanded as a sum of terms and the most important terms selected. In practice there is only enough data for terms containing a maximum of two residues, terms with higher numbers of residues would produce parameters of zero or close to zero. One of the most interesting things about Robson's method is that it allows this separation and evaluation of the different infor-

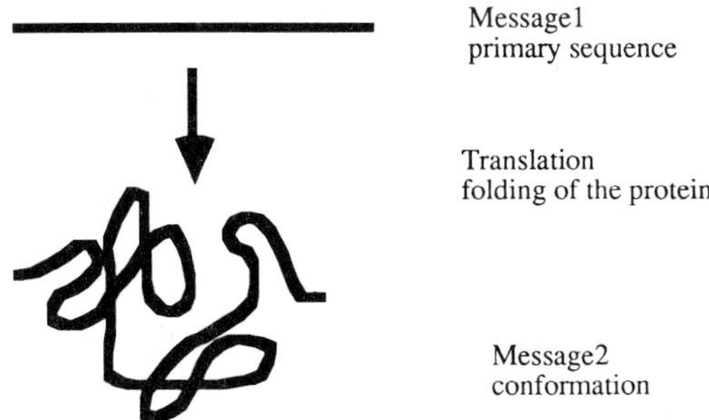

Figure 1. Robson's method: biological and informational interpretations.

mation terms. The terms containing no greater than two residues can be split into three types.

(a) Information a residue carries about its own secondary structure—intraresidue information.
(b) Information a residue carries about another residues secondary structure which does not depend on the other residues type—directional information.
(c) Information a residue carries about another residues secondary structure which depends on the other residues type.

Gibrat *et al.* (12) improved Robson's method by recalculating all the information terms using the larger database that was then available. This larger database also allowed investigation of the use of side chain/side chain interactions. This was found to improve overall prediction accuracy.
Advantages:

- the method is robust and theoretically sound
- the interpretation of protein folding as message translation opens up a new conceptual way of looking at the problem
- the method allows the separation and evaluation of the different types of information involved in the folding of the protein: intraresidue, directional, pair interaction, etc.

Disadvantages:

- the reasons for designation of secondary structure are not directly related to chemical and physical theory
- it treats as a 'black box' the folding of the protein, a process which ultimately needs to be understood

2.3 Nearest neighbour methods

In nearest neighbour statistical methods, the secondary structure of a new primary sequence is classified to be the same as that of the closest primary sequence to it of known secondary structure (in some defined space). This is based on the reasonable hypothesis that similar primary sequences will have similar secondary structures. However, this hypothesis is not always true, as secondary structure formation is also based on long-range interactions. Studies (13–15) have shown many examples where local sequence similarity does not imply a structural relationship. This nearest neighbour prediction method is simple to understand, with the main complication being the definition of what it means for two sequences to be similar. In practice the algorithm is usually extended to include not only the nearest neighbour, but the *k*-nearest neighbours. A number of different *k*-nearest neighbour approaches have been proposed. Levin *et al.* (16) used their own similarity matrix, a sequence

of 7 and $k = 1$. Nishikawa and Ooi (17) used the Kubota homology matching algorithm, a sequence of 8 and a variable k. Sweet (18) used the standard Dayhoff homology matching algorithm, a sequence of 12, and $k = 15$. Salzberg and Cost (19) used a Stanfil and Waltz similarity matrix, a sequence of 19, and a variable k. Zhang *et al.* (20) used their own similarity matrix, a sequence of 13, and $k = 25$. Yi and Lander (8) used a complicated similarity measure based on the Bowie environmental scoring scheme, a sequence of 13, and $k = 50$. It should be noted that these values of k are larger than is usual to use with this algorithm.
Advantages:

- predicting secondary structure on the basis of similarities is simple and easy to understand
- a large amount of information is preserved (e.g. ϕ and ψ angles) (18)
- the rules could be reversed for protein engineering

Disadvantages:

- there is little use of chemical or physical theory
- there is no explanation of the prediction
- short identical sequences often have differing secondary structure
- the method is computationally expensive for prediction

2.4 Neural networks

A neural net learning system is a network of non-linear processing units that have adjustable connection strengths. There are many different types of neural networks: conjugate gradients, cascade correlation, Hopfield nets, Boltzmann machines, etc. (21). However, almost all work applied to secondary structure prediction has been based on feed-forward networks (multilayer perceptrons), using the back-propagation learning rule (22). Learning consists of altering the weights of connections between the units in response to a teaching signal which provides information about the correct classification in input terms (primary and secondary structure). The goal is to find a good input–output mapping which can then be used to predict the test set.

A large number of authors have applied neural networks to secondary structure prediction: refs 9, 23–30, and many more. These different approaches differ mainly in the size of the window of amino acids and the data set of proteins they use. A typical approach is that of Qian and Sejnowski (23), they used a net with one hidden layer, a window size of 13, and 21 inputs for each position (*Figure 2*).
Advantages:

- neural network prediction methods are readily available
- neural network methods are often successful in practice

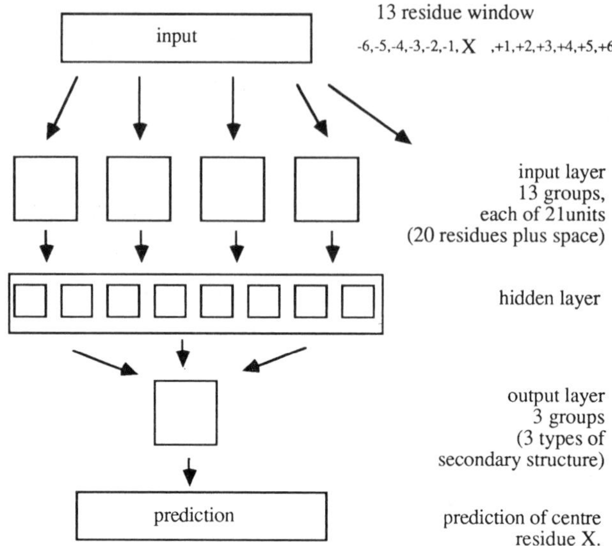

Figure 2. Neural network structure.

Disadvantages:

- there is little use of chemical or physical theory
- the method has very poor explanatory power—a Hinton diagram means nothing to a protein chemist
- they are statistically rather poorly characterized

2.5 The Lim method

Lim (31) developed a theory of formation of α-helices and β-sheets based on the packing of polypeptide chains in globular proteins. This theory was used to produce stereochemical rules for predicting secondary structure based on patterns of residue types (10). The theory was developed by a logical and stereochemical analysis of the packing of molecular models and provided many important insights into protein structure. The actual prediction rules developed are quite complicated but computer implementations now also exist.

As an example of the insight provided by Lim, he noted for α-helices: that to make contact with the main body of a protein an α-helix needs a hydrophobic side. This occurs if the hydrophobic residues face internally and can pack closely together (linked). This is achieved by overlapping the following positions: linked hydrophobic pairs (position i and $i + 4$), terminal pairs (position i and $i + 3$), and hydrophobic triplets (position i, $i + 1$, and $i + 4$) or (i, $i + 3$, and $i + 4$).

Advantages:

- the method introduced a new and important theory of protein formation
- the rules have a clear basis in protein chemical theory and so have good explanatory power

Disadvantages:

- the database of known structures is not used in a systematic way
- the rules are often complex and difficult to understand
- no qualitative comparison is made between the rules

2.6 The Cohen method

The prediction method of Cohen and co-workers, is based on using a rule-based formalism to encode structural knowledge about proteins (11). The rules are simply lists of generalized amino acid sequences or patterns that are associated through physical and chemical theory with specific secondary structures. An implementation of these rules is available (32).

A typical rule is that for a strong hydrophobic diamond pattern without aromatics:

$$[S, *, *, S, S, *, *, S] \rightarrow \alpha\text{-helix} \qquad S = I, L, \text{or } V.$$

Advantages:

- the rules have good explanatory power
- attention is paid to use of the results of secondary structure prediction in a system which uses hierarchical structure constraints to determine conformation

Disadvantages:

- the database is not used in a systematic way to find the rules
- the rules give the impression of being overfitted
- no quantitative comparison is made of the rules

2.7 Machine learning methods

The first full-scale application of machine learning to secondary structure prediction occurred with the work of King and Sternberg (33). They developed a learning algorithm PROMIS that was specifically designed for the problem. Background knowledge about the physico-chemical properties was encoded as a generalization hierarchy of classes derived from the work of Taylor (34). A typical rule is:

[*, (tiny or (small and polar) or p), *, tiny or
(polar minus aromatic) or p)] → coil.

Since the application of PROMIS, more powerful machine learning methods have been developed known as inductive logic programming (ILP) methods (see Appendix). One ILP program, GOLEM, has been successfully applied to the α/α domain subset of protein structures (35). ILP methods have the advantage over other machine learning methods for secondary structure prediction in that they are specifically designed to learn structural relationships between objects (technically they use predicate logic). The application of GOLEM involved three stages, an initial prediction and two smoothing stages.

Advantages:

- they use the database of known structures
- they can exploit physico-chemical knowledge
- they have been successfully applied
- they can generate more comprehensible rules that statistical methods

Disadvantages:

- no state of the art method is available that can predict protein sequences from all domain classes
- the rules generated are not as easily comprehensible as the rules generated by human experts

3. Evaluating prediction methods

How good are secondary structure prediction methods? Which of the many secondary structure prediction methods is best? At present there can be no unequivocal answers to these questions, as a number of methodological problems exist that preclude a definitive answer. Because of these problems, the reader is hereby warned that it is dangerous to read a paper about a secondary structure prediction method, notice its headline accuracy in the abstract, and assume that the method will produce the same accuracy on the protein you are interested in, and that the method is necessarily better than another prediction method with a lower headline accuracy.

It is probably fair to asses the current state of the art in protein secondary structure prediction as follows: if all you know about the protein is its primary structure, the best methods will be able to predict its secondary structure with about 65% accuracy; if a number of homologous sequences are available then the accuracy rises to around 70%; and if the structural domain of the protein is known, the accuracy can rise to about 80% for some domain types.

To understand the methodological problems surrounding the proper assessment of prediction methods, it is necessary to understand something about how secondary structure prediction methods are formed. Typically what happens is this: a database of proteins of known primary and secondary structure is selected, this database is divided into a training and testing set, the prediction method (GOR, neural net, etc.) learns how to predict using the training proteins, and then the prediction method is tested using the test proteins to give a final headline accuracy. A number of problems arise in comparing the accuracy of different prediction methods formed in this way: the use of different data sets of proteins, the inclusion of homologous proteins, different definitions and types of secondary structure, poor measures of success, and difficulties in accurately estimating the measures of success.

3.1 The use of different proteins

The most obvious problem is that different proteins have been used to train and test the different methods. The Brookhaven Data Bank (36) contains the crystallographic structure of several hundred proteins. From these proteins the different workers have selected widely different subsets. The choice of proteins used has a direct bearing on the accuracy of prediction. For example, in analysing the work of Qian and Sejnowski (23): if the three proteins with the highest accuracy of prediction are excluded from the test set, the accuracy is 62.9%; if the three proteins with the lowest accuracy of prediction are excluded from the test set, the accuracy increases is 67.2%. This variation of ~ 4% may not appear very much, but it is the level of difference between the best accuracies claimed today and those ten years ago.

3.2 The use of homologous proteins

Every protein used to test a prediction method data should be non-homologous with every other protein, i.e. the test proteins should be as statistically independent of each other as possible. This condition is aimed to ensure that the sample is unbiased and that the secondary structure prediction method is not specific to known proteins, and is therefore not overfitted. It is now common practise to test for sequence homology between the proteins, with the cut-off normally put around 30% sequence identity. This is fine as far as it goes, but it ignores the problem of structural homology, e.g. a number of globins share little sequence homology. The accuracies of most secondary structure predictions are based on the use of several globins, resulting in a biased prediction of accuracy. Structural homology tests should also be used to select proteins (37).

3.3 The use of different definitions of secondary structure

Just as different authors have used different proteins, they have also used different definitions of secondary structure. There is no generally accepted unambiguous and physically meaningful definition of the various secondary structure types. The Brookhaven Data Bank uses the subjective crystallographer designation of secondary structure. Objective computer algorithms have been developed to assign secondary structure from conformation (38,39). These algorithms are in effect procedural definitions of secondary structure; taking as input the conformation of a protein (in atomic coordinate form) and producing as output a designated secondary structure type for each residue in the protein. The definitions are based on the presence of hydrogen bonding, the position of C^{α} atoms, and the ϕ and ψ angles. It has been found that these different definitions can disagree markedly, especially at the ends of secondary structure elements.

3.4 The use of different types of secondary structure

A related problem to the designation of secondary structure is that there is no agreement about how many types of secondary structure to predict. The most common practice is to carry out a three-state prediction, this involves predicting: α-helices, β-strands, and coil (where coil is a dustbin class for all other possible secondary structure types). One ambiguity in this is whether to class 3_{10} helices with α-helices or with turns. It is also quite common to predict four-states: α-helices, β-strands, turn (where turns are well-defined turn regions), and coil (where coils are regions of undefined secondary structure). It is obviously difficult to compare a three and a four-state prediction, four-state prediction being more difficult; it is probably most difficult to discriminate between turns and coils than between any other classes.

3.5 Measurements of success

Most current measures of prediction success are based on examining the residue by residue predictions of the different algorithms (*Table 1*).

The most commonly used prediction measure is accuracy (success rate) (*Table 2*)—it is sometimes referred to as Q_3 accuracy. This measure is the most easily understood. However, it is natural to expect of a measure of

Table 1. Confusion matrix

		PREDICTED yes	PREDICTED no	Total
ACTUAL	yes	*a*	*b*	a + b
ACTUAL	no	*c*	*d*	c + d
ACTUAL	Total	a + c	b + d	*N*

a, positive correct prediction (prediction); *d*, negative correct prediction (non-prediction); *b*, underprediction (an error of omission); *c*, overprediction (an error of commission); *N*, total number of data items.

Table 2. Three measures of association between predicted and actual structure

Measure	Definition	1	2	3
Accuracy	$\frac{(a+d)}{N}$	good	poor	poor
Matthews correlation	$\frac{(ad)-(bc)}{\sqrt{(a+b)(a+c)(b+d)(c+d)}}$	poor	good	poor
Weight of evidence	$\log\left(\frac{ad}{cb}\right)$	poor	poor	good

Column 1 refers to the intuitive ease of understanding of the measure. Column 2 refers to the property of having the value 0 when no association exists and the value 1 when complete association exists. Column 3 refers to invariance when the relative frequency with which data are sampled from the different states is varied.

association, that the limits of variation should be known, and that it should take the value 0 when there is no association and 1 when there is complete association. This is not the case for accuracy, with the lower limit not being 0. A sounder statistical, but more complex measure, is (Matthew's) correlation r (*Table 2*) (40). This measure is also becoming commonly used for protein secondary structure prediction. It has the value of: –1 with complete negative association, 0 with no association, and 1 with complete association. It is closely related to chi square: $\psi^2 = N^*r$. One problem with correlation is that it is not invariant when the relative frequency with which data are sampled from the different states is varied. For example: if the value of c and d were divided by two it would be desirable for the measurement of association to remain the same. One measure that does this is weight of evidence (*Table 2*). It is closely related to correlation. It may be of value in comparing relative success of prediction for different types of protein.

It could be argued that accuracy, Matthew's correlation, and weight of evidence do not take into account the main point of secondary structure prediction, which is to produce useful results to feed into model-based studies of protein structure. This is because they consider all mistakes in prediction to be equally bad. For a model-based study, it is far less costly to make a mistake of predicting a residue to be a coil at the very start of a helix (where secondary structure may not even be well defined) than it is to make a mistake in the middle of the helix. This can be taken into account by assigning a cost to making a mistake in prediction at each residue position. In the case of an α-helix, the cost would be low at the start, increase towards the middle, and then decrease towards the end (a correct prediction has no cost). A prediction method would aim to minimize the cost of a prediction.

In measuring the success of a prediction method it is customary to add all the predictions for different proteins (with differing lengths) together and form the mean as a final score for the prediction method. This is misleading, as the natural unit for secondary structure prediction is the whole protein and not individual residues. It would be better to give the mean and variance of the prediction method based on whole proteins not residues.

3.6 Difficulties in estimating the measures of success

Even if two prediction methods used exactly the same data, and definition of secondary structure, it would still be difficult to identify what differences in measures of success could be said to be statistically reliable. This is because protein sequences are strongly autocorrelated, e.g. if one residue is in a helix the next one is also likely to be in the helix. This makes statistical analysis difficult. An efficient method of estimating success is the use of resampling techniques such as cross-validation (41). Use of such a technique helps ensure that a quoted prediction accuracy is reliable. Ideally the form of cross-validation known as leave one out should be used, although using this makes problems with homologous proteins in the database particularly severe.

4. A practical example

A practical example is presented to illustrate the practice of secondary structure prediction. The aim of this example is expository. It is not intended as a comparative trial of different prediction methods; that would need many more proteins for statistical validity. The example is intended to show the current state of the art in secondary structure prediction, to illustrate what can be achieved, and what is not possible. The protein used for the example was adaC O6 Methyl G.DNA Methyltranferase from *E. coli* (the sequence and secondary structure were kindly provided by P. Moody of the University of Leeds). This protein has recently been crystallized, but had not been reported in the literature (*Figure 3*); its structure was therefore not widely known, allowing a blind trial of different prediction methods to be carried out. The secondary structure of the protein was assigned using QUANTA which implements the Kabsch and Sander algorithm (42). This assignment was edited slightly to remove isolated residues assigned to be α-helix or β-strand, and to extend helices where the capping residues are three or five turns.

4.1 The trial procedure

The procedure of the trial was as follows. The primary sequence of the protein was sent out to a number of leading workers in the field, and they were asked to make a secondary structure prediction using whatever method they thought most suitable. In addition to the primary sequence they were also sent an alignment involving eight other homologous proteins which they were free to use or ignore as they saw fit. The alignment came from the pre-processing of the Rost and Sander secondary structure prediction algorithm

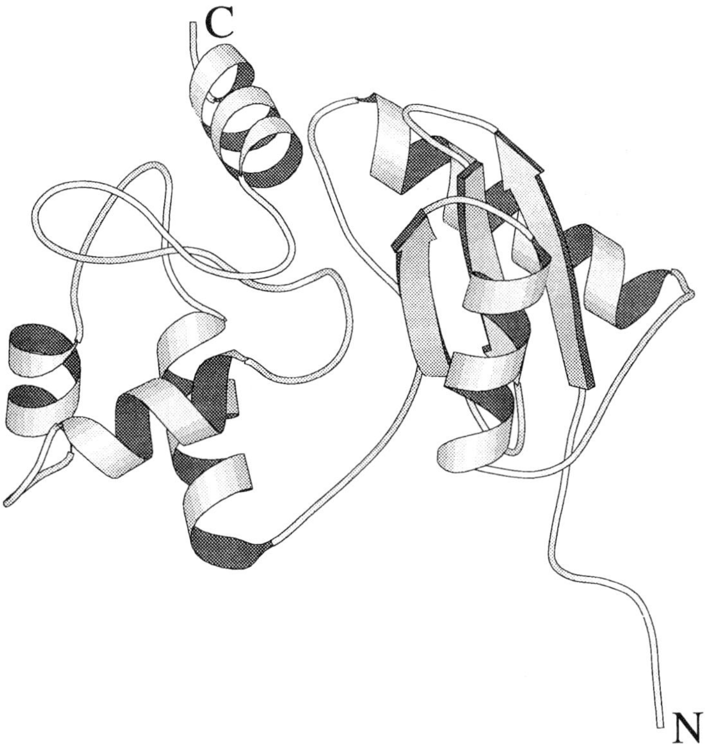

Figure 3. The tertiary structure of adaC O6 Methyl G.DNA Methyltranferase from *E. coli* (displayed using Molscript) (48).

and involved: ada_ecoli (99% sequence identity), ada_salty (74%), dat1_bacsu (40%), ogt_ecoli (39%), adab_bacsu (35%), mgmt_mouse (35%), mgmt_rat (33%), mgmt_crigr (31%).

4.2 Prediction methods applied

Ten secondary structure predictions were made of the adaC sequence.

(a) The Rost and Sander (Sander) method (9) was applied automatically using electronic mail (see below). This method is based on a complicated series of neural networks. It uses information from the alignment of homologous sequences and it smoothes its prediction to avoid predictions such as helices consisting of single residues. Although its claim to have 'better than 70% accuracy' is still unproven, it success is well validated and it is the prediction method that all the others have to beat. One important innovation of the Rost and Sander method is the ability to use this method by e-mail, making it completely objective (send the word *help* to `PredictProtein@Embl-Heidelberg.de` for detailed instructions).

(b) The COMBINE method of Biou *et al.* (43) (Garnier) was applied by J. Garnier, one of the authors of the method. This method is a combination of the GOR III (an update of the GOR method) (7), a nearest neighbour method (16), and a method that uses patterns of hydrophilic/hydrophobic patterns. COMBINE was applied to an alignment of the eight homologous sequences using CLUSTAL.

(c) The method of Sasagawa and Tajima (Sasagawa) (30) was applied by its authors. This is a neural network with two modules, one for predicting α-helix, the other for β-strand. Both modules are three layer nets trained by back-propagation.

(d) The method of Zvelebil *et al.* (Zvelebil) (44) was applied by M. J. E. Sternberg, one of its authors (this author was aware of the α + β nature of the domain). The Zvelebil method is an extension of the GOR method to include information from homologous sequences.

(e) The WASP method (Barton) (45) was applied by G. Barton, one of its authors. The basic prediction was refined using conserved hydrophobic residues using the authors subjective knowledge of protein structures.

(f) The method of Wako and Blundell (Blundell) (46) was applied by T. Blundell's laboratory. This is a statistical method that utilizes hydrophobicity patterns and capping rules to define boundaries of secondary structure regions.

(g) A number of methods from the Leeds package of secondary structure prediction algorithms were applied by J. Shirazi. The 'vanilla' GOR method (GOR) (7), the Chou–Fasman method (Chou–Fasman) (6), and an algorithmic formulation of the Lim prediction rules (Lim) (10). In addition, the Leeds package allows a joint prediction to be made using the results of: Robson, Chou–Fasman, Lim, and a few other prediction methods (Leeds).

4.3 Results

The protein formed an α + β domain which is the most difficult type to predict. Despite this, the best prediction methods do well, with that of Rost and Sander achieving an accuracy of 79% (*Table 3*). It is interesting that the first 40 residues appear to be less well predicted than the remaining residues, this may be due to some edge effect. For the remaining sequence the remaining secondary structure elements are almost all correctly predicted by the different methods (the only area of dispute is around residue 90) (*Figure 4*).

This main difference between the best method, Sander, and the other methods, is its success at predicting β-strands. It has a C_b correlation of 0.633 (W_b = 3.61), the next best method is Le1 with a C_b correlation of 0.380 (W_b = 2.51). Its main errors are wrong prediction of an α-helix at the start of the sequence, and it predicts the last strand to longer than it is. The relative suc-

Table 3. The results of the 11 secondary structure prediction methods

Name	Q_3	C_a	C_b	C_c	W_a	W_b	W_c
Sander	0.794	0.752	0.633	0.612	3.99	3.61	2.88
Sasagawa	0.578	0.293	0.104	0.280	1.310	1.77	1.29
Barton	0.639	0.498	0.344	0.400	2.25	1.91	1.70
Zvelebil	0.622	0.478	0.239	0.456	2.17	1.34	2.00
Blundell	0.639	0.508	−0.078	0.582	2.34	0.79	2.76
Chou–Fasman	0.528	0.236	0.327	0.269	1.08	2.08	1.25
Garnier	0.728	0.709	0.283	0.485	3.68	2.14	2.24
GOR	0.511	0.323	0.220	0.247	1.54	1.31	1.26
Leeds	0.683	0.553	0.380	0.468	2.74	2.51	2.07
Lim	0.711	0.576	0.290	0.605	2.92	1.91	2.92

Q_3 is the accuracy of the prediction, C_a is the correlation of α-helix structure, C_b is the correlation of β-strand structure, C_c is the correlation of coil structure, W_a is the weight of evidence of α-helix structure, W_b is the weight of evidence of β-strand structure, and W_a is the weight of evidence of coil structure.

cess of the Sander method is probably not due to chance, as it is the only prediction method that incorporates two features that are important for prediction success: it includes information from homologous proteins, as does the second best prediction method (Garnier), and it smoothes its prediction to avoid predictions such as residues in isolated α-helices or β-strands. The Garnier method does not smooth its prediction, and it is clear from its prediction that it would probably do better if it did.

Considering the other methods. The GOR method needs to be updated, with the recalculation of the information terms using the most up to date database; this will also improve the Zvelebil and Barton methods. The value of human insight is shown by the success of the, now very old, Lim rules. In the Blundell method the prediction of β-strands needs to be improved. The Sas method would be improved if its output was smoothed. The results for Chou–Fasman method confirm other results showing that it is now outdated. The success of Leeds and Garnier show the usefulness of using several methods to make a joint prediction.

5. Recommendations

As there is as yet no one clear best prediction method, it is sensible to apply a few different prediction methods to your sequence. Regions where the prediction methods agree are more likely to be correctly predicted than other areas. I recommend that you start with the following methods: the Rost and Sander method, as it is quite successful and readily available by e-mail; the Leeds package as it is also quite successful, incorporates a number of interesting methods, and is easily available; the COMBINE method (Garnier) could also be usefully applied.

```
....,....1....,....2....,....3....,....4....,....5....,....6
MTAKQFRHGGENLAVRYALADCELGRCLVAESERGICAILLGDDDATLISELQQMFPAAD
--------------EEEEEEEE--EEEEEEEE---EEEEEEE--HHHHHHHHHHH-----
--HHHHH-----EEEEEEEEEE---EEEEE-----EEEEE-----HHHHHHHHHH-----   Sander
------------HHHH-H---HHH--------HH-HHHH---------HHH-HH------   Sasagawa
------------EEEEEEEE----EEEEEE-----HHHHHH-------HHHHHHHH----   Barton
---------------EEEEEH---HHHHHHHHHHHHEEEE---------HHHH-------   Zvelebil
----------HHHHHHHHHHHHHHH-HHHHHHHH----------HHHHHHHHH-------   Blundell
-HHHHHHH--HHHHHH-HHHHHHHHHHHHHHHHHHEEEEHHHHHHHHHHHHHHHHHHHHH   Chou-Fasman
---HHH------HE--E---------EEE----------------HH---HHHHH-----   Garnier
HHHHHHHH--HHHEEHHHHH-HH--HHEE-------EEEEHHHH-HHHHHHHHHHHHHH-   GOR
-HHHH-----HHHHHHHHHHHHH-------------EEEEE----HHHHHHHHHHH----   Leeds
----------HHHHHHHHHHHHH--EEEEEE-HHHHHHHHH------HHHHHHHH-----   Lim

....,....7....,....8....,....9....,....10...,....11...,....1
NAPADLMFQQHVREVIASLNQRDTPLTLPLDIRGTAFQQQVWQALRTIPCGETVSYQQLA
-----HHHHHHHHHHHHHHH---------------HHHHHHHHHHHH------EEHHHHH
------HHHHHHHHHHHHH----------------HHHHHHHHHHHHH---HHHHHHHHH   Sander
---HH-H-HHH----HH-----------------HHHHHH---------------H----   Sasagawa
--HHHHHHHHHH---HHHHHH---EEEEEEEE-----HHHHHHHHHHH----EEEHHHHH   Barton
---HHHHHHHHHHHHHHHH------EEEEE------EEEE-HHHEEE----EEEEHHHHH   Zvelebil
------HHHHHHHHHHHHHH---------------HHHHHHHHHHH-----HHHHHHHHH   Blundell
HHHHHHHHHHHHHHH----------------------HHHHHHHHH------EEEEHHHH   Chou-Fasman
-----HHHHHHHHHHHHHH-----HE---H------HHHHHHHHHH--------HHHHHH   Garnier
-HHHHHHHHHHHHHHHEH--HHH-------HEH-HHHHHHHHHHHEEEE----E-EE-HH   GOR
---HHHHHHHHHHHHHHH-----------------HHHHHHHHHHH-------EEEHHHH   Leeds
-----EEHHHHHHHHHHHHHHH----------HHHHHHHHHHHHHHHH-------HHHHH   Lim

2...,....13...,....14...,....15...,....16...,....17...,....1
NAIGKPKAVRAVASACAANKLAIVIPCHRVVRGDGSLSGYRWGVSRKAQLLRREAENEER
HHHH----HHHHHHHHH------------EE------------HHHHHHHHHH-------
HHH---HHHHHHHHH------EEEEEEEEEEE--------------HHHHHHHH------   Sander
------H----HHHHHHHH---EH------E------------H-HHHHHHH-H------   Sasagawa
HHH----EEEE---------EEEEEEHHHHHHHH-----------HHHHHHHHH------   Barton
EH---HHHHHHHHHHH---EEEEEEEEEEEE----EEEE------EHHHHHHHH------   Zvelebil
HHHH---HHHHHHHHHHHH--EEEEEE--EEEEEE--------EEEEEEEEEEEEE----   Blundell
HH--HHHHHHHHHHHHHHHHHHEEEEEEEEE----------------HHHHHHHHHHHHH   Chou-Fasman
HHH---HHHHHH----------EEE----EE--------------HHHHHHHH-------   Garnier
HHHHHH-HHHHHHHHHHHHHHHHHH--EEEEE-----EEEEEEEEEHHHHHHHHHHHHHH   GOR
HHH----HHHHHHHHHHHHHHHEEE--EEEE---------------HHHHHHHHHHHHHH   Leeds
HHH-----HHHHHHHHHHHHEEEE---EEEEE-------------HHHHHHHHH------   Lim
```

Figure 4. The primary and secondary structure of adaC O6 Methyl G.DNA Methyltranferase from *E. coli*, along with the 11 secondary structure predictions. The primary structure is represented using the single letter code: a residue in an α-helix by 'H', in a β-strand by 'E', and turn and coil by '–'.

Information from homologous proteins should be used if it is available. This can either involve using a prediction method that does this implicitly (e.g. the method of Rost and Sander), or by aligning the sequences yourself, using your favourite prediction method on each sequence, and generalizing the result in some way. Regions with well-defined secondary structure (e.g. an α-helix or β-strand) generally vary less in sequence than coil regions. The sequences should also be examined for the conservation of patterns of conservation of hydrophobic residues.

The prediction should be smoothed to remove impossible or unlikely predictions, e.g. an isolated residue predicted to be in an α-helix. This can be done either implicitly in a prediction method (e.g. the method of Rost and Sander), or by hand using your knowledge of the protein. If the prediction method produces a probability of each secondary structure type, then this can prove useful in smoothing the prediction.

6. Conclusions

The prediction of protein secondary structure is the most general method of obtaining some structural information from any newly-determined sequence and a major step in the general protein folding problem. A vast variety of prediction methods exist. These can be divided into three main groups: the statistical/empirical methods that predict structure on the basis of statistical trends in the data; physico-chemical methods that predict structure on the basis of knowledge about the physical and chemical basis of protein structure; and the machine learning methods that try to combine the best features of the statistical and the physico-chemical methods.

There is no clear single best prediction method, and a number of important methodological problems exist that make the proper evaluation of prediction methods difficult. The current state of the art for secondary structure prediction is: given only a single sequence, about 65% accuracy; with homologous sequences about 70% accuracy; and if the structural domain of the protein is known, the accuracy can rise to about 80% for some domain types. These values have risen slowly but steadily over the years, and there are good reasons to hope that they will increase in the future in line with the increasing sizes of databases. Currently however, the results of any secondary structure prediction method should be treated with a healthy degree of scepticism.

To illustrate the practice of secondary structure prediction, state of the art methods were applied to predict the secondary structure of adaC O6 Methyl G.DNA Methyltranferase from *E. coli* (a protein whose structure has recently been solved but as of yet is not widely known), homologous sequences were available but the domain type was not known. This produced the quite impressive result that the best prediction method had an accuracy of 79%, with almost all the secondary structure elements correctly placed (one leading helix missed).

It is recommended that in attempting a secondary structure prediction: a number of different methods be used; homologous sequences should be used if available; the prediction should be smoothed; and finally, some common sense should be applied.

Acknowledgements

The primary structure of the C-terminal of the ada protein is that published by Demple *et al.* (1985). *Proc. Natl. Acad. Sci. USA,* **82**, 2688. It is from *E.*

coli B. The entry in the EMBL/SWISS-PROT database picked up by Rost and Sanger is for *E. coli* K12. The three-dimensional structure of the adaC protein (of *E. coli* B) was determined by Peter Moody of Leeds University in collaboration with Madeleine H. Moore, Jacqueline M. Gulbis, Eleanor J. Dodson, and Bruce Demple. The work was funded by the Yorkshire Cancer Research Campaign. The following people should be thanked for being brave and kind enough to submit predictions: Hioshi Wako and Tom L. Blundell, Fumiyoshi Sasagawa, Michael J. E. Sternberg, Geoff Barton, Jean Garnier, Burkhard Rost and Chris Sander. Jack Shirazi should also be thanked for running the Leeds package.

References

1. Pearl, L. H. and Taylor, W. R. (1987). *Nature*, **329**, 351.
2. Bowie, J. U., Lüthy, R., and Eisenberg, D. (1991). *Science*, **253**, 164.
3. Luthy, R., Bowie, J. U., and Eisenberg, D. (1992). *Nature*, **356**, 83.
4. Taylor, W. R. and Green, N. M. (1989). *Eur. J. Biochem.*, **179**, 241.
5. Cohen, F. E., Kosen, P. A., Kuntz, I. D., Epstein, L. B., Ciardelli, T. L., and Smith, K. (1986). *Science*, **234**, 349.
6. Chou, P. Y. and Fasman, G. D. (1974). *Biochemistry*, **13**, 222.
7. Garnier, J., Osguthorpe, D. J., and Robson, B. (1978). *J. Mol. Biol.*, **120**, 97.
8. Yi, T. and Lander, E. S. (1993). *J. Mol. Biol.*, **232**, 1117.
9. Rost, B. and Sander, C. (1993). *J. Mol. Biol.*, **232**, 584.
10. Lim, V. I. (1974). *J. Mol. Biol.*, **88**, 873.
11. Cohen, F. E., Abarbanel, R. M., Kuntz, I. D., and Fletterick, R. J. (1986). *Science*, **234**, 349.
12. Gibrat, J. F., Garnier, J., and Robson, B. (1987). *J. Mol. Biol.*, **198**, 425.
13. Kabsch, W. and Sander, C. (1985). *Nature (London)*, **317**, 207.
14. Argos, P. (1987). *J. Mol. Biol.*, **197**, 331.
15. Sternberg, M. J. E. and Islam, S. A. (1990). *Protein Eng.*, **4**, 125.
16. Levin, J. M., Robson, B., and Garnier, J. (1986). *FEBS Lett.*, **205**, 303.
17. Nishikawa, K. and Ooi, T. (1986). *Biochim. Biophys. Acta*, **871**, 45.
18. Sweet, R. M. (1986). *Biopolymers*, **25**, 1565.
19. Salzburg, S. and Cost, S. (1992). *J. Mol. Biol.*, **227**, 371.
20. Zhang, X., Mesirov, J. P., and Waltz, D. J. (1992). *J. Mol. Biol.*, **225**, 1049.
21. Simpson, P.F. (1990). *Artificial neural systems.* Pergamon Press.
22. Rumelhart, D. E., Hinton, G. E., and Williams, R. J. (1986). *Nature*, **323**, 533.
23. Qian, N. and Sejnowski, T. J. (1988). *J. Mol. Biol.*, **202**, 865.
24. Bohr, H., Bohr, J., Brunak, S., Cotterill, R. M. J., Lautrup, B., Norskov, L., *et al.* (1988). *FEBS Lett.*, **241**, 223.
25. Holley, L. H. and Karplus, M. (1989). *Proc. Natl. Acad. Sci. USA*, **86**, 152.
26. McGregor, M. J., Flores, T. P., and Sternberg, M. J. E. (1989). *Protein Eng.*, **2**, 521.
27. McGregor, M. J., Flores, T. P., and Sternberg, M. J. E. (1990). *Protein Eng.*, **3**, 459.
28. Kneller, D. G., Cohen, F. E., and Langridge, R. (1990). *J. Mol. Biol.*, **214**, 171.
29. Stolorz, P., Lapedes, A., and Xia, Y. (1992). *J. Mol. Biol.*, **225**, 363.

30. Sasagawa, F. and Tajima, T. (1993). *Comput. Appl. Biosci.*, **9**, 147.
31. Lim, V. I. (1974). *J. Mol. Biol.*, **80**, 857.
32. Presnell, S. R., Cohen, B. I., and Cohen, F. E. (1993). *Comput. Appl. Biosci.*, **9**, 373.
33. King, R. D. and Sternberg, M. J. E. (1990). *J. Mol. Biol.*, **216**, 441.
34. Taylor, W. R. (1986). *J. Theor. Biol.*, **119**, 205.
35. Muggleton, S., King, R. D., and Sternberg, M. J. E. (1992). *Protein Eng.*, **5**, 647.
36. Bernstein, F. C., Koetzle, T. F., Williams, G., Meyer, D., Brice, M. D., Rodgers, J. R., *et al.* (1977). *J. Mol. Biol.*, **112**, 535.
37. Orengo, C. A., Flores, T. P., Taylor, W. R., and Thornton, J. M. (1993). *Protein Eng.*, **6**, 485.
38. Kabsch, W. and Sander, C. (1983). *FEBS Lett.*, **155**, 179.
39. Sklener, H., Etchebest, C., and Lavery, R. (1989). *Proteins*, **6**, 46.
40. Kendall, M. and Stuart, A. (1977). *The advanced theory of statistics*. Griffen and Company, London.
41. Weiss, S. M. and Kulikowski, C. A. (1991). *Computer systems that learn*. Morgan Kaufmann, San Mateo.
42. Kabsch, W. and Sander, C. (1983). *Biopolymers*, **22**, 2577.
43. Biou, V., Gibrat, J. F., Levin, J. M., Robson, B., and Garnier, J. (1988). *Protein Eng.*, **2**, 185.
44. Zvelebil, M. J. J. M., Barton, G. J., Taylor, W. R., and Sternberg, M. J. E. (1987). *J. Mol. Biol.*, **195**, 957.
45. Boscott, P. E., Barton, G. J., and Richards, W. G. (1993). *Protein Eng.*, **6**, 261.
46. Wako, H. and Blundell, T. L. (1994). *J. Mol. Biol.*, **238**, 693.
47. Srinivasen, A., Muggleton, S., and Bain, M. (1992). In *International workshop on inductive logic programming*. Institute for New Generation Computer Technology, Tokyo.
48. Kraulis, P. J. (1991). *J. Appl. Crystallogr.*, **24**, 946.

Appendix 1 Inductive logic programming (ILP)

The use of the ILP program GOLEM is illustrated by working through the following simple problem. Consider the problem of learning a prediction rule for the following primary and secondary structures (*Table 4*). In ILP, the examples and rules learnt from them are represented in a logic programming language, e.g. PROLOG. The input *observations* to GOLEM are the secondary structure positions. The positions with α-helix structure (positive examples) are given in a file as follows:
alpha(6), alpha(5), alpha(7), alpha(8), alpha(9), alpha(10).
The positions with coil structure (negative examples) are given in another file:

alpha(1), alpha(2), alpha(3), alpha(4), alpha(5), alpha(11), alpha(12), alpha(13), alpha(14), alpha(15).

The input *background knowledge* is given in a third file. This background knowledge is represented by the predicates: position, hydrophobic, not_pro,

Table 4. Illustrative protein secondary structure prediction problem

No.	**2°**	**1°**	**Class**
1	_	A	not_pro and hydrophobic
2	_	P	pro
3	_	P	pro
4	_	D	not_pro
5	_	P	pro
6	H	D	**not_pro**
7	H	M	**not_pro and hydrophobic**
8	H	D	not_pro
9	H	T	**not_pro**
10	H	M	**not_pro and hydrophobic**
11	_	N	not_pro
12	_	P	pro
13	_	Q	not_pro
14	_	P	pro
15	_	K	not_pro

No, residue number; 1°, primary structure; 2°, secondary structure (H, α-helix; —, coil); class, classification of the residue (classes in bold are from the pairs of random examples selected by GOLEM). The classes hydrophobic {H, W, Y, F, M, L, I, V, C, A, G, T, K}, pro {P}, and not_pro {A, C, D, E, F, G, H, I. K. L, M, N, O, Q, R, S, T, V, W, Y} are used.

and succ. The position predicate represents the primary structure, e.g. position(2,P). states that the residue at position 2 is a proline. The hydrophobic predicate represents the class of hydrophobic residues, e.g. hydrophobic(V). states that valine is hydrophobic. The not_pro predicate represents the class of residues that are not proline, e.g. not_pro(V). states that valine is not proline. The succ predicate represents positional (arithmetic) information, e.g. succ(2,3). states that position 2 is one place closer to the N-terminal than position 3.

When told 'to induce', the GOLEM algorithm randomly picks a sample of pairs of examples, e.g. the pairs: (6,7) and (9,10). For each of these pairs GOLEM constructs a rule (hypothesis) in PROLOG where the body of the rule states all the common properties of the two examples. Both examples produce the same rule:
alpha(Pos1):-

position(Pos1, Residue1), not_pro(Residue1), succ(Pos1, Pos2),
position(Pos2, Residue2), not_pro(Residue2), hydrophobic(Residue2).

A residue position is an α-helix if:

it is not proline and
it is followed by a residue that is not proline and hydrophobic

Experimentation is carried out to test the rule. It is found that the rule covers two positive examples and no negative examples. GOLEM then tries

to expand the rule by extending the best pairs to triples, e.g. (6,7,8). This produces the rule that has all common properties of the triplet:
alpha(Pos1):-

position(Pos1, Residue1), not_pro(Residue1), succ(Pos1, Pos2),

position(Pos2, Residue2), not_pro(Residue2).

A residue position is an α-helix if:

- it is not proline and
- it is followed by a residue that is not proline

This rule covers all five positive examples and no negative examples. GOLEM therefore concludes that this rule is the best that can be found.

Many of the features of GOLEM are not demonstrated in this simple problem. The background knowledge contains only the simple relation succ, and so the problem could have been dealt with using much simpler learning algorithms. It is also not usual to learn from a set containing all possible examples. The examples are usually broken up into training and testing examples to check how accurate the rules are on unseen examples. In addition, real sets of examples usually contain errors. GOLEM has a sophisticated method of avoiding overfitting and dealing with noise (47).

5

Prediction of transmembrane protein topology

GUNNAR VON HEIJNE

1. Introduction

Membrane proteins fulfil many critical cellular functions, and make up a large fraction of all proteins. In fact, it has recently been estimated that as many as 35–40% of all yeast genes may code for integral membrane proteins (1,2). Yet, only a handful of high resolution membrane protein three-dimensional structures are known, making it even more important to extract as much structural information as possible from the amino acid sequences. This chapter will focus on the topology prediction problem: how to identify transmembrane segments and how to predict the overall orientation of the protein in the membrane.

2. Structural classes: helix bundles and β-barrels

So far as is known, integral membrane proteins are built according to one of two architectural principles: as helix bundles or β-barrels, *Figure 1*. In the helix bundle proteins, the membrane spanning segments are formed by long stretches of predominantly apolar amino acids that fold into transmembrane α-helices. Typical examples are the photosynthetic reaction centre (3) and bacteriorhodopsin (4) which have 11 and 7 transmembrane helices, respectively. Helix bundle proteins are found in nearly all cellular membranes, and serve as cell surface receptors, ion channels, active and passive transporters, and membrane-bound biosynthetic or proteolytic enzymes.

β-Barrels are large antiparallel sheets rolled into a cylindrical structure where the strands have apolar amino acids that project from the outer surface of the barrel at every second position in the sequence. A number of porins from the outer membrane of Gram-negative bacteria have been shown to form β-barrels (5,6). Porins form passive but selective diffusion pores for small solutes across the outer bacterial membrane, and may also exist in the outer membrane of mitochondria.

So far, no mixed helix bundle/β-barrel structures have been unequivocally

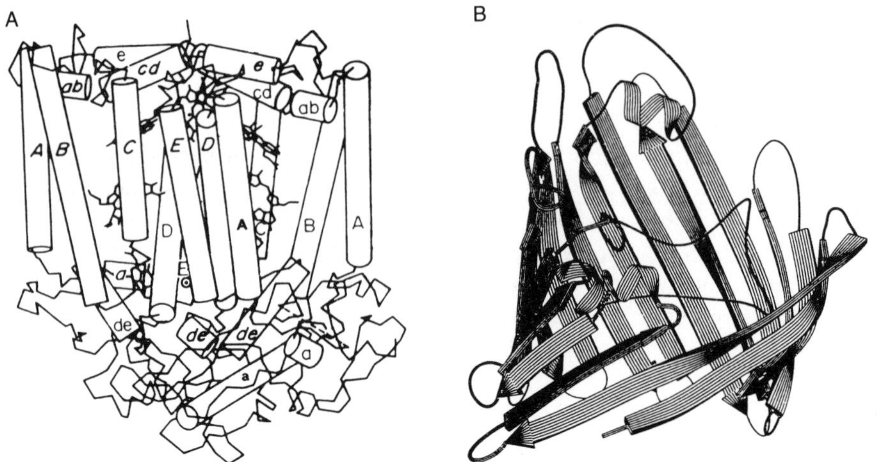

Figure 1. Helix bundle and β-barrel proteins. (A) The *Rb. sphaeroides* photosynthetic reaction centre (reproduced from ref. 41 with permission). (B) The *R. capsulatus* porin (reproduced from ref. 5 with permission).

shown to exist, though recent electron microscopy data on the acetylcholine receptor have been interpreted as a central five-helix bundle surrounded by a large outer ring of β-strands (7).

3. Membrane insertion: setting-up the topology

3.1 Helix bundle proteins

An important characteristic of helix bundle proteins from membrane systems as diverse as the bacterial inner membrane, the eukaryotic plasma membrane, the mitochondrial inner membrane, and the chloroplast thylakoid membrane is the so-called positive inside rule (8), namely the observation that positively charged residues (Arg and Lys) tend to be much more frequent in non-translocated as compared to translocated parts of the protein. It has also been shown experimentally both in *E. coli* and in mammalian cells that positively charged residues in regions flanking the apolar transmembrane segments are very potent topological determinants, whereas negatively charged residues have a much weaker influence on the topology (8). In *E. coli*, the available data suggest that membrane insertion is in most cases a locally determined process where individual 'helical hairpins' composed of two neighbouring apolar segments and a connecting loop insert as units. A possible exception is when the translocated loop is very long (> 60–70 residues), in which case the N-terminal apolar segment serves as a signal sequence that triggers translocation of the loop through the so-called *sec*-machinery (9). Long loops that are translocated by the *sec*-machinery can contain any number of positively charged residues, and hence do not conform to the positive inside rule (10).

3.2 β-Barrel proteins

β-Barrel proteins are first translocated through the inner bacterial membrane by the *sec*-machinery, transiently reside in the periplasmic space, and then insert into the outer membrane (11). Many porins form very stable oligomers in the membrane, and membrane insertion may be dependent on oligomerization (12). Not much is known about the membrane insertion process, and no equivalent of the positive inside rule has been discovered.

4. Prediction of transmembrane helices and membrane topology

The basic facts that can be used to predict the location of transmembrane helices and their orientations are:

(a) Transmembrane helices are predominantly apolar and 15–30 residues long.
(b) Short (< 60 residues) loops and tails follow the positive inside rule.
(c) Long (> 60 residues) translocated and non-translocated loops and tails have statistically detectable differences in amino acid composition not related to the positive inside rule (13).
(d) In *E. coli* proteins and possibly also in eukaryotic proteins, N-terminal tails follow the positive inside rule irrespective of length (our unpublished observation).
(e) In eukaryotic but not prokaryotic proteins, the net charge difference across the most N-terminal transmembrane segment (counting both positively and negatively charged residues and including 15 flanking residues on each end) correlates with its orientation (the positive end remains cytoplasmic) (14,15).

4.1 Hydrophobicity plots

In all current prediction schemes, candidate transmembrane helices are first identified by some kind of hydrophobicity plot, i.e. a graph showing the local hydrophobicity of the amino acid sequence as a function of position. To make a hydrophobicity plot, one has to choose a hydrophobicity scale and an averaging procedure.

Hydrophobicity scales come in all sizes and shapes, *Table 1*, and many attempts to find the best scale have been made. However, 'best' means different things to different authors: the best scale may be the one that is best at predicting the number of transmembrane segments without worrying about

Table 1. Some hydrophobicity scales specifically designed for membrane protein structure prediction

Aa	KD	GES	Argos	vH	PG	Edel21	Eis
Ala	1.8	1.6	1.6	0.267	13.85	0.4397	0.25
Cys	2.5	2.0	1.2	1.806	15.37	1.150	0.04
Asp	−3.5	−9.2	0.1	−2.303	11.61	−2.588	−0.72
Glu	−3.5	−8.2	0.2	−2.442	11.38	−1.270	−0.62
Phe	2.8	3.7	2.0	0.427	13.93	0.4345	0.61
Gly	−0.4	1.0	0.6	0.160	13.34	−0.8634	0.16
His	−3.2	−3.0	0.3	−2.189	13.82	0.0268	−0.40
Ile	4.5	3.1	1.7	0.971	15.28	1.546	0.73
Lys	−3.9	−8.8	0.2	−2.996	11.58	−1.502	−1.10
Leu	3.8	2.8	2.9	0.623	14.13	1.517	0.53
Met	1.9	3.4	3.0	0.136	13.86	1.746	0.26
Asn	−3.5	−4.8	0.3	−1.988	13.02	−1.414	−0.64
Pro	−1.6	−0.2	0.8	−0.451	12.35	−1.721	−0.07
Gln	−3.5	−4.1	0.5	−1.814	12.61	−1.656	−0.69
Arg	−4.5	−12.3	0.5	−2.749	13.10	−0.7010	−1.80
Ser	−0.8	0.6	0.8	−0.119	13.39	−0.3841	−0.26
Thr	−0.7	1.2	0.9	−0.083	12.70	−0.0078	−0.18
Val	4.2	2.6	1.1	0.721	14.56	0.5056	0.54
Trp	−0.9	1.9	1.1	−0.875	15.48	−0.0638	0.37
Tyr	−1.3	−0.7	0.7	−0.386	13.88	−0.4585	0.02

Positive values indicate hydrophobic residues. References: KD (38), GES (39), Argos (40), vH (26), PG (22), Edel21 (20), Eis (34).

their precise ends, or best at predicting the number of membrane embedded residues, or best at predicting the precise points at which transmembrane helices start and end (these points may be some distance away from the surface of the lipid bilayer). The latter two criteria are difficult to apply since a test set of high resolution three-dimensional structures is required. To date, there are only four helical structures known to sufficient resolution: the photosynthetic reaction centre with 11 transmembrane helices (3), bacteriorhodopsin with 7 transmembrane helices (4), and the coat proteins from phages Pf1 and M13, each with one transmembrane helix (16–19). As lipids or detergent molecules are not visible in these structures, it is also difficult to decide where the lipid headgroup region is located relative to the protein. Not surprisingly, different authors come to different conclusions as to which is the optimal scale (20–24), and the best advice seems to be to try a couple of scales and look for consistently predicted candidate transmembrane segments.

Averaging often helps to bring out the hydrophobic regions more clearly, though some authors prefer to use non-averaged profiles and predict transmembrane segments on the basis of rules such as 'search for a continuous sequence of 20–24 points above the average line with a maximum of two non-adjacent exceptions' (22). Averaging windows can be rectangular or of some other shape (triangular, trapezoid, etc.). Triangular or trapezoid windows produce somewhat smoother profiles than rectangular ones, and may also be preferred on the grounds that they more accurately represent the gradual transition from the lipophilic centre of the bilayer to the aqueous surroundings (25,26). Typical window sizes are 15–25 residues (corresponding to the number of residues required to span the membrane in a helical conformation), though smaller windows are used when the aim is to accurately predict the ends of the helices.

No published method based on hydrophobicity analysis is 100% accurate in identifying the correct number of transmembrane helices in proteins of known topology, and most methods are rather poor at predicting the precise helical ends (22). When large samples of proteins are analysed, one always finds a considerable overlap between the distributions of peak hydrophobicities of bona fide transmembrane segments and of non-membrane segments (27), and candidate segments with hydrophobicities in the overlap region cannot be predicted as transmembrane with any certainty.

4.2 Charge bias analysis

A significant improvement in the prediction of transmembrane topology can be obtained by incorporation of the positive inside rule. According to this rule, all acceptable topologies must have the loops with high numbers of Arg and Lys residues on the cytoplasmic side. This idea was first incorporated in the TOP-PRED algorithm (15,26), which is described in *Protocol 1*. TOP-PRED is available in Macintosh and Unix version from the author.

Protocol 1. The TOP-PRED algorithm

1. Construct a hydrophobicity plot using the GES scale and a trapezoid averaging window.
2. Identify all 'certain' transmembrane helices defined as those peaks with a hydrophobicity value above an upper cut-off C1.
3. Identify all 'putative' transmembrane helices defined as those peaks with a hydrophobicity value below the upper cut-off C1 but above a lower cut-off C2.
4. Construct all possible topologies that include all the certain transmembrane segments and either include or exclude the putative segments.
5. For each possible topology, calculate the difference Δ_+ in the number of Arg + Lys residues between the two sides of the structure. Exclude all loops (and the C-terminal tail) with lengths > 60 residues from the calculation, but include the N-terminal tail irrespective of its length.
6. Choose the topology with the largest Δ_+ as the best prediction.

Protocol 1. *Continued*

7. If many loops have lengths > 60 residues, their overall amino acid compositions may provide some guidance as to their extra- or intracellular location.

An example is shown in *Figure 2.*

With an appropriate choice of parameters, the TOP-PRED method works surprisingly well for bacterial inner membrane proteins, where at least 90% of the 35 or so known topologies are correctly predicted. Expressed in terms of transmembrane segments, at least 99% are correctly identified to within, say, ± 5–10 residues. Bacterial proteins generally have a very clear charge bias that is equally strong for N-terminal and C-terminal loops, and also have very few long loops and tails. They are thus ideally suited for this kind of approach. Eukaryotic proteins, on the other hand, often have very long extramembraneous domains and sometimes have rather weakly hydrophobic transmembrane segments; in addition, the charge bias tends to become less evident for the more C-terminal loops (15). This makes topology prediction more difficult, although some guidance can be obtained from the net charge difference across the most N-terminal transmembrane segment, as well as from the overall amino acid compositions of long (> 60 residues) loops.

A computationally more advanced method based on the same general ideas has recently been developed by Jones *et al.* (28), who report that 64 out of 83 known topologies (a mixture of prokaryotic and eukaryotic proteins) are correctly predicted.

4.3 Prediction of helix ends

For modelling purposes, two kinds of helix 'ends' need to be identified: the points at which the transmembrane helix leaves the lipid bilayer, and the physical ends of the helical stretch. The appearance of charged, polar, or aromatic residues may signal the point of exit from the bilayer (8,29,30). In addition, since, in the membrane embedded parts of the transmembrane helices, the more hydrophobic residues tend to face the lipid and also tend to be more variable between homologous sequences, while in helical segments exposed to the aqueous phase the more polar residues tend to be more variable and face the surrounding medium, one would expect that the point of exit from the bilayer would be signalled by a transition from a situation where the more hydrophobic and more variable sides of the helix coincide to a situation where the more polar and more variable sides coincide. This in fact appears to be the case (31).

There is no published method for specifically identifying the physical ends of transmembrane helices, though it would seem reasonable to use secondary structure prediction methods developed for globular proteins to look for helical N- or C-caps and segments of high turn probability (32,33). With the very

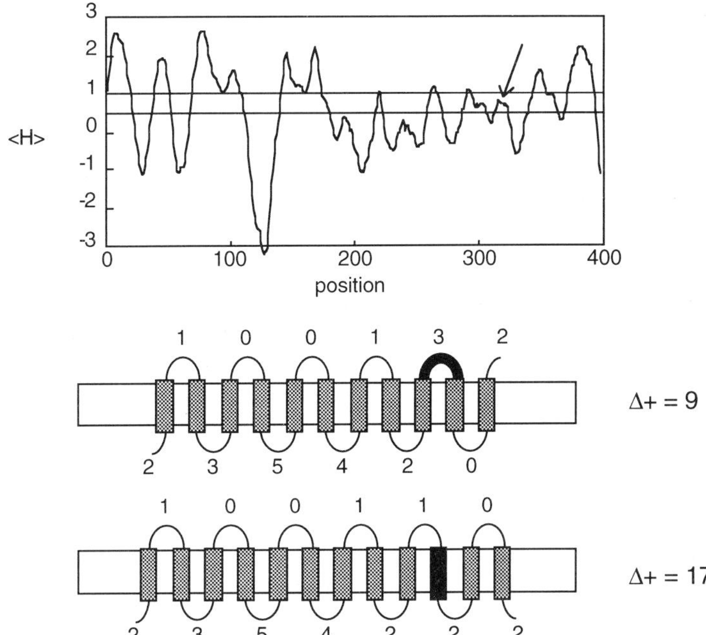

Figure 2. TOP-PRED prediction of the topology of the *E. coli* LacY protein. The hydrophobicity plot (top) is shown with the two cut-offs C1 and C2 marked. A peak classified as a 'putative' transmembrane helix is indicated by the *arrow*. The bottom panels show the two possible topologies, one excluding and one including the 'putative' helix. The bottom one is chosen as the best prediction on the basis of its greater charge bias; this is also the experimentally determined topology (reproduced from ref. 26 with permission).

small database of known three-dimensional structures presently available, it would be difficult if not impossible to assess the accuracy of such methods.

4.4 Hydrophobic moment analysis

An additional parameter that is useful when analysing helix bundle proteins is the hydrophobic moment (34). This measures the amphiphilicity of a helical segment (or any other periodic conformation such as a β-strand) as viewed along the axis of a helical wheel plot, and is calculated as:

$$\mu_H = \frac{1}{N}\left\{\left[\sum_{n=1}^{N} H_n \sin(\partial n)\right]^2\right\} + \left\{\left[\sum_{n=1}^{N} H_n \cos(\partial n)\right]^2\right\}^{1/2}$$

where N is the number of residues in the segment, H_n is the hydrophobicity of residue n, and δ is the angle between two successive residues ($\delta = 100°$ for an ideal α-helix and $\delta = 180°$ for a fully extended β-strand). The hydrophobic moment can be used to quantitate the degree of amphiphilicity of a helix, and typically varies between 0 and 0.4 (measured over 11 residues using the Eis scale) for a transmembrane segment and between 0.5 and 1.0 for a surface-seeking helix such as found in lytic peptides. The direction of the hydrophobic moment vector can also be used to predict the face of a transmembrane helix that will be exposed to the lipid environment.

5. Prediction of transmembrane β-strands

The three-dimensional structure of three different outer membrane porins are known, and have prompted the prediction of transmembrane β-strands in other porins based on sequence alignments (35,36). A more generally applicable method (37) is based on the fact that, in the transmembrane strands, every second residue is hydrophobic and faces the lipid (the inward facing residues can be either hydrophobic or hydrophilic). In addition, almost every strand in the known structures is flanked by aromatic residues (Phe, Tyr, or Trp). To identify transmembrane strands, a 'sided hydrophobicity' profile is calculated as:

$$H_s(i) = \frac{1}{4}[h(i-2) + h(i) + h(i+2) + h(i+4)]$$

with h-values taken from the hydrophobicity scale of Eisenberg (34). When residue $i - 2$ or $i + 4$ is aromatic its hydrophobicity value is increased to 1.6; this will bias the prediction toward strands with aromatic residues at the ends. When applied to known structures, at least 15 of the 16 transmembrane strands can be identified by this method, with only two or three overpredictions. If observed high sequence variability or high turn probability are used to filter the prediction, most of the overpredictions can be removed.

Acknowledgements

Supported by grants from the Swedish Natural Sciences Research Council (NFR), the Swedish Technical Research Council (TFR), and the Swedish National Board for Industrial and Technical Development (NUTEK).

References

1. Goffeau, A., Nakai, K., Slominski, P., and Risler, J. L. (1993). *FEBS Lett.*, **325**, 112.
2. Goffeau, A., Slonimski, P., Nakai, K., and Risler, J. L. (1993). *Yeast*, **9**, 691.
3. Deisenhofer, J. and Michel, H. (1991). *Annu. Rev. Biophys. Biophys. Chem.*, **20**, 247.
4. Henderson, R., Baldwin, J. M., Ceska, T. A., Zemlin, F., Beckmann, E., and Downing, K. H. (1990). *J. Mol. Biol.*, **213**, 899.
5. Weiss, M. S., Abele, U., Weckesser, J., Welte, W., Schiltz, E., and Schulz, G. E. (1991). *Science*, **254**, 1627.

6. Cowan, S. W., Schirmer, T., Rummel, G., Steiert, M., Ghosh, R., Pauptit, R. A., *et al.* (1992). *Nature*, **358**, 727.
7. Unwin, N. (1993). *J. Mol. Biol.*, **229**, 1101.
8. von Heijne, G. (1994). *Annu. Rev. Biophys. Biomol. Struct.*, **23**, 167.
9. Andersson, H. and von Heijne, G. (1993). *EMBO J.*, **12**, 683.
10. von Heijne, G. and Gavel, Y. (1988). *Eur. J. Biochem.*, **174**, 671.
11. Bosch, D., Voorhout, W., and Tommassen, J. (1988). *J. Biol. Chem.*, **263**, 9952.
12. Struyve, M., Moons, M., and Tommassen, J. (1991). *J. Mol. Biol.*, **218**, 141.
13. Nakashima, H. and Nishikawa, K. (1992). *FEBS Lett.*, **303**, 141.
14. Hartmann, E., Rapoport, T. A., and Lodish, H. F. (1989). *Proc. Natl. Acad. Sci. USA*, **86**, 5786.
15. Sipos, L. and von Heijne, G. (1993). *Eur. J. Biochem.*, **213**, 1333.
16. Shon, K. J., Kim, Y. G., Colnago, L. A., and Opella, S. J. (1991). *Science*, **252**, 1303.
17. Nambudripad, R., Stark, W., Opella, S. J., and Makowski, L. (1991). *Science*, **252**, 1305.
18. McDonnell, P. A., Shon, K., Kim, Y., and Opella, S. J. (1993). *J. Mol. Biol.*, **233**, 447.
19. Henry, G. D. and Sykes, B. D. (1992). *Biochemistry*, **31**, 5284.
20. Edelman, J. (1993). *J. Mol. Biol.*, **232**, 165.
21. Turner, R. J. and Weiner, J. H. (1993). *Biochim. Biophys. Acta*, **1202**, 161.
22. Ponnuswamy, P. K. and Gromiha, M. M. (1993). *Int. J. Pept. Protein Res.*, **42**, 326.
23. Fasman, G. D. and Gilbert, W. A. (1990). *Trends Biochem. Sci.*, **15**, 89.
24. Cornette, J. L., Cease, K. B., Margalit, H., Spouge, J. L., Berzofsky, J. A., and De Lisi, C. (1987). *J. Mol. Biol.*, **195**, 659.
25. Claverie, J.-M. and Daulmiere, C. (1991). *CABIOS*, **7**, 113.
26. von Heijne, G. (1992). *J. Mol. Biol.*, **225**, 487.
27. von Heijne, G. (1986). *EMBO J.*, **5**, 3021.
28. Jones, D. T., Taylor, W. R., and Thornton, J. M. (1994). *Biochemistry*, **33**, 3038.
29. Ballesteros, J. A. and Weinstein, H. (1992). *Biophys. J.*, **62**, 107.
30. Landolt-Marticorena, C., Williams, K. A., Deber, C. M., and Reithmeier, R. A. F. (1993). *J. Mol. Biol.*, **229**, 602.
31. Donnelly, D. and Cogdell, R. J. (1993). *Protein Eng.*, **6**, 629.
32. Presta, L. G. and Rose, G. D. (1988). *Science*, **240**, 1632.
33. Juretic, D., Lee, B., Trinajstic, N., and Williams, R. W. (1993). *Biopolymers*, **33**, 255.
34. Eisenberg, D., Schwarz, E., Komaromy, M., and Wall, R. (1984). *J. Mol. Biol.*, **179**, 125.
35. Jeanteur, D., Lakey, J. H., and Pattus, F. (1991). *Mol. Microbiol.*, **5**, 2153.
36. Welte, W., Weiss, M. S., Nestel, U., Weckesser, J., Schiltz, E., and Schulz, G. E. (1991). *Biochim. Biophys. Acta*, **1080**, 271.
37. Schirmer, T. and Cowan, S. W. (1993). *Protein Sci.*, **2**, 1361.
38. Kyte, J. and Doolittle, R. F. (1982). *J. Mol. Biol.*, **157**, 105.
39. Engelman, D. M., Steitz, T. A., and Goldman, A. (1986). *Annu. Rev. Biophys. Biophys. Chem.*, **15**, 321.
40. Argos, P., Rao, J. K. M., and Hargrave, P. A. (1982). *Eur. J. Biochem.*, **128**, 565.
41. Rees, D. C., Komiya, H., Yeates, T. O., Allen, J. P., and Feher, G. (1989). *Annu. Rev. Biochem.*, **58**, 607.

6

Comparative modelling of proteins

N. SRINIVASAN, K. GURUPRASAD, and TOM L. BLUNDELL

1. Introduction

The prediction of protein three-dimensional (3D) structure from amino acid sequence is most successful when the structures of one or more homologues are known. Structural information can then be extrapolated to the new sequence and a 3D model may be derived, well before the structure of the new protein is determined by X-ray crystallography or NMR. This approach is most appropriately known as comparative modelling, but is also referred to as homology modelling, or knowledge-based modelling.

The attractions of comparative modelling are several:

(a) Proteins have a limited number of folds and the structure of a new protein family can resemble a known fold even though there are no significant similarities in sequences.

(b) The number of proteins of known amino acid sequence is many times more than that of proteins of known 3D structure and the difference is widening despite significant developments in crystallography and NMR.

(c) As the models built using this approach are reasonably accurate they form a rational basis for explaining experimental observations in the light of tertiary structure and redesigning proteins with a view to improving their performance.

(d) A model can be used as a starting point in the determination of protein structure using NMR and X-ray crystallography.

1.1 Early modelling studies

The first modelling studies, based on knowledge of homologous or other proteins with a common fold, were carried out in the late 1960s and early 1970s and relied upon the construction of wire or plastic models (1). Only later were interactive computer graphics exploited (2). The first models were constructed by taking the existing coordinates of a single known structure, and altering those side chains not identical in the protein to be modelled. This approach to protein modelling is still employed today with considerable success, especially when the proteins are similar. When the sequences are more dissimilar (i.e. less than 30% sequence identity), models constructed on this basis can have significant problems. Most obviously, the backbone of the model closely resembles that of the structure employed in the modelling, deletions and especially insertions are not handled well, and there are often clashes between side chains. When relationships are very distant, there is little hope of producing a reliable model.

1.1.1 α-Lactalbumin

Browne *et al.* (1) published the first report on comparative modelling. They modelled bovine α-lactalbumin on the known three-dimensional structure of hen egg white lysozyme. These proteins contain identical patterns of disulfide bonds and the sequence identity between them is 39%. Apart from replacements of side chains, only deletions needed to be modelled; there are no insertions in α-lactalbumin. Using their procedures developed for the refinement of X-ray structures Warme *et al.* (3) produced a model for α-lactalbumin on the basis of the crystal structure of lysozyme. When the X-ray structure of α-lactalbumin was determined, Acharya *et al.* (4) reported that the two independent models were generally correct except for the carboxyl-terminal regions of the models where they differed from each other and from the crystal structure.

1.1.2 α-Lytic protease

McLachlan and Shotton (5) modelled α-lytic proteinase of the fungus *Myxobacter* 495 on the basis of the structures of the mammalian chymotrypsin and elastase. As the sequence identity between the sequence to be modelled and the known structures was of the order of 18%, modelling was a difficult task. Moreover, an alignment between the sequences showed a large number of gaps, including five which were between 6 and 19 residues in length.

Subsequently, James and co-workers (6,7) determined the crystal structure of α-lytic proteinase and compared the X-ray structure with the model. Although they found that portions of both domains were built correctly, misalignment of the sequences led to local errors. For residues 214–220 a four residue offset disrupted the active site and Ser 214, an invariant residue in serine proteinases, was replaced by an asparagine. As a result of their comparison, James and co-workers suggested that protein structure predictions should be used to establish homology but not structural features, especially of catalytic or substrate binding sites. In retrospect this can be seen to be a consequence of the difficulty not only in aligning the sequences but also in modelling the variable, mainly loop regions.

1.1.3 Insulin-related structures

The X-ray structure of the first polypeptide hormone, porcine pancreatic insulin (8), formed the basis for modelling other insulins and many important related hormones including insulin-like growth factors (9). A recent NMR study of the structure of human insulin-like growth factor by Campbell and co-workers (10) confirmed the general tertiary structure, but found that the C peptide was disordered in solution. Bedarker *et al.* (11) and Issacs *et al.* (2) modelled porcine relaxin, which is distantly related to insulin but has a conserved pattern of three disulfide bridges and two invariant glycines. Although the sequence identity did not extend to the core, this was found to be conserved as hydrophobic, giving support to the proposal that relaxin adopted an insulin-like fold. Recent X-ray analysis of human relaxin by Kossiakof and co-workers (12) confirmed the insulin-like fold.

1.1.4 Mammalian serine proteinases

Greer (13) was the first to model variable regions on the basis of equivalent regions from homologous proteins of known structure. He superimposed the structures of trypsin, elastase, and chymotrypsin and found many equivalent C^{α} atoms within 1.0 Å of one another; all of the remaining positions corresponded to solvent exposed loop regions where the insertions/deletions were located. When the model sequence (heavy chain of haptoglobin) was aligned with the basis structures, the insertions and deletions relative to the known structures also corresponded to the loops. Greer's strategy was to build the main chain of both the spatially conserved and variable loop regions from fragments of each of the three known structures. The construction of the loop regions, however, was not as straightforward as that for the structurally conserved regions. Although deletions of one or two residues were easily accommodated and similar length loops were extrapolated from one of the homologous structures, one long loop was only partially modelled. Side chains were modelled according to the conformation found at the equivalent positions for those identical side chains in the known structures. Greer (14) subsequently applied this approach to the modelling of a number of different serine proteinases. A comparison of the model of trypsin and the crystal structure by James and co-workers (15) showed that the regions involved in substrate binding were inaccurately modelled.

1.1.5 Renin

Structural analysis of renin and renin–inhibitor complexes was stimulated by their involvement in release of the hormone angiotensin from angiotensinogen and the role of angiotensin in blood pressure regulation. The first models for renin (16–19) were constructed using the three-dimensional structure of the distantly related fungal proteinases, as no refined, high resolution structure of a mammalian enzyme was available. Later structures were constructed using the structure of pepsin and chymosin as those became available in the refined forms (20,21). Inaccuracies in the models arose from the differences in the arrangement of helices and strands between the mammalian and fungal aspartic proteinases, as well as the rather different variable regions that are found in renins. Nevertheless, the catalytic and active site cleft in general was well modelled and the models have been widely exploited in the design of antihypertensives. A more detailed retrospective analysis of a renin model is dealt in Section 3.1.

2. Different approaches to comparative modelling

Since the mid 1980s a large number of models of other proteins have been constructed and reported in the literature. Many of the models were built using rule-based procedures encoded as computer programs. A list of models reported in the literature is given in Johnson *et al.* (22) with full references.

Figure 1 shows a typical flow chart for protein modelling which includes learning from existing protein structures and applying the knowledge available to model a novel protein (23). Some of the stages are not described in detail in this chapter as they are discussed in other chapters of this book and in Johnson *et al.* (22).

The first step in the comparative modelling is to identify the known 3D structures that will form the basis for the structure of the unknown. This may be trivial if the sequence identity between the model sequence and the homologues of known structure is high (generally more than 35%). In situations where sequence identity is low, the recognition of the fold is a very difficult task. Many approaches to fold recognition have been proposed (24–30) as discussed in Chapter 8.

Once the homologues of known three-dimensional structure have been identified, the next non-trivial task is the alignment of the sequence to be modelled with those of the known structure. This is at the heart of comparative modelling and is the most frequent and serious source of errors. The correct sequence alignment is straightforward if the percentage identities between the compared sequences is high (say, more than 45%). Sequence alignment becomes extremely difficult when the sequence identity is low (25% or below).

Automated and rule-based modelling procedures have been developed in recent years in order to minimize subjective manual decisions. Such modelling techniques fall into two classes:

- the assembly of rigid fragments from homologous and other proteins of known structure
- the use of restraints such as interatomic distances to construct models that have the best agreement with homologues of known structure

2.1 Modelling—fragment based

Many approaches depend on the assembly of rigid fragments from known protein structures. Local main chain and side chain conformers from equivalent fragments in known structures are extrapolated to the sequence of the

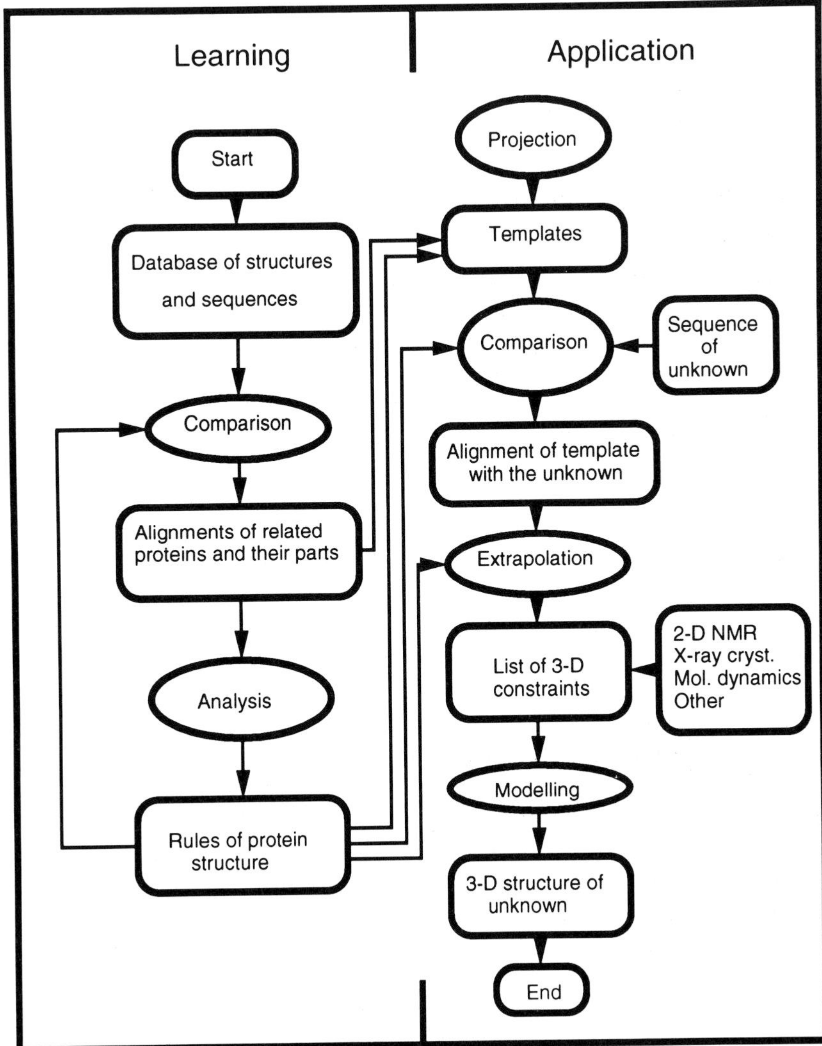

Figure 1. A flow chart for rule-based modelling of proteins. The learning process comprises of the derivation of rules from the comparisons of sequences and three-dimensional structures, and application involves extrapolation of the rules to the sequence of a new protein and the construction of a three-dimensional model. Figure used with permission (23).

unknown. Jones and Thirup (31) were the first to demonstrate that a protein structure can be built from a combination of segments from other proteins, and this was supported by the subsequent studies by Unger *et al.* (32), Claessens *et al.* (33), and Levitt (34). The comparative modelling software COMPOSER (*Figure 2*) developed originally by Blundell and co-workers

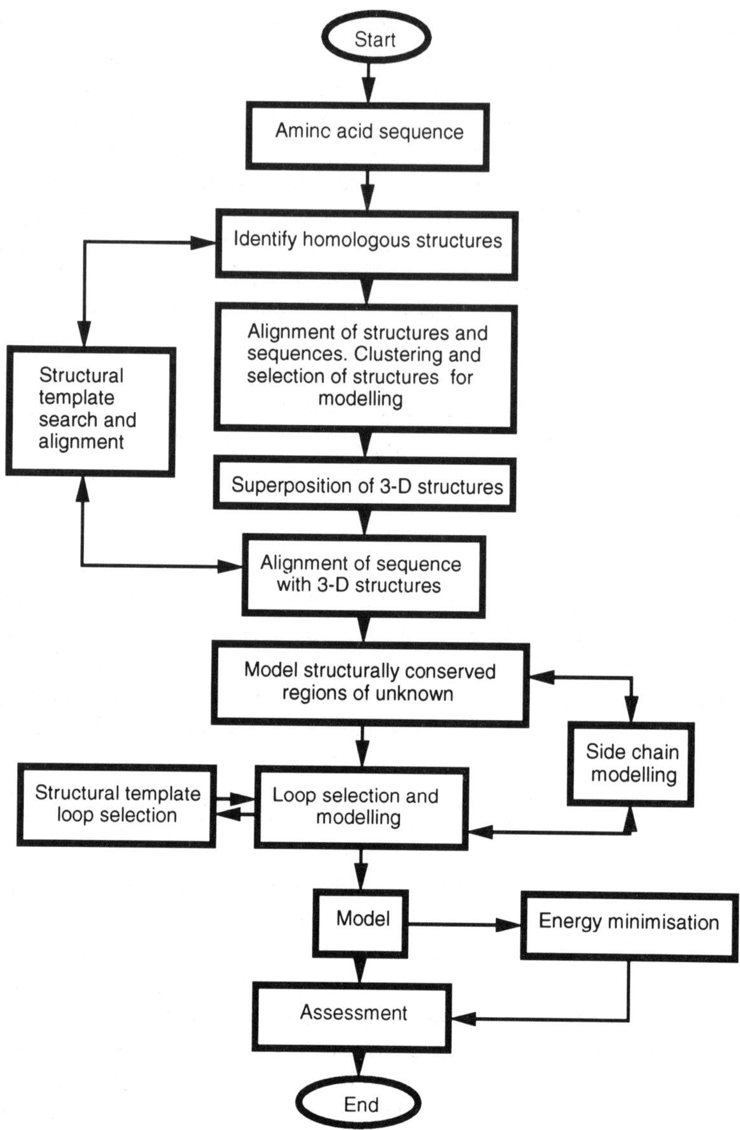

Figure 2. A flow chart for the comparative modelling procedure, COMPOSER. Various steps in COMPOSER and the tools used are indicated. Figure used with permission (22).

(35,36) and incorporated in the software SYBYL (Tripos Inc., St. Louis) also depends on the assembly of rigid fragments.

2.1.1 Modelling conserved regions

Comparative protein modelling emphasizes that accuracy is likely to be greatest when the segments of the model are selected from homologues. The first step is to assemble a set of rigid fragments which may represent the 'framework'. We define the framework of a family of related structures as the mean structure (or weighted mean structure) of the superimposed structurally conserved regions (SCRs), which are often helices and extended strands (37). There are many sophisticated superposition procedures (see review in ref. 22). In the comparative modelling approach encoded in COMPOSER (35,36) the framework is identified by simultaneous rigid body superposition of homologues of known structure (37). The mean of topologically equivalent C^{α} atoms, defined by a distance cut-off (typically 2.5 Å), constitutes the framework. The contributions of each of the homologues is best weighted, preferably by the square of sequence identity with the unknown (38). For every SCR in the unknown, the corresponding SCR (in a known structure) with the greatest sequence similarity is superposed on the framework. *Figure 3* illustrates the superposition of subtilisin BPN′, proteinase K, and thermitase, which have been used in modelling subtilisin Carlsberg.

2.1.2 Modelling variable regions

Comparisons of structures of homologous proteins reveal that, while secondary structural elements tend to be well-conserved, loop regions accommodate most replacements, deletions, and insertions. Hence, the loop regions have proven particularly difficult to model accurately. However, the accurate modelling of loops may be crucial, as they are often involved in recognition processes, for example, the hypervariable regions of immunoglobulins. Loops are also known to be more mobile and disordered than the core regions. In some crystal structures the mean positions of the loops are undetermined due to dynamic or static disorder, although many loop structures are well determined in high resolution protein structures.

Most approaches to modelling variable regions involve a search for fragments of suitable length and end-to-end distances, together with a check that the modelled loop does not clash with the rest of the protein (*Figure 4A*). The identified segment is usually fitted to the anchor regions (the ends of the intervening regions in the model that are mainly helices and strands). Subtle 'tweaking' of single bonds is sometimes permitted to achieve the best superposition. The selection of the correct conformer can be improved by considering the rms difference (rmsd) in the overlap (anchor) regions and the sequence similarity between the identified segment and the one to be modelled. For short fragments, simple geometry-based criteria are sufficient to reproduce native conformations; but, for long segments, especially those with glycyl residues, the sequence similarity must be used as a constraint in searching for a suitable fragment. *Figure 3* also shows the loops built connecting framework regions of subtilisin Carlsberg.

A retrospective analysis of COMPOSER-built protein models whose crystal structures are available confirms that errors are greatest in the loops (38). The main reasons are that they vary in length, sequence, and conformation,

(a) Subtilisin, *Bacillus amyloliquefaciens* (1sbt)

(d) Fitted structures (1sbt, 1tcc, 2prk)

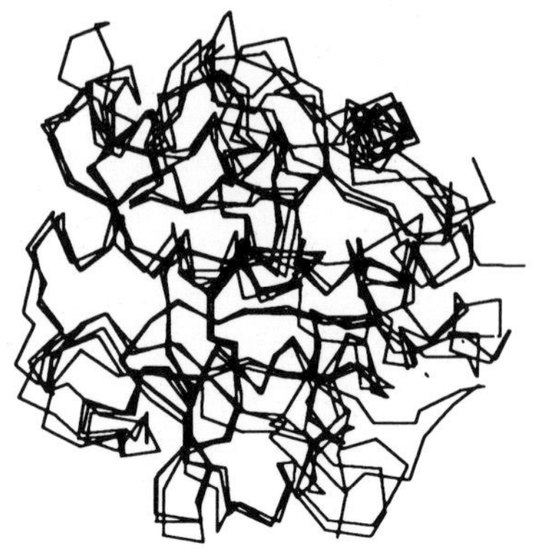

(b) Thermitase, *Thermoactinomyces vulgaris* (1tcc)

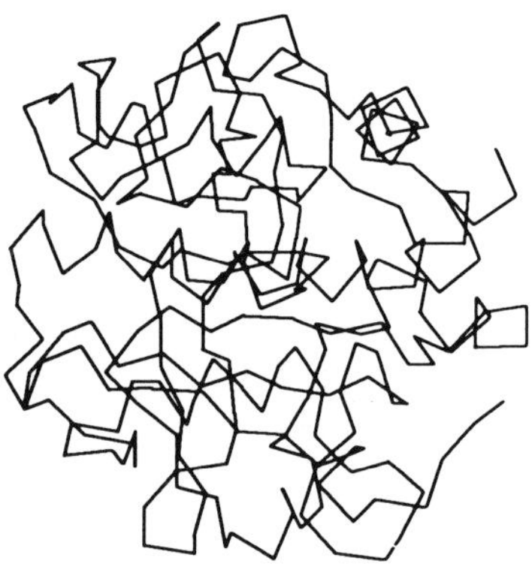

(e) Structurally conserved regions (1sbt, 1tcc, 2prk)

(c) Proteinase K, *Tritirachium album limber* (2prk)

(f) Average framework of the model

Figure 3. COMPOSER modelling of subtilisin Carlsberg (38) using the known structures of (a) subtilisin BPN′ (Brookhaven code: 1sbt), (b) thermitase (1tec), and (c) proteinase K (2prk). The multiple superposition of known structures (37) is shown in (d) and the structurally conserved regions (SCRs) in (e). The framework which represents the mean or

(g) Superposition of model with X-ray structure (1sbc)

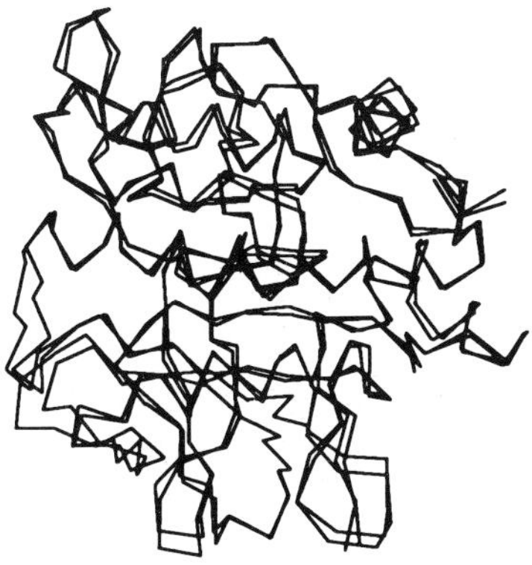

(h) Model – main chain and side chains

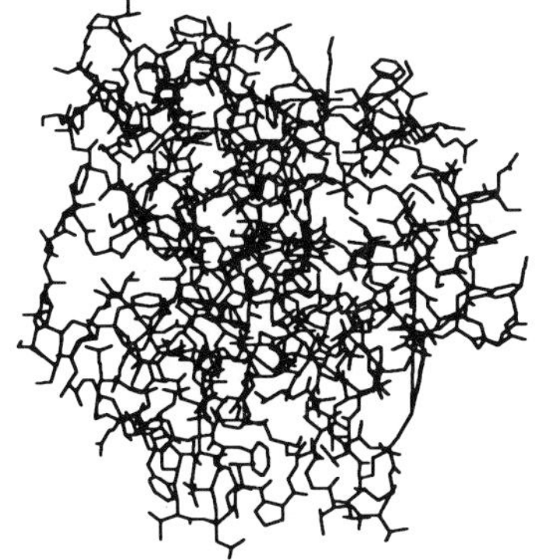

(i) X-ray structure (1sbc) – main chain and side chains

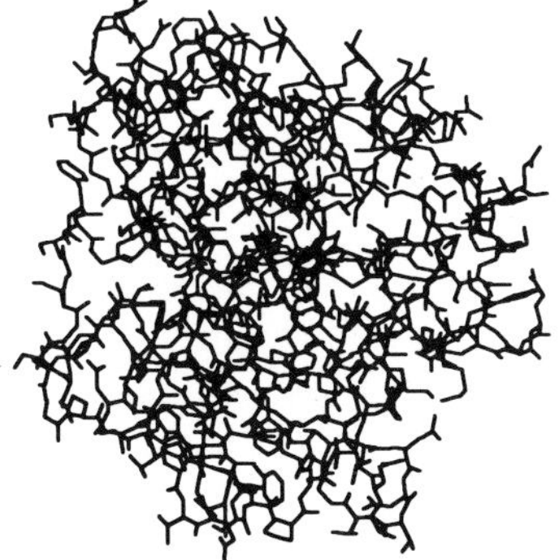

weighted mean of the SCRs is shown in (f). The superposition of C^{α} positions in the modelled and X-ray structure (1sbc) is shown in (g); (h) and (i) correspond to the model and X-ray structures respectively with all non-hydrogen atoms displayed. Figure used with permission (22).

even among the proteins in a given family. They undergo large movements about the mean position in the family as they are usually at the molecular surface rather than in a tightly packed hydrophobic region. The temperature factors are usually high and the electron density poorly defined.

i. Minimizing errors in loop modelling

Topham *et al.* (39) have developed a sophisticated procedure which ranks candidate loops by matching them with structural templates. The templates reflect the amino acid substitutions that are compatible with the local structural environment for each amino acid, defined in terms of main chain conformation, solvent accessibility, hydrogen bonding, disulfide bonding, and *cis* peptide conformations (24).

If a loop is modelled on the equivalent segment in a homologous protein, it provides a more accurate structure than when it is modelled on an unrelated protein. If no segment of the same length is found in homologous proteins, a loop of slightly different length but similar sequence may be useful. Sibanda and Thornton (40) show that, in homologous proteins, β-hairpin conformations are often conserved for most of the loop, even if there are insertions and deletions. This observation is exploited in the collar extension approach, in which use is made of a loop from a homologue that differs by one or two residues at the most and has greater than 40% sequence identity with the loop to be modelled. The framework is extended to include similar regions and the rest of the loop is modelled on a fragment obtained from a search of all known protein structures. This approach minimizes the error in the loop region (38). (*Figure 4B*). A recent exhaustive compilation of families of loops linking contiguous regular secondary structural elements and the templates derived (40a) have been shown to be effective in predicting loop conformational class in comparative modelling (40b).

2.1.3 Modelling side chains

The proper conformations of side chains are important in the packing of amino acids. In modelling side chains of homologous proteins, preferred side chain torsion angles, close packing, disulfide cross-links, hydrogen bonds, ion pairs, and other electrostatic interactions must be considered.

i. Conformations of side chains

The side chain torsion angles, N–C^{α}–C^{β}–C^{γ} (χ_1), C^{α}–C^{β}–C^{γ}–C^{δ} (χ_2) etc. of amino acid residues in proteins prefer values that correspond to the three staggered orientations about a single bond, namely, +60° (*gauche minus* or *g*−), −60° (*gauche plus* or *g*+), and 180° (*trans* or *t*). This has been confirmed by many analyses using unrelated protein structures. Of the three values, torsion angles corresponding to *t* and *g*+ are preferred for χ_1, while *t* is preferred for χ_2. When the C^{γ} atom is trigonal as in Asp, Phe, Tyr, Trp, Asn, or His, preferred values of χ_2 are either +90° or −90°.

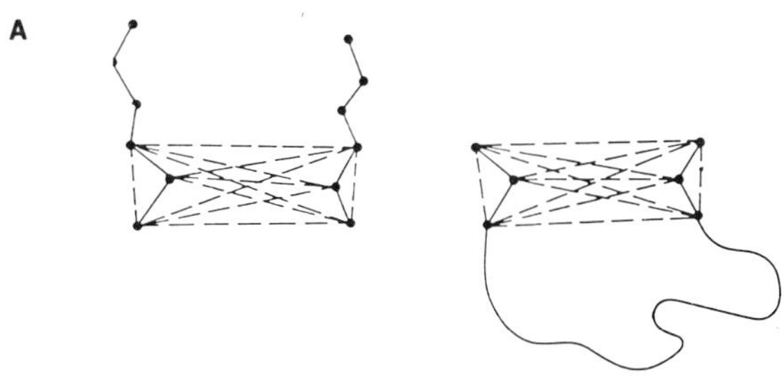

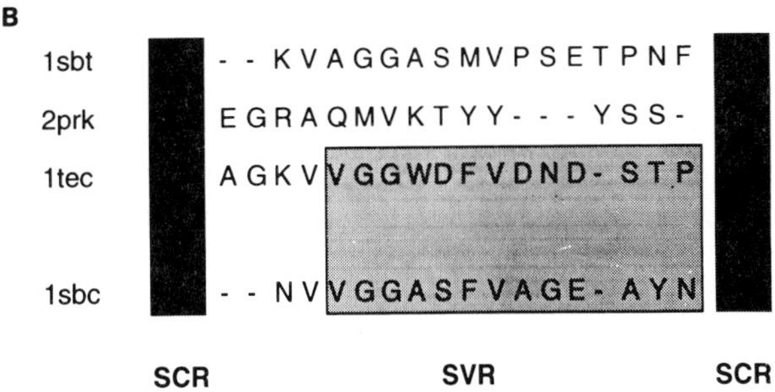

Figure 4. Modelling of structurally variable regions (SVRs). (A) The two figures at the top demonstrate the ring closure procedure. All possible distances between two sets of C^{α} atoms (three in each) in the anchor region of the structurally conserved regions (SCRs) flanking the loop are used to search for possible loops from known 3D structures (top *left*). The segment selected from the database search is melded on to the SCRs (top *right*). (B) The collar extension approach which uses information from the equivalent loop of a homologue is shown at the bottom. Subtilisin Carlsberg (1sbc), which is modelled on other members of the subtilisin family (1sbt, 2prk, and 1tec), but the loop has no *exact* equivalent in the known structures. One of the SCRs (darkly shaded) of thermitase (1tec) is extended (lightly shaded) in order to model most of the residues in the loop of subtilisin Carlsberg. Note the sequence similarity in the extended regions. The final two residues of the loop are modelled based on a search for fragments with compatible end-to-end and anchor distances from a database of known 3D structures. Figure used with permission (22).

The preferred side chain torsion angle distribution depends on the secondary structure of the polypeptide (40c). Analysis of high resolution protein structures shows that the preferred conformation of χ_1 for residues in an α-helix is *t*, although polar residues with short side chains have a preferred χ_1 value of $g+$. Rotamer libraries have been constructed by Ponder and Richards (41) and others that indicate the preferred side chain torsions for all the amino acids. A rotamer library derived by Dunbrack and Karplus (42) has been used to correlate (ϕ,ψ) values with side chain torsion angle probabilities. Well-defined rules for side chain building have been derived; these depend on the main chain torsion angles and the side chain orientation in the known homologues.

ii. Side chain modelling

Johnson *et al.* (22) have reviewed approaches to side chain modelling, using simulated annealing, Monte Carlo simulation, and knowledge-based approaches. These methods use many different approaches to limit the search of side chain conformers and simultaneously to predict the positions of side chains by considering residue–residue interactions. However, in a comparative modelling exercise, modelling of a side chain is best performed by extrapolating from the side chain positions in the topological equivalent positions of the homologous structures. Variation in relative positions of the backbones should also be considered while modelling a side chain position as suggested by the mutagenesis work of Matthews and co-workers (43).

Schiffer *et al.* (44) use energetics to identify the correct side chain conformation in comparative modelling. The initial orientations of side chains are obtained from a homologue. The solvation energy term was found to be essential in predicting the structure of surface residues, whereas a simpler molecular mechanics calculation is sufficient to model correctly those in the core. Wilson *et al.* (45) have proposed a similar approach using 'energetics' as a tool to position side chains in comparative modelling. They use a library of rotamers and explore an average of five to six different conformations per residue. In their modelling of α-lytic protease and other proteins, they have shown that the correct prediction could be obtained for a great majority of the side chains.

COMPOSER employs a fast and objective way of modelling side chains. The side chains are modelled using rules defining the probabilities of orientation of side chains in the equivalent position in homologues and rules defining preferred conformations in various secondary structures (36). Where the side chain orientation is not expected to be conserved in a topologically equivalent position in the known homologues, the general rules concerning side chain conformation in helices and sheets are used. *Figure 5* shows examples of side chain modelling in COMPOSER. Energy minimization, molecular dynamics, or restraint-based comparative modelling can then be applied to improve the model.

2.2 Modelling—restraint based

Although comparative modelling using assembly of fragments has proved remarkably successful, procedures for modelling by satisfaction of distance or

Figure 5. Side chain modelling in COMPOSER. (A) Topologically equivalent side chains, Trp 113 of subtilisin BPN′ (1sbt), Phe 113 of proteinase K (2prk), and Tyr 121 of thermitase (1tec) are shown superposed (*left*) and the superposition of modelled (thin lines) and X-ray (thick lines) conformations of Trp 112 in the equivalent position of subtilisin Carlsberg (1sbc) is shown (*right*). (B) Similar to (A) for Trp 104 of 1sbt, Tyr 104 of 2prk, and Trp 112 of 1tec (*left*) and Tyr 103 of 1sbc, modelled and crystal orientation (*right*). Note that the accuracy in the side chain modelling is better where the side chain orienta-

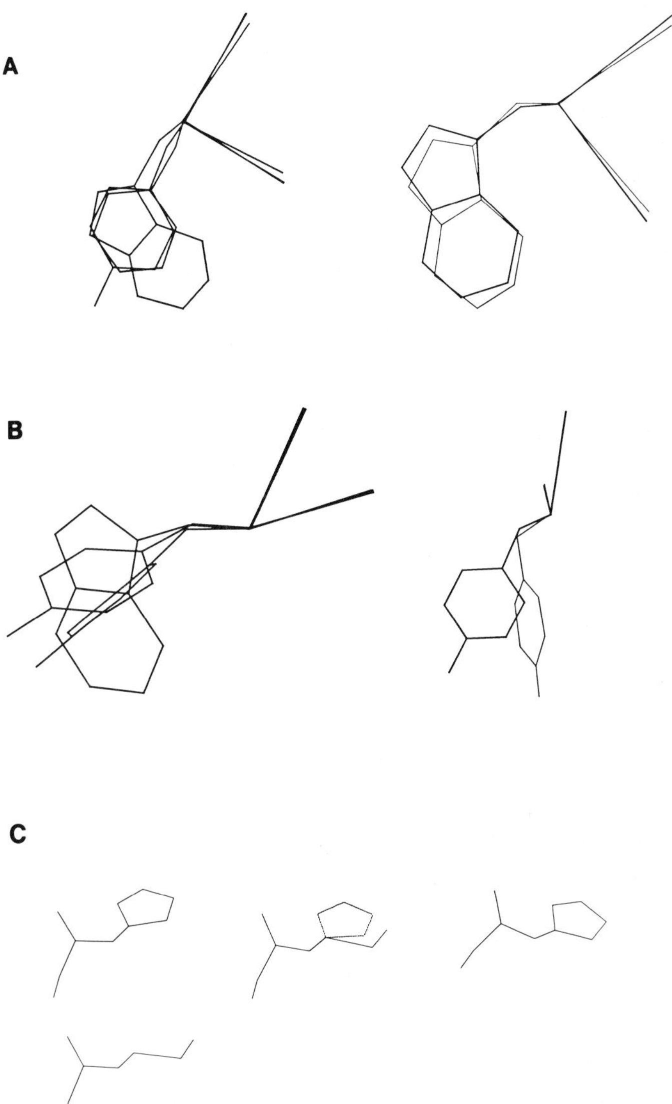

tions are conserved in the structures used to build the model. (C) Illustration of how a side chain is substituted. In this example, His 82 of myoglobin is being built. The two plots on the *left* show His in the most probable conformation and the Met from which it is being built. The middle plot shows C^{β}, C^{γ}, and $C^{\delta 2}$ of His are least squares fitted to the C^{β}, C^{γ}, and $S^{\delta 2}$ of Met with weights of 1.0, 1.0, and 0.01 respectively. This ensures that the plane of His and the C^{β} and C^{γ} positions of His are defined. The plot on the right shows that the S^{δ} and C^{ε} of Met are discarded to leave the modelled His residue. (A and B are reproduced with permission from ref. 22 and C is reproduced with permission from ref. 46.)

other restraints may also be used to advantage, especially where the homologues are distantly related. Such procedures derive restraints, such as interatomic distances, from homologous protein structures. These restraints involve main chain/main chain, main chain/side chain, and side chain/side chain distances, and are analogous to the distance information derived from nuclear Overhauser effects in multidimensional NMR experiments. Using distance geometry, an ensemble of models satisfying input restraints is built.

2.2.1 Methods that use distance restraints

Havel and Snow (47) proposed a method which derives distance and chirality restraints based on alignment with homologous known structure(s). Distance geometry procedures are used to derive an ensemble of structures compatible with input restraints. The model of the silver pheasant ovomucoid third domain has a rmsd of 0.78 Å (for C^{α} atoms) with the basis structure of the Japanese quail ovomucoid third domain (JQOM3). The rmsd between the model and its corresponding experimental structure (SPOM3) was 2.36 Å over the C^{α} atoms and 2.60 Å for the backbone atoms. This difference is comparable to that between the experimental structures of SPOM3 and JQOM3, which is 2.28 Å for the C^{α} atoms. Thus, the method has produced a model which is very close to the template structure rather than the experimental structure of SPOM3. Most errors originated in the amino-terminal segment, and if this segment is ignored, the rmsd is 1.1 Å, which is smaller than the rmsd between X-ray structures of SPOM3 and JQOM3. Similar methods have been proposed by Srinivasan *et al.* (48), Brocklehurst and Perham (49), Fujiyoshi-Yoneda *et al.* (50), and Friedrichs *et al.* (51); the last two methods exploit molecular dynamics.

Taylor (52,53) has used multiple sequence alignment and distance restraints in packing of α-helices to build three-dimensional models starting from idealized folds. His method can be extended to comparative modelling.

2.2.2 A method based on structural feature restraints

Šali and Blundell (54) have developed a comparative modelling procedure (MODELLER) which arrives at a three-dimensional model by optimally satisfying restraints extrapolated from known structures which are homologous to the model sequence. These restraints are expressed as probability density functions (*pdfs*) for the features to be restrained. The features are structural properties at residue positions and relationships between residues; they include solvent accessibility, secondary structure, and hydrogen bonding. For example, the probabilities for the main chain conformation of an equivalent residue in a similar protein are expressed as a function of similarity between the two sequences. Several such *pdfs* were obtained from the correlations between structural features in families of homologous proteins which have been aligned on the basis of their 3D structures. The *pdfs* restrain C^{α}–C^{α} distances, main chain N–O distances, and main chain and side chain dihedral angles. A smoothing procedure (adapted from ref. 55) is used in the derivation of these relationships to minimize the problem of a sparse database. The 3D model of a protein is obtained by optimization of the molecular *pdf* such that the model violates the input restraints as little as possible. The molecular *pdf* is derived as a combination of *pdfs* restraining individual spatial features of the whole molecule. The optimization procedure is a variable target function method that applies the conjugate gradients algorithm to positions of all non-hydrogen atoms. The method is automated and is illustrated by:

(a) The model of a domain of endothiapepsin constructed on the basis of other aspartate proteinase domains (*Figure 6*).
(b) The modelling of trypsin on elastase and tonin which have about 40% sequence identity with trypsin.

In the first case, the rmsd with the crystal structure was 0.76 Å although only C^{α}–C^{α} restraints were used. In the second example, when compared with the crystal structure, the best of 11 very similar models has an rmsd of 0.7 Å for 195 topologically equivalent C^{α} atoms.

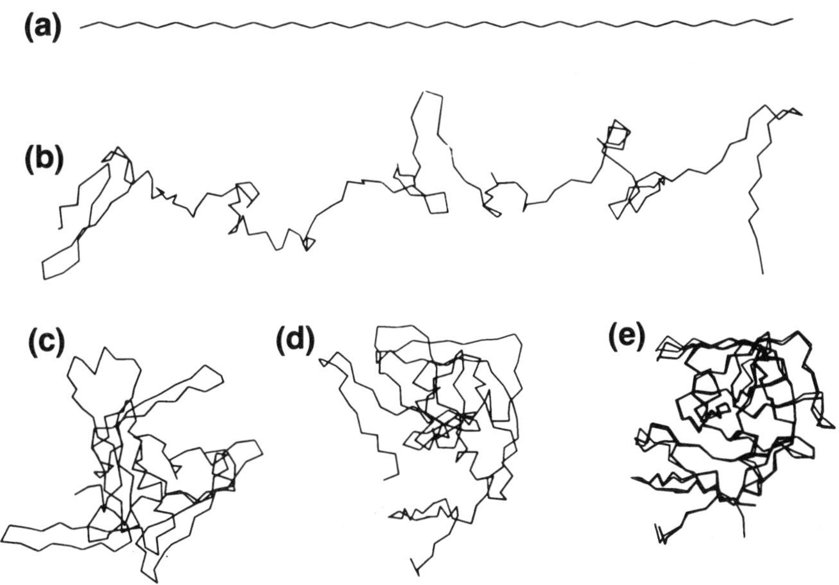

Figure 6. Generation of a model of endothiapepsin using MODELLER on the basis of several homologous aspartate proteinases and from protein structures in general and expressed as distance constraints on atomic positions. (a) The initial extended main chain (C^{α} trace), (b) local constraints are considered first, (c) and (d) are intermediate structures that arise as global constraints are applied, and (e) is the final model compared with its crystal structure. Figure used with permission (23).

2.3 Energy minimization and molecular dynamics

Energy minimization of a model is necessary in order to relieve short contacts and to rectify bad geometry that may be present in the model. Inconsistencies in geometry are possible particularly in the 'anchor' regions where a loop is melded with the core of the protein. The purpose of energy minimization for a model built using a comparative modelling procedure is to rectify such inconsistencies; energy minimization will not significantly alter the structure.

Before starting energy minimization it is important to inspect the structure to check stereochemistry and to make sure that no serious problems exist. Major problems in stereochemistry of the model should be rectified first using interactive graphics. The model can then be subjected to energy minimization. A suggested minimization strategy for comparative protein models is given below.

Force fields such as CHARMM (56), AMBER (57), or TRIPOS (SYBYL software—Tripos Inc.) are used. If major problems persist in a local region of the model (e.g. in 'anchor' regions), a simulated annealing protocol focused on the problematic region can be adopted. Subsequently, the whole model can be considered for energy minimization. At the beginning the electrostatic term in the force field need not be considered, as the problems with the models are expected to be due to short contacts and bad geometry alone' omitting electrostatic terms will also speed up the minimization process. When the structure to be modelled is close to a homologue of known structure, the backbone of the model can be fixed and side chains alone can be allowed to move during the first few rounds of energy minimization, as most of the short contacts are expected to involve side chain atoms. This strategy will also ensure minimum perturbation to the backbone structure. For every round of energy minimization, initially about ten cycles of fast optimization procedures such as Simplex can be used; this will bring the model close to a local minimum in the energy profile. Simplex can be followed by procedures such as Powell torsional gradient minimization to optimize further the system towards convergence. When most problems with the stereochemistry have been rectified, the electrostatic term can be invoked in the force field; this will optimize the geometry of the hydrogen bonds in the structure. Once the side chain positions are energy refined, all the atoms (including backbone atoms) in the model can be allowed to move and further rounds of energy minimization can be employed. The energy minimization process can be terminated once all the inconsistencies in the geometry are rectified and short contacts are relieved. The absolute value of the total energy is not important although success of the energy minimization procedure may be implied by a large negative energy value.

In order to study the flexibility over various local regions of a model, and to explore other local minima, molecular dynamics can be employed. This is particularly valuable when distance restraints derived from NMR nuclear Overhauser effect are available; such restraints can be included in a molecular dynamics simulation in order to optimize the model structures while satisfying distance restraints.

2.4 Outlook

The overall accuracy of the models derived by restraint-based modelling is of the order obtained in assembly of rigid fragments approach for proteins that are similar to those of known structure. However, the former method is likely to model a residue in a position interacting favourably with another residue. This is not guaranteed in the latter approach because when modelling a side chain, its interaction with surroundings is not automatically considered. But the time taken to model a protein is normally much more in restraint-based modelling than in COMPOSER-like approaches. Hence, it might be appropriate to use both the methods in a given modelling exercise when the protein to be modelled is distantly related to the homologues. First, using COMPOSER one can get an approximate model and subsequently the interatomic interactions could be optimized using MODELLER and energy minimization.

3. Accuracy of models and evaluation of models

3.1 Retrospective assessment of protein models

Retrospective analyses of a number of three-dimensional models built using COMPOSER have been reported (21,38,58, K. Guruprasad and T. L. Blundell, unpublished results). *Table 1* shows comparisons between modelled structures and their corresponding crystal structures (both for the framework regions and all C^{α} atoms), as well as the expected error in the crystal structures based on their resolution. The smaller errors between the model and its X-ray structure over the framework region show that the models resemble the X-ray structure of the modelled protein more than the X-ray structure of the closest homologue.

The rmsd (all C^{α} atoms) for superpositions between the modelled and X-ray structures lie in the range of 0.7–1.7 Å for most models. A representative example of errors along the polypeptide of the model of hen egg white lysozyme compared with its crystal structure is shown in *Figure 7A*, in which SCRs and SVRs are indicated. The stereo diagrams in *Figure 8* illustrate superposition of a few modelled and corresponding X-ray structures. The errors are generally greatest in the loop regions, particularly if they are built on an unrelated protein, but these errors can be minimized when it is possible to extend the structurally conserved regions on the basis of a loop of similar sequence but different length in a homologue (*Figure 4B*). This has resulted in more accurate models for the variable regions in myoglobin, subtilisin Carlsberg (38), and in human cathepsin D (59).

Table 1. Known structures built using COMPOSER, their accuracy and comparison of their X-ray structure with homologues

PDB code of the protein modelled	Closest homologue (A)		Homologue with best resolution		Homologue with least resolution		rmsd (Å) between X-ray str. of unknown and A (equivalences)	rmsd (Å) between the model and X-ray str. (equivalences)	
	Code	**% seq. sim.**	**Code**	**Exp. error (Å)[a]**	**Code**	**Exp. error (Å)[a]**		**Framework**	**Whole chain**
5MBN	2HHB	25	1ECA	0.18	2LHB	0.40	1.37 (138)	0.83 (107)	0.99 (146)
1MPP	2APR	30	2APR	0.33	1YPA	0.96	1.85 (282)	1.71 (282)	3.47 (354)
1DTX	1AAP	36	5PTI	0.03	1AAP	0.21	0.91 (56)	0.72 (54)	0.85 (56)
2PTN	4CHA	44	3EST	0.27	3RP2	0.36	1.52 (219)	0.84 (177)	1.15 (223)
1BBC	1PPA	47	1BP2	0.29	1PP2	0.59	1.30 (118)	1.10 (94)	1.56 (124)
1LZT	1LZ1	60	1LZ1	0.21	1ALC	0.29	0.76 (129)	0.65 (115)	0.73 (129)
1AZU	2AZA	62	1PAZ	0.23	2AZA	0.55	0.88 (123)	0.68 (50)	0.90 (126)
1SBC	1SBT	70	2PRK	0.21	1SBT	0.59	0.73 (272)	0.67 (213)	1.07 (274)

Reproduced with permission (38).

[a] Estimated experimental error is derived by extrapolation of rmsd for solvent inaccessible regions versus the resolution of independently determined structures given in Figure 4 of ref. 76.

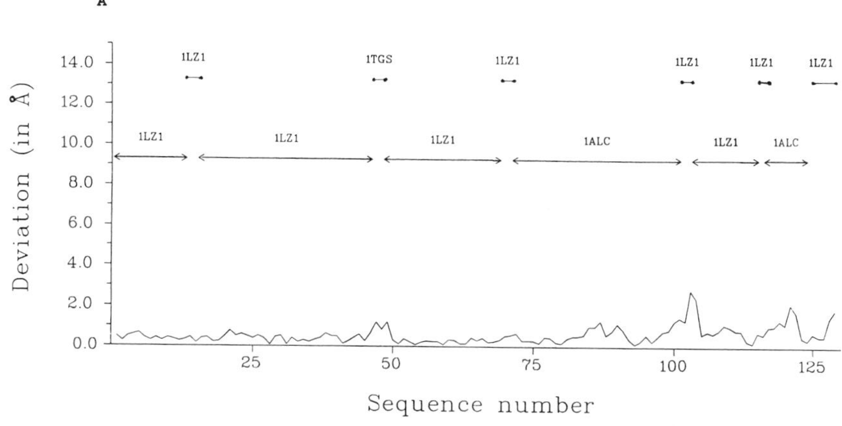

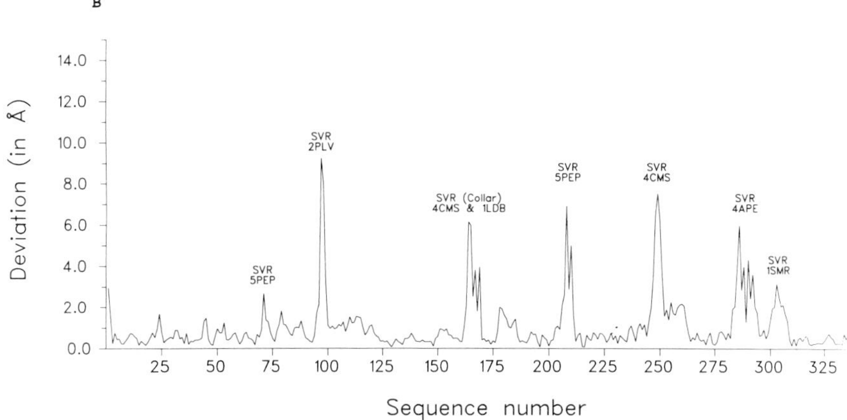

Figure 7. (A) The deviation of the modelled hen egg white lysozyme from the X-ray structure (1lzt) upon superposition is plotted against residue number at every C^{α} position. Proteins which contribute to modelling various regions (SVRs at the top and SCRs at the bottom) of lysozyme are indicated. Figure used with permission (38). (B) Same as in (A), but for the model of human cathepsin D (59). The peaks correspond to significant errors in the model. Whether these regions are in the SCRs or SVRs is indicated. The four letter code of the protein structures on which these regions were modelled are also indicated.

The error in the core region can be reduced by generating a framework in which contributions from the known homologues are weighted on the basis of their sequence similarities with the unknown. This is particularly important if the percentage sequence identities of the unknown with the known structures vary greatly. For example, the azurin from *P. aeruginosa* was modelled on the azurin from *A. denitrificans*, pseudozaurin from *A. faecalis* and the plastocyanins from *Poplus Algra* and *E. prolifara*. The percentage sequence

A

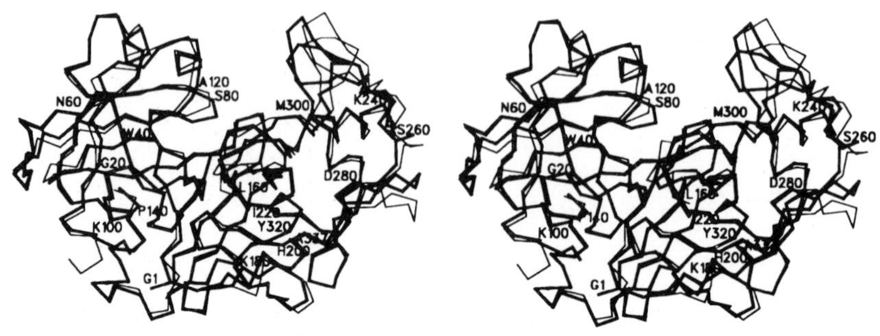
N60
A120
S80
M300
K240
S260
W40
G20
D280
L160
P140
I220
K100
Y320
H200
K180
G1

B

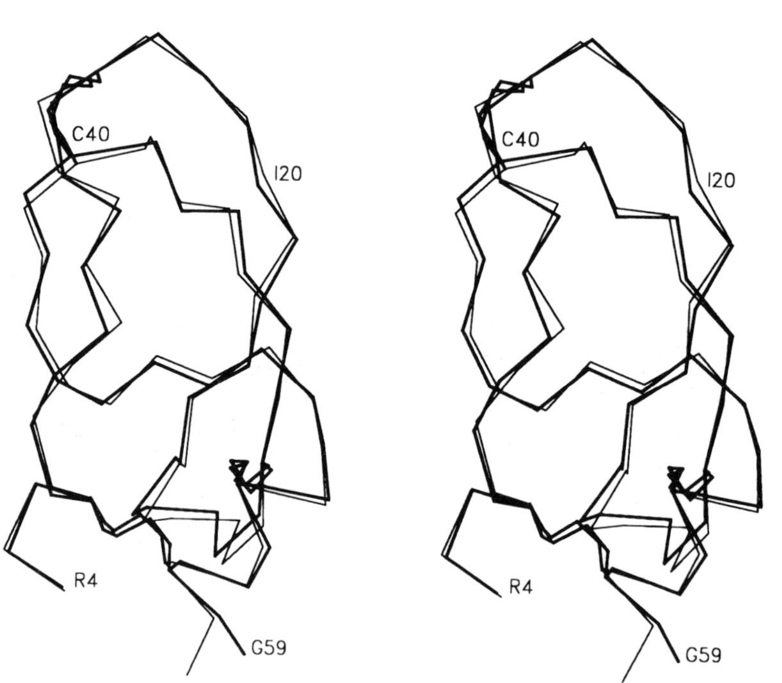
C40
I20
R4
G59

C

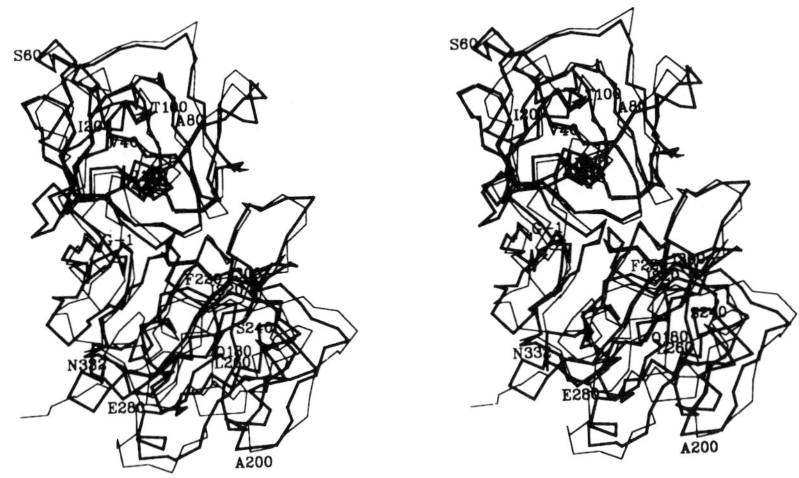

D

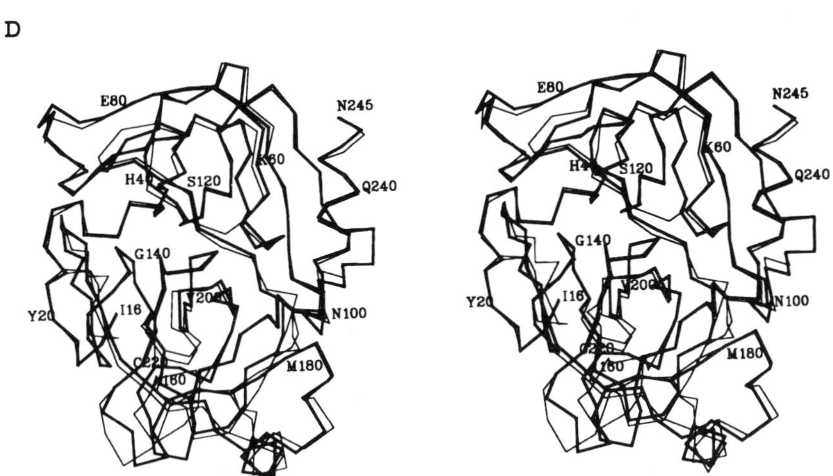

Figure 8. Superposition of C$^{\alpha}$ traces of X-ray (thick line) and modelled (thin line) structures. Every twentieth residue is marked (A) Cathepsin D (1lyb), (B) α-dendrotoxin (1dtx), (C) mucor pepsin, and (D) trypsin (2ptn). Figures 8B–D used with permission (38).

identities of the modelled protein with these homologues is 62%, 12%, 20%, and 30% respectively. A model constructed from equal contributions from these homologues differs from the true structure over the framework region by an rmsd of 1.43 Å. The error is reduced to 0.68 Å if the homologues are weighted using the square of their sequence identity to the unknown (38).

Frazao *et al.* (21) have analysed a model of human renin constructed using the known structures of three fungal proteinase (rhizopuspepsin, endothiapepsin, and penicillopepsin) and two mammalian enzymes (porcine pepsin and calf chymosin) before the crystal structures of human and mouse renins were determined (60). A comparison of the X-ray and modelled human renin structures shows that the 280 C^α atoms forming the framework have a rmsd of 0.84 Å, which is smaller than the rmsd between the X-ray structure of human renin and the structures used in the model building with COMPOSER. Some of the errors, for example from the flap that covers the active site, arise from the fact that the model was constructed on the basis of the unliganded forms of pepsin and chymosin, whilst the crystal structure of renin has a bound inhibitor (60). Errors in the relative orientation of both domains in the renin model are difficult to avoid in view of the rigid body shifts that occur in the aspartic proteinases. Although the catalytic residues are well modelled, there are some problems in modelling the specificity pockets, as a result of alterations to well-defined loops in the other proteinases. Other features of renin, such as the conformation of a large loop containing several prolines including two in the *cis* conformation, were not modelled correctly. However, an increase in the number of different structures for a family increases the accuracy of the models.

The model of human cathepsin D (59) was built prior to the determination of the crystal structure (61,62). We used the crystal structure of Baldwin *et al.* (61) which is at a higher resolution (2.5 Å) for the evaluation of the model. The rms deviation between the model and the crystal structure is 0.87 Å for 300 of the 338 C^α positions. The overall rms deviation for all 338 residue positions is 1.7 Å reflecting that it is a good model. The deviation in C^α positions is plotted against the residue number in *Figure 7B*. There are seven peaks corresponding to significant errors in the model. However, the peaks are sharp indicating that the regions of error are small. All these regions are present within the SVRs of the model that are exposed to the solvent. These errors, however, did not affect the conclusions drawn about the specificity of the enzyme as the SVRs are far from the active site. it is also clear from the figure that segments modelled from within the homologous protein family have smaller errors compared with those modelled from an unrelated protein. The peak around position 96 corresponds to an insertion unique to human cathepsin D amongst the aspartic proteinase family; this was modelled on an unrelated protein (polio virus—PD. code: 2plv). Interestingly this region is also not defined in the crystal structure. The errors indicated in this region in *Figure 7B* are the glycines at positions 98 and 99, which were modelled from polio virus. The SVR around 160 was modelled using the collar extension approach and this reduced both the size and magnitude of error of the segment that was modelled from an unrelated protein. The proline-rich loop around 310, unique to renins and cathepsin D, was initially modelled on the equivalent segment from pepsin. Subsequent to the determination of the crystal structure of mouse renin (1smr) this loop was remodelled on mouse renin in which the local sequence is similar. Retrospectively, it can be seen that the conformation of this region agrees with the crystal structure, although it is spatially displaced, partly due to the rigid body shift between the N and C domains which is often observed in the aspartic proteinases.

The superimposition of the modelled and X-ray structures is shown in *Figure 8A*. The predictions made using the three-dimensional model of cathepsin D proved to be useful in confirming some of the structure–activity relationships using site-directed mutagenesis experiments (63).

Problems due to unique features in the unknown (often at loops) have also been noted by many others who compared modelled and crystal structures. Weber (64) has shown that the substrate binding site of HIV protease was modelled accurately and could be used to design inhibitors.

3.2 Estimation of errors in protein models

An assessment of likely errors is required for every model. This can be performed at different levels of structural organization, namely, to:

- check the correctness of the overall fold
- detect errors over localized regions in an otherwise correctly folded sequence
- check stereochemical parameters like bond lengths, bond angles, and to identify short contacts

3.2.1 Detection of gross misfolds and errors in local regions

Assessment of the compatibility of a sequence with a given fold is a challenging task, particularly if the model is built on a protein structure which is not obviously similar to the model sequence. Several attempts have been made to identify structural descriptors which discriminate between incorrectly and correctly folded structures. Novotny *et al.* (65) constructed incorrect models deliberately; they built the sequence of haemerythrin, which is predominantly α-helical, on to a domain of immunoglobulin, which is predominantly β-sheet and vice versa (*Figure 9A*). After energy minimization, they compared these incorrect models with native protein folds and found that the side chains could be readily accommodated into the incorrect folds. However, they found that with the incorrect folds more non-polar groups were exposed to the surroundings and more polar side chains were buried within the core than expected from experimentally defined protein structures. If these solvent

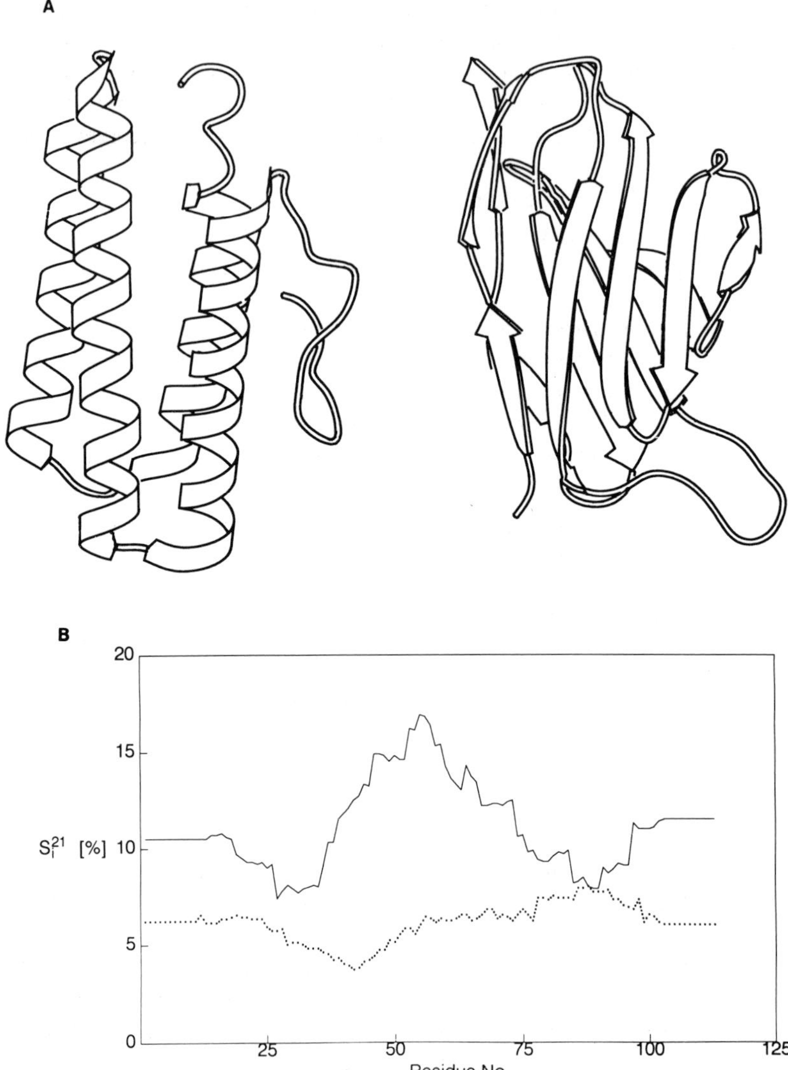

Figure 9. Identification of an incorrect fold. (A) The sequence of an immunoglobulin variable domain, which is predominantly β-sheet (top *right*), is threaded on to the fold of haemerythrin, which is a four-helical bundle (top *left*). The scores for the compatibility of the residues to their structural environments were calculated using the method of Topham *et al.* (73). (B) Profiles for the correct (solid line) and incorrect (dotted line) folds calculated using a 21 residue moving window are shown at the bottom. Figure used with permission (22).

effects are not taken into consideration, the empirical energy difference between the folded and misfolded forms is not significant. Subsequently, Novotny *et al.* (66) reinvestigated their 'incorrect' models, using a conformational sampling program (CONGEN) to model side chains. They found that several features could help distinguish between correct and incorrect folds:

- non-polar side chains exposed to the solvent
- buried ionizable groups
- empirical free energy functions that incorporate solvent effects

Bryant and Amzel (67) and Sander and co-workers (68) analysed known protein structures and known misfolds and their conclusions are similar to those obtained by Novotny and co-workers.

Sippl and co-workers (69,70) have used potentials of mean force, derived for interactions between C^{β} atoms in 3D structures, to calculate the conformational energy of different folds for a given amino acid sequence (71). Their procedure can identify native protein folds among a large number of incorrect models, although the discrimination may not be satisfactory for proteins with large prosthetic groups or iron–sulfur clusters. They have also found a linear relationship between the solvation free energy of folding and protein size, for known crystal structures. The misfolded structures were found to show a higher solvation free energy than predicted. This approach can discriminate between correctly folded and misfolded structures; but in some cases the misfolded structures have values close to the predicted value and hence need very careful analysis.

Lüthy (72) have extended their earlier work for generating 3D profiles and developed a method to assess 3D models. The compatibility of a sequence to a 3D fold is tested by calculating scores for the association of a residue with its structural environment (solvent accessibility and local secondary structure). The incorrect models can be identified by examining the profile scores calculated over a 21 residue window moving along the polypeptide chain. Topham *et al.* (73) have calculated propensity tables from a database of aligned homologous proteins and used a greater range of local environments to probe the errors in local regions. Using a smaller moving window frame (five residues) they assessed a model of the dust mite allergen Der pI. The detection of overall misfold of the sequence of immunoglobulin threaded on to haemerythrin is shown in *Figure 9B*.

If many sequences that are homologous to the modelled protein are available, substitution tables can be used to assess the model. Overington *et al.* (24) align the sequence of the model with all similar proteins in a sequence data bank. The residue substitution pattern derived from this alignment is viewed in the light of the structural environment at every residue position in the model and this substitution pattern is compared with the one derived from the alignment of known homologous tertiary structures. The errors in local regions in a correctly folded structure are more difficult to recognize. Work in progress is particularly aimed at detecting errors in local regions in a correctly folded sequence, using a combination of environment-dependent amino acid substitution and propensity tables (R. Sowdhamini, N. Srinivasan, C. M. Topham, J. P. Overington, and T. L. Blundell, unpublished results).

3.2.2 Stereochemical quality

Bond lengths and bond angles need to be precise in order to make reliable use of a protein model in studies of molecular recognition. Thornton and co-workers (75) have analysed the (ϕ,ψ) values, peptide planarity, bond lengths and bond angles, hydrogen bond geometry, and side chain conformations of known protein structures as a function of crystallographic resolution. From these data they arrived at expected values of these parameters depending upon the resolution (for X-ray structures) of the 3D structure. For example, a model whose accuracy is comparable to that of a typical 2 Å resolution crystal structure is expected to have more than 80% of the (ϕ,ψ) values within the allowed regions in the Ramachandran map. Of course statistics of this kind are a necessary but not a sufficient indicator of such accuracy. The program PROCHECK (75) gives visual and quantitative information about various structural features and assessment of the model.

3.2.3 Outlook

A positive health check by one method does not imply that the model is fully correct, although by using many methods working on various different principles it may be possible to uncover a large number of the inaccuracies in a model. Local inaccuracies are most difficult to identify. With the present state of art, the final evaluation of any model can only be made when a experimental structure analysis has itself been completed!

4. Use of protein models and summary comments

The wealth of sequences available, together with the interest in protein structure as a basis for design, has created interest in comparative protein modelling techniques in many laboratories. The wide availability of molecular graphics workstations, with user-friendly interfaces, has further contributed to the popularity of the approach.

In this chapter we have described progress from a partly subjective, mainly interactive exercise carried out by experts, to a more automated and rule-based approach, which exploits our extensive knowledge of protein structure and interactions. We emphasize the importance of correct sequence alignment of the protein to be modelled and its homologues. Incorrect alignment is the source of the most serious errors. We have shown how knowledge of the structures of proteins in general can be used to assist in the construction of useful models.

Such models have a value that is operationally defined. In some cases a model that defines the approximate disposition of groups relative to a ligand binding site may be of value in suggesting a mechanism or an explanation for specificity. Approximate models may also be useful in the early stages of novel ligand design. However, for many purposes, especially in drug design, a more precise model will be required in order to assess the likely effects of small changes in the chemistry of the ligand or the sequence of a protein. Precise models can now be constructed in favourable circumstances, most importantly where there are homologues of known structure. Most proteins, of interest in design of drugs, pesticides, and vaccines, will be present in very low copy numbers in the organism. This means that quantities of the protein will be not directly available for experimental studies and that the sequence will be defined from cloning and sequencing the cDNA. This will provide a route for expression of the protein for crystallization or solution NMR studies. But, while these experiments are being undertaken, models prove to be useful.

Protein models have played an important role in many design processes. In the design of inhibitors, for example, of HIV proteinase for AIDS antivirals or human renin for antihypertensives, models were exploited for several years before X-ray crystal structures of target enzyme complexes were available. For HIV proteinase the relationship with aspartic proteinases of known structure was distant, and the models were a very rough guide. For renin the models were quite precise and allowed useful elaboration of lead compounds. This included suggestions of cyclization to decrease flexibility, of decreases in size to improve oral availability, of removal of peptide bonds to decrease proteolysis, of addition of groups to improve lipid solubility, and of modification of groups to improve the specificity of ligand binding.

Models are also useful in protein engineering. They suggest sites where mutations might be introduced effectively. They have provided a guide to the construction of useful chimeric molecules where parts of one protein are grafted on to another. This has been particularly effective in the production of humanized antibodies or chimeric growth factors, for example hybrid neurotrophic factors comprised of fragments of nerve growth factor, brain derived neurotrophic factor (BDNF), and neurotrophic factor 3 (NT3).

The increasing interest on the part of medicinal chemists, molecular biologists, pharmacologists, plant biotechnologists, and many others, who have little direct knowledge of protein structure, underlines the importance of procedures that use, in an automatic way, rules about and knowledge of protein structure. Such approaches are certain to play an important role, alongside experimental techniques, in design of novel proteins and ligands that recognize them.

Acknowledgements

We thank Dr Mark Johnson and Dr R. Sowdhamini for compiling some of the information detailed in this chapter, during the preparation of an exhaustive review on modelling (22). N. S. and K. G. are supported by Tripos Inc. and Parke-Davis respectively.

Note added in proof: The readers are urged to go through some of the papers that have appeared since this chapter was prepared: loop modelling (77–78), side chain modelling (79–82), assessment of protein models (83–86), and general review on comparative modelling (87).

References

1. Browne, W. J., North, A. C. T., Philips, D. C., Drew, K., Vanaman, T. C., and Hill, R. L. (1969). *J. Mol. Biol.*, **42**, 65.
2. Issacs, N., James, R., Niall, H., Bryant-Greenwood, G., Dodson, G., Evans, A., *et al.* (1978). *Nature*, **271**, 278.
3. Warme, P. K., Momany, F. A., Rumball, S. V., Tuttle, R. W., and Scheraga, H. A. (1974). *Biochemistry*, **13**, 768.
4. Acharya, K. R., Stuart, D. I., Walker, N. P. C., Lewis, M., and Phillips, D. C. (1989). *J. Mol. Biol.*, **208**, 99.
5. McLachlan, A. D. and Shotton, D. M. (1971). *Nature New Biol. (London)*, **229**, 202.
6. Brayer, G. D., Delbaere, L. J. T., and James, M. N. G. (1979). *J. Mol. Biol.*, **131**, 743.
7. Delbaere, L. T. J., Brayer, G. D., and James, M. N. G. (1979). *Nature*, **279**, 165.
8. Blundell, T. L., Dodson, G. G., Hodgkin, D. C., and Mercola, D. A. (1972). *Adv. Protein Chem.*, **26**, 279.
9. Blundell, T. L., Bedarkar, S., Rinderknecht, E., and Humbel, R. E. (1978). *Proc. Natl. Acad. Sci. USA*, **75**, 180.
10. Cooke, R. M., Harvey, T. S., and Campbell, I. D. (1991). *Biochemistry*, **30**, 5484.
11. Bedarkar, B., Turnell, W. G., Schwabe, C., and Blundell, T. L. (1977). *Nature*, **270**, 449.
12. Eigenbrot, C., Randal, M., Quan, C., Burnier, J., O'Connell, L., Rinderknecht, E., *et al.* (1991). *J. Mol. Biol.*, **221**, 15.
13. Greer, J. (1980). *Proc. Natl. Acad. Sci. USA*, **77**, 3393.
14. Greer, J. (1981). *J. Mol. Biol.*, **153**, 1027.
15. Read, R. J., Brayer, G. D., Jurášek, L., and James, M. N. G. (1984). *Biochemistry*, **23**, 6570.
16. Blundell, T. L., Sibanda, B. L., and Pearl, L. (1983). *Nature*, **304**, 273.
17. Sibanda, B. L., Blundell, T., Hobart, P. M., Fogliand, M., Bindra, J. S., Dominy, B. W., *et al.* (1984). *FEBS Lett.*, **174**, 102.
18. Carlson, W., Karplus, M., and Haber, E. (1985). *Hypertension*, **7**, 13.
19. Akahane, K., Nakagawa, S., and Umeyama, H. (1985). *Hypertension*, **7**, 3.
20. Hutchins, C. and Greer, J. (1991). *Crit. Rev. Biochem. Mol. Biol.*, **26**, 77.
21. Frazao, C., Topham, C., Dhanaraj, V., and Blundell, T. L. (1994). *Pure Appl. Chem.*, **66**, 43.
22. Johnson, M. S., Srinivasan, N., Sowdhamini, R., and Blundell, T. L. (1994). *Crit. Rev. Biochem. Mol. Biol.*, **29**, 1.

23. Šali, A., Overington, J. P., Johnson, M. S., and Blundell, T. L. (1990). *Trends Biochem. Sci.*, **15**, 235.
24. Overington, J. P., Johnson, M. S., Šali, A., and Blundell, T. L. (1990). *Proc. R. Soc. Lond.*, **B241**, 146.
25. Johnson, M. S., Overington, J. P., and Blundell, T. L. (1993). *J. Mol. Biol.*, **231**, 735.
26. Maiorov, N. V. and Crippen, G. M. (1992). *J. Mol. Biol.*, **227**, 876.
27. Bowie, J. U., Lüthy, R., and Eisenberg, D. (1991). *Science*, **253**, 164.
28. Sippl, M. J. and Weitckus, S. (1992). *Proteins*, **13**, 258.
29. Godzik, A., Kolinski, A., and Skolnick, J. (1992). *J. Mol. Biol.*, **227**, 227.
30. Jones, D. T., Taylor, W. R., and Thornton, J. M. (1992). *Nature*, **358**, 86.
31. Jones, T. H. and Thirup, S. (1986). *EMBO J.*, **5**, 819.
32. Unger, R., Harel, D., Wherland, S., and Sussman, J. L. (1989). *Proteins*, **5**, 335.
33. Claessens, M., Cutsen, E. V., Lasters, I., and Wodak, S. (1989). *Protein Eng.*, **4**, 335.
34. Levitt, M. (1992). *J. Mol. Biol.*, **226**, 507.
35. Blundell, T. L., Sibanda, B. L., Sternberg, M. J. E., and Thornton, J. M. (1987). *Nature*, **326**, 347.
36. Blundell, T. L., Carney, D., Gardner, S., Hayes, F., Howlin, B., Hubbard, T., *et al.* (1988). *Eur. J. Biochem.*, **172**, 513.
37. Sutcliffe, M. J., Haneef, I., Carney, D., and Blundell, T. L. (1987). *Protein Eng.*, **1**, 377.
38. Srinivasan, N. and Blundell, T. L. (1993). *Protein Eng.*, **6**, 501.
39. Topham, C. M., McLeod, A., Eisenmenger, F., Overington, J. P., Johnson, M. S., and Blundell, T. L. (1993). *J. Mol. Biol.*, **229**, 194.
40. Sibanda, B. L. and Thornton, J. M. (1993). *J. Mol. Biol.*, **229**, 428.
40a. Donate, L. E., Rufino, S. D., Canard, L. H. J., and Blundell, T. L. (1996). *Protein Sci.*, (submitted).
40b. Rufino, S. D., Donate, L. E., Canard, L. H. J., and Blundell, T. L. (1996). *J. Mol. Biol.*, (submitted).
40c. McGregor, M. J., Islam, S. A., and Sternberg, M. J. E. (1987). *J. Mol. Biol.*, **198**, 295.
41. Ponder, J. W. and Richards, F. M. (1987). *J. Mol. Biol.*, **193**, 775.
42. Dunbrack, R. L. and Karplus, M. (1993). *J. Mol. Biol.*, **230**, 543.
43. Baldwin, E. P., Hajiseyedjavadi, O., Baase, W. A., and Matthews, B. W. (1993). *Science*, **262**, 1715.
44. Schiffer, C. A., Caldwell, J. W., Kollman, P. A., and Stroud, R. M. (1990). *Proteins*, **8**, 30.
45. Wilson, C., Gregoret, L. M., and Agard, D. A. (1993). *J. Mol. Biol.*, **229**, 996.
46. Sutcliffe, M. J., Hayes, F. R. F., and Blundell, T. L. (1987). *Protein Eng.*, **1**, 385.
47. Havel, T. F. and Snow, M. E. (1991). *J. Mol. Biol.*, **217**, 1.
48. Srinivasan, S., March, C. J., and Sudarsanam, S. (1993). *Protein Sci.*, **2**, 277.
49. Brocklehurst, S. M. and Perham, R. N. (1993). *Protein Sci.*, **2**, 626.
50. Fujiyoshi-Yoneda, T., Yoneda, S., Kitamura, K., Amisaki, T., Ikeda, K., Inoue, M., *et al.* (1991). *Protein Eng.*, **4**, 443.
51. Friedrichs, M. S., Goldstein, R. A., and Wolynes, B. G. (1991). *J. Mol. Biol.*, **222**, 1013.
52. Taylor, W. R. (1991). *Protein Eng.*, **4**, 853.
53. Taylor, W. R. (1993). *Protein Eng.*, **6**, 593.

54. Šali, A. and Blundell, T. L. (1993). *J. Mol. Biol.*, **234**, 779.
55. Sippl, M. J. (1990). *J. Mol. Biol.*, **213**, 859.
56. Brooks, B. R., Bruccoleri, R. E., Olafson, B. D., States, D. J., Swaminathan, S., and Karplus, M. (1983). *J. Comput. Chem.*, **4**, 187.
57. Weiner, S. J., Kollman, P. A., Case, D. A., Singh, U. C., Ghio, C., Alagona, G., *et al.* (1984). *J. Am. Chem. Soc.*, **106**, 765.
58. Topham, C. M., Thomas, P. Overington, J. P., Johnson, M. S., Eisenmenger, F., and Blundell, T. L. (1991). *Biochem. Soc. Symp.*, **57**, 1.
59. Scarborough, P. E., Guruprasad, K., Topham, C., Richo, G. R., Conner, G. E., Blundell, T. L., *et al.* (1993). *Protein Sci.*, **2**, 264.
60. Dhanaraj, V., Dealwis, C. G., Frazao, C., Badasso, M., Sibanda, B. L., Tickle, I. J., *et al.* (1992). *Nature*, **357**, 466.
61. Baldwin, E. T., Bhat, T. N., Gulnik, S., Hosur, M. V., Sowder, R. C., Cachau, R. E., *et al.* (1993). *Proc. Natl. Acad. Sci. USA*, **90**, 6796.
62. Metcalf, P. and Fusek, M. (1993). *EMBO J.*, **12**, 1293.
63. Scarborough, P. E. and Dunn, B. M. (1994). *Protein Eng.*, **7**, 495.
64. Weber, I. T. (1990). *Proteins*, **7**, 172.
65. Novotny, J., Bruccoleri, R. E., and Karplus, M. (1984). *J. Mol. Biol.*, **177**, 787.
66. Novotny, J., Rashin, J. J., and Bruccoleri, R. E. (1988). *Proteins*, **4**, 19.
67. Bryant, S. H. and Amzel, L. M. (1987). *Int. J. Pept. Protein Res.*, **29**, 46.
68. Baumann, G., Frommel, C., and Sander, C. (1989). *Protein Eng.*, **2**, 329.
69. Hendlich, M., Lackner, P., Weitckus, S., Floeckner, H., Froschauer, R., Gottsbacher, K., *et al.* (1990). *J. Mol. Biol.*, **216**, 167.
70. Sippl, M. J. (1993). *Proteins*, **17**, 355.
71. Sippl, M. J. (1993). *J. Comput. Aid. Mol. Des.*, **7**, 473.
72. Lüthy, R., Bowie, J. U., and Eisenberg, D. (1992). *Nature*, **356**, 83.
73. Topham, C. M., Srinivasan, N., Thorpe, C. J., Overington, J. P., and Kalsheker, N. A. (1994). *Protein Eng.*, **7**, 869.
74. Morris, A. L., MacArthur, M. W., Hutchinson, E. G., and Thornton, J. M. (1992). *Proteins*, **12**, 345.
75. Laskowski, P. A., MacArthur, M. W., Moss, D. S., and Thornton, J. M. (1993). *J. Appl. Crystallogr.*, **26**, 283.
76. Hubbard, T. J. P. and Blundell, T. L. (1987). *Protein Eng.*, **1**, 159.
77. Rosenbach, D. and Rosenfeld, R. (1995). *Protein Sci.*, **4**, 496.
78. Sudarsanam, S., Dubose, R. F., March, C. J., and Srinivasan, S. (1995). *Protein Sci.*, **4**, 1412.
79. Laughton, C. A. (1994). *J. Mol. Biol.*, **235**, 1088.
80. Kono, H. and Doi, J. (1994). *Proteins*, **19**, 244.
81. Tanimura, R., Kidera, A., and Nakamura, H. (1994). *Protein Sci.*, **3**, 2358.
82. Hwang, J. K. and Liao, W. F. (1995). *Protein Eng.*, **8**, 363.
83. Church, W. B., Palmer, A., Wathey, J. C., and Kitson, D. H. (1995). *Proteins*, **23**, 422.
84. Samudrala, R., Pederson, J. T., Zhou, H. B., Luo, R., Fidelis, K., and Moult, J. (1995). *Proteins*, **23**, 327.
85. Šali, A., Potterton, L., Yuan, F., Vanvlijmen, H., and Karplus, M. (1995). *Proteins*, **23**, 318.
86. Harrison, R. W., Chatterjee, D., and Weber, I. T. (1995). *Proteins*, **23**, 463.
87. Šali, A. (1995). *Curr. Opin. Struct. Biol.*, **6**, 437.

7

Antibody combining sites: structure and prediction

ANTHONY R. REES, STEPHEN J. SEARLE, ANDREW H. HENRY, NICHOLAS WHITELEGG, and JAN PEDERSEN

1. Introduction

Antibodies possess a vast repertoire of specificity and affinity. To understand the molecular basis of antibody function requires a knowledge of the structures of free and antigen-bound antibodies, determined by X-ray crystallography or NMR, or ideally both. Although the number of antibody structures is increasing rapidly—at the time of writing 48 structures have been deposited in the Brookhaven Protein Database (PDB) (1)—the rate of structure determination will never match that of sequence acquisition. Thus, to explore fully the structural variability of antibody combining sites, encoded in immunoglobulin variable region sequences, effective structural prediction methods have become essential. Such models will have increasing utility where the structural effects of point mutations within the antibody combining site need to be assessed, where minimal perturbation strategies for the grafting of complementarity determining regions (CDRs) during the process of 'humanization' are required, and even where the residues to be targeted for random mutation in the gene library methods now available (e.g. phage libraries) are selected in a more rational manner on the basis of their likely accessibility to antigen. In time, when the antibody sequence/structure relationship is fully understood, *de novo* design of antibodies will also become possible (2).

This chapter outlines the relevant structural features of antibody variable regions and provides a step by step methodological approach to CDR modelling embodied in the computer program AbM (3–5).

2. Antibody structure

2.1 General features of antibodies

The four chain structure of antibodies, depicted in *Figure 1* for an IgG was established by Porter (6) and Edelman (7) in the late 1950s. The manner in which these heavy and light chains were organized into three-dimensional

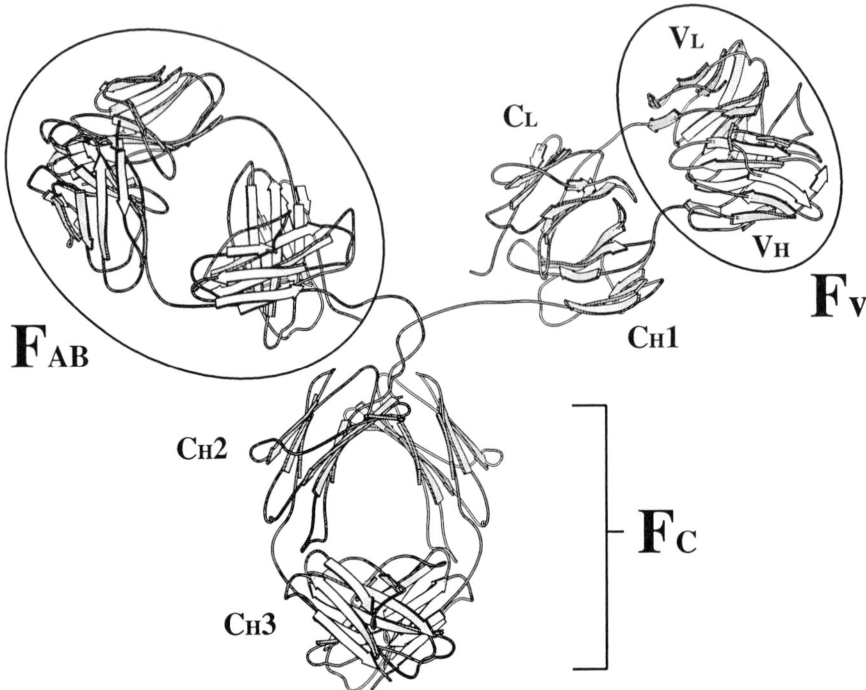

Figure 1. A cartoon of an IgG molecule showing the way in which the heavy and light chains fold to form a multidomain protein. The Fab region consists of one variable-type and one constant-type domain in each chain, non-covalently associated to give the structure V_L+V_H for the Fv region and C_L+C_H1 for the constant region. The Fc region is formed from the heavy chain only and consists of two constant regions $(C_H2)_2$ and $(C_H3)_2$. The segment of polypeptide between the C_H1 and C_H2 regions is known as the 'hinge' region. The angle between the Fab and Fc regions, dictated by the hinge behaviour, differs between and even within antibodies (12). This is illustrated in the figure by showing two rather different orientations of each Fab with respect to the Fc.

domains was demonstrated when the first Fab (8) and Fc (9) X-ray structures were determined. The quaternary organization of these domains became clear when structures of intact IgG molecules became available (10,11). These first two structures were of hinge deleted IgGs (see *Figure 1*), the effect of which was to constrain the Fab arms so that overall the antibody was T-shaped. Recently, the structure of a 'normal' IgG, determined by Harris *et al.* (12), has suggested that the Fab arms are somewhat independent of each other in terms of their orientation with respect to the Fc region, behaviour that may be important when the two combining sites are undergoing dynamic binding to separate molecules of the cognate antigen.

The crystallographic information currently available for antibodies and their Fab fragments, either alone or complexed with antigen, is summarized in *Table 1*.

Table 1. List of antibodies and their fragments for which X-ray coordinates are currently available from the Brookhaven database of protein structures (PDB)

PDB code	Name	Resolution (Å) /R-factor (%)	Notes
1 baf	ANO2	2.9/19.5	Anti-DNP
1bbd	8F5	2.8/19.0	Anti-HRV2
1bbj	B72.3	3.1	Anti-cancer antigen
1cgs	1cgs	2.6/21.8	Anti-hapten
1dbb	DB3	2.7/21.0	Progesterone complex
1dfb*	3D6	2.7/17.7	Anti-HIVgp41
1eap	17e8	2.5/18.6	Abzyme complex
1fai*	R19.9	2.7/18.9	Anti-arsenate
1fgv	H52	1.9/18.0	Humanized anti-CD18
1fig	1F7	3.0/22.0	Abzyme complex
1for	Fab17–1A	2.75/17.4	Anti-rhinovirus
1frg	26–9	2.8/19.0	Anti-flu HA complex
1fvc	Hu–4D5	2.2/18.3	Humanized 4D5
1fvd	4D5	2.5/17.9	Anti-HER2
1ggi	50.1	2.8/18.8	Anti-HIVgp120 peptide complex
1gig	HC19	2.3/19.5	Anti-flu HA
1hil	17/9	2.0/19.0	Anti-flu HA peptide complex
1igf	B13I2	2.8/18.0	Anti-peptide Fab′
1igi	26–10	2.7/17.7	Anti-digoxin
1igm	hulgM	2.3/20.1	Human IgM Fv
1ind	CHA255	2.2/18.8	Anti-hapten complex
1jel	JE142	2.8/19.3	Anti-protein complex
1jhl	D11.15	2.4/21.4	Anti-HEL comlex
1mam	YST9.1	2.5/21.5	
1mcp*	McPC603	2.7/22.5	Anti-PC
1mfa	SE155–4	1.7/16.6	Anti-CHO complex
1nbv	BV04–01	2.0/24.6	Anti-ssDNA
1ncb	NC41	2.5/16.5	Anti-flu NA complex
1tet	TE33	2.3/14.8	Anti-cholera peptide complex
1vfa*	D1.3	1.8/15.8	Anti-lysozyme
2cgr	2cgr	2.2/21.4	Anti-hapten complex
2fb4*	Kol	1.9/18.9	Myeloma Fab
2fbj*	J539	1.95/19.4	Anti-galactan
2gfb	CNJ206	3.0/21.3	Abzyme
2hfl*	HyHEL5	2.54/24.5	Anti-lysozyme
2hfm*	HyHEL10	3.0/24.6	Anti-lysozyme
4fab*	4–4–20	2.7/21.5	Anti-fluorescein complex
6fab*	36–71	1.9/20.9	Anti-phenylarsenate
7fab*	New	2.0/16.9	
8fab	Hil	1.8/17.3	Human myeloma
1ivl	M29B	2.17/17.5	LC dimer
1lvd	LEN	1.8/24.0	LC dimer
1mcw	MCW/Weir	3.5/17.0	Hybrid LC dimer
1rei	REI	2.0/	LC dimer
1wtl	WAT	1.9/15.7	LC dimer
2bjl	Loc	2.8/21.6	LC dimer
2mcg	MCG	2.0/18.7	LC dimer
2rhe	RHE	1.6/14.9	LC dimer

Resolution and R-factors are given where known. Where the structure is a complex this is stated. References for each of the structures can be obtained directly from the PDB. Antibodies used for the mean β-barrel calculations described in the main text are marked with an asterisk. An additional antibody, Gloop 2, not in the PDB was also used (Jeffrey, P., Taylor, G. L. and Rees, A. R., unpublished data)

2.2 The antibody fold

The characteristic folding pattern of immunoglobulin domains was first seen in the Fab structure determined by Poljak and colleagues (8). It consisted of two antiparallel β-sheets packed to form a β-sheet 'sandwich', the two halves of the sandwich being held together by a disulfide bond (*Figure 2a*). This pattern, now known as the immunoglobulin fold, is highly conserved and can be found with slight variations in a large superfamily of immunoglobulin-like molecules (see ref. 13 and references therein). The topology of this fold, commonly showing a 'Greek key' motif, is shown in *Figure 2b*. Within antibodies two similar but distinct patterns are seen, one characteristic of the variable domains (V_L for light chain; V_H for heavy chain) and containing the antigen recognition, or complementarity determining regions (CDRs), and one for the constant domains (C_L for light chain; C_H1 for heavy chain).

From *Figure 2* the positions of the three CDRs with respect to the core β-sheet framework can also be seen. Each CDR is a loop connecting two β-strands and has a fixed orientation on the framework depending on its length and other characteristic sequence motifs (see later). Due to their close proximity to one another, mutational changes within a CDR, introduced by protein engineering methods, may be propagated to adjacent CDRs, or to the framework itself. A structure or model to guide such changes is therefore most important.

2.3 Variable region domain packing: formation of the framework

In the variable region (Fv) structure, the V_L and V_H domains associate non covalently, with an average rotation one to the other of about 180° about the pseudo-diad axis (see *Figure 3*). This generates a β-barrel structure in which four of the strands from each domain form the barrel framework while the CDRs (three from each domain) form a contiguous surface at the N-terminal end. Although the framework sequences are well conserved between antibodies (*Table 2*), structurally the β-sheets can vary in their orientation with respect to one another by as much as 30° ± 18° (14,15).

At the V_H–V_L interface, close complementarity is particularly important. *Table 3* shows the distribution of amino acids found in human and mouse V region at each of the interfacial residue positions, from which it can be seen that many positions are conserved with high fidelity. There are variations however and, when procedures that result in random pairings of V_L and V_H domain (e.g. gene libraries) are used, incompatibilities across the interface may be encountered resulting in abortive pairing. Of course, heavy chain and light chain CDRs are also brought into contact by this dimer formation and may either add to or subtract from the binding energy.

Although the three-dimensional structures of Fv regions are well conserved, they are not identical in all antibodies, and care should be taken when modelling new sequences on existing X-ray structures. We have carried out an analysis of 12 antibody structures (* in *Table 1*) to determine where these differences occur. The β-strands comprising the V region β-barrel were defined according to their secondary structure and sequence conservation, rather than by excluded surface area as described by Chothia *et al.* (17). The barrels of 11 structures were then fitted on to the 12th (selected at random) and mean coordinates were calculated. The 12 structures were then fitted on to these mean coordinates and new mean coordinates calculated. This was repeated until the mean coordinate set converted (~10 cycles). The variance for the mean coordinates was then calculated for each N, C^{α}, C atom set in the barrel strands. This average barrel was fitted to the surface of a hyperboloid:

$$\frac{X^2}{A^2} + \frac{Y^2}{B^2} - \frac{Z^2}{C^2} = 1$$

where the parameters A, B, and C are taken from Novotny *et al.* (18,19). The strands of the mean barrel are shown in *Figure 4a*. The root mean square (rms) derivation of the fit to the theoretical hyperboloid was 2.10 Å. Using this mean barrel, the deviations of ten of the Fv region barrels were calculated. These data are plotted in *Figure 4b* and *4c* for the four strands of V_L and V_H, respectively, and show rms deviation as a function of N, C^{α}, C position along the strand. This analysis has pin-pointed 'hot-spots' where residues deviate markedly in their backbone conformation from one antibody to another. The most disordered residues (> 3σ from mean coordinates) are found in strands F and G of the heavy chain. Strand G is particularly variable with only two residues being retained during the fitting procedure. Interestingly, these two strands form the take-off points for CDR H3, the most conformationally variable of the six CDRs. Although some other strands show variability at their ends (e.g. the C-terminus of strand C of the light chain), the residues involved lie at the variable region–constant region interface and are unlikely to influence CDR conformation. To summarize, of the eight strands in the Fv barrel, only parts of each strand exhibit high positional conservation. The mean barrel calculated on the basis of these structurally conserved positions can be used to fit new sequences with high confidence (see later).

2.4 CDR conformation

The combining site of an antibody is derived from the juxtaposition of six interstrand loops (or CDRs), three derived from the heavy chain (H1, H2, and H3) and three from the light chain (L1, L2, and L3). Four of the six are encoded in the V_L and V_H germline V region genes (L1, L2, H1, and H2)

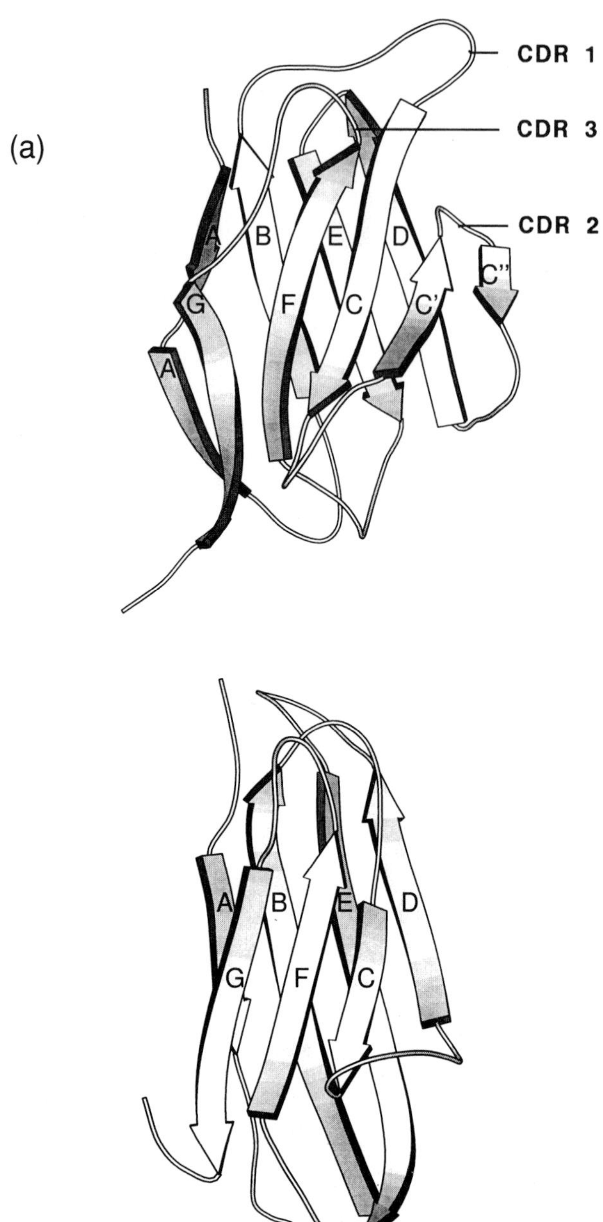

Figure 2. (a) Detail of the domain structure of immunoglobulins. Upper: the variable (V) domain, consisting of nine β-strands (A–G, C′, and C″) packed to form a β-sheet sandwich. The loops connecting the strands form the complementarity determining regions (CDRs) and are shown generically as CDR1, CDR2, and CDR3. Lower: the constant (C) domain, consisting of only seven β-strands packed in a similar manner to the V domain, but lacking the two additional strands C′ and C″. (b) The relative positions and hydrogen bonding pattern (= = =) of the β-strands in immunoglobulin constant and variable domains shown as a two-dimensional topology diagram. The type of connectivity seen in these types of domain is known as a 'Greek key' motif.

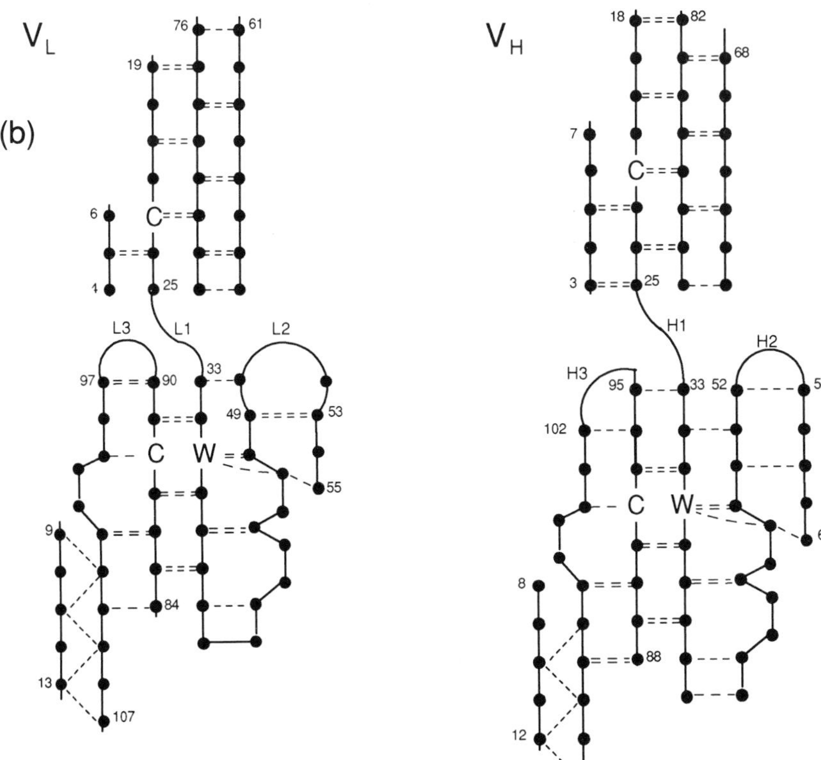

V_L
(b)
76
61
19
6
C
1
25
L3
L1
L2
97
90
33
49
53
C
W
55
9
84
13
107
V_H
18
82
68
7
C
3
25
H1
H2
H3
95
33
52
56
102
C
W
60
8
88
12
112

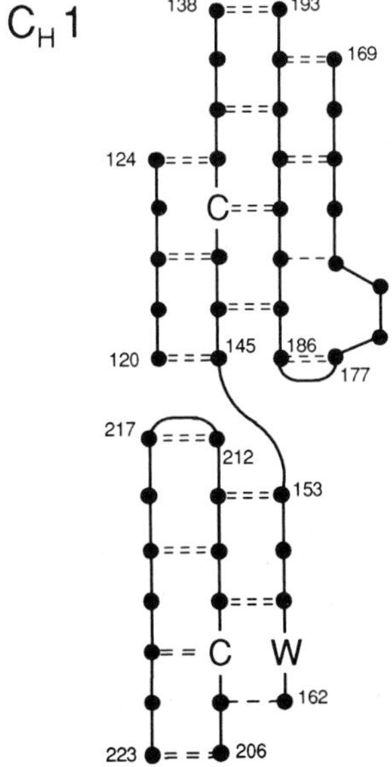

C_H1
138
193
169
124
C
120
145
186
177
217
212
153
C
W
162
223
206

(a)

(b)

Figure 3. The pairing of V_L and V_H domains to form an F_V region. Plan (a) and side (b) views are shown to indicate the tight, non-covalent packing of aromatic and other hydrophobic residues at the V_L–V_H interface (V_H is right-hand domain). Close inspection will reveal the manner in which the two conserved tryptophan residues, one from each

while the remaining two are formed at the junctions of the V genes and their respective modifying gene segments (+J for light chain and +DJ for heavy chain). Despite considerable variability in these CDR sequences from one antibody to another (20), there is some structural similarity for many of the CDRs. This was analysed in our laboratory in 1985 and 1986 (21,22) when six antibody structures were compared and the high conservation of H1, H2, and L2 conformations noted.

The present database of antibody structures (see *Table 1*) contains combining sites with various combinations of CDR length and sequence. The length ranges (seen in the sequence database) are:

CDR	**No. of residues**
L1	10–17
L2	7 (highly conserved)
L3	7–11
H1	5,7
H2	9–12 (defined as in ref. 50)
H3	4–27

The structures of these CDRs often vary considerably with length. Four of the CDRs (L2, L3, H2, and H3) are hairpin loops. The preferred conformations of such loops have been extensively studied for other proteins (23,24) but the high variability in length of the antibody loops has complicated the development of a rigorous structural classification. For example, while H3 is by definition a hairpin loop since it interconnects two adjacent β-strands, its high length variation produces a large number of backbone conformations, the relationship between which has not been fully elucidated (but see later).

The most comprehensive structure classification of CDRs has been described by Chothia, Lesk, and co-workers (25–27) for both murine and human sequences. A full structural description of these 'canonical' families (for CDRs L1, L2, L3, H1, and H2) is outside the scope of this article but a summary of their classification is shown in *Table 4*. Briefly, the method assigns CDRs to a limited set of conformers based on the presence of residues that influence hydrogen bonding, packing, or preferred torsion angles. Should such a residue or residues be mutated by the antibody engineer then, of course, it is likely that a conformational change will result—a further reason why molecular models to guide the engineering is desirable.

While this method is useful for CDRs where exact obedience to the 'rules' is seen, further development is necessary in two areas. First, there are CDRs that are as yet structurally unclassified. That is to say, they lack the appropriate key residues in those positions that define a particular canonical class. The second issue relates to the question of whether all permutations of CDRs will be allowed. A particular canonical conformation in one V_L or V_H domain may be disallowed because of overriding energetic factors associated

Table 2. Antibody variable domain sequences of light chains (L) and heavy chains (H) from known structures

Structure	LFR1	L1	LFR2	L2	LFR3	L3	LFR4
glb2	DIQMTQSPSSLSASLGERVSLTC	RASQEISG------YLS	WLQQKP DGTI KRLIY	AASTLDS	GVPKRFSGRRSGSDYSLTISSLESED FADYYC	LQYLS--YPLT	FGAGT KLELKRA
2hfl	DIVLTQSPAIMSASPGEKVTMTC	SASSSVN-------YMY	WYQQKS GTSP KRWIY	DTSKLAS	GVPVRFSGSGSGTSYSLTISSMETED AAEYYC	QQWGR--NP-T	FGGGT KLEIKRA
3hfm	DIVLTQSPATLSVTPGNSVSLSC	RASQSIGN------NLH	WYQQKS HESP RLLIK	YASQSIS	GIPSRFSGSGSGTDFTLSINSVETED FGMYFC	QQSNS--WPYT	FGGGT KLEIKRA
2fbj	EIVLTQSPAITAASLGQKVTITC	SASSSVS-------SLH	WYQQKS GTSP KPWIY	EISKLAS	GVPARFSGSGSGTSYSLTINTMEAED AAIYYC	QQWTY--PLIT	FGAGT KLELKRA
2fb4	QSVLTQPPSASG-TPGQRVTISC	SGTSSNIG----SSTVN	WYQQLP GMAP KLLIY	RDAMRPS	GVPDRFSGSKSGASASLAIGGLQSED ETDYYC	AAWDVSLNAYV	FGTGT KVTVLGQ
1fb4	ESVLTQPPSASG-TPGQRVTISC	TGTSSNIG----SITVN	WYQQLP GMAP KLLIY	RDAMRPS	GVPTRFSGSKSGTSASLAISGLEAED ESDYYC	ASWNSSDNSYV	FGTGT KVTVLGQ
2mcp	DIVMTQSPSSLSVSAGERVTMSC	KSSQSLLNSGNQKNFLA	WYQQKP GQPP KLLIY	GASTRES	GVPDRFTGSGSGTDFTLTISSVQAED LAVYYC	QNDHS--YPLT	FGAGT KLEIKRA
1rei	DIQMTQSPSSLSASVGDRVTITC	QASQDII------KYLN	WYQQTP GKAP KLLIY	EASNLQA	GVPSRFSGSGSGTDYTFTISSLQPED IATYYC	QQYQS--LPYT	FGQGT KLQIT--
2rhe	ESVLTQPPSASG-TPGQRVTISC	TGSATDIG----SNSVI	WYQQVP GKAP KLLIY	YNDLLPS	GVSDRFSASKSGTSASLAISGLESED EADYYC	AAWNDSLDEPG	FGGGT KLTVLGQ
4fab	DVVMTQTPLSLPVSLGDQASISC	RSSQSLVHS-QGNTYLR	WYLQKP GQSP KVLIY	KVSNRFS	GVPDRFSGSGSGTDFTLKISRVEAED LGVYFC	SQSTH--VPWT	FGGGT KLEIKRA
2f19	DIQMTQTTSSLSASLGDRVTISC	RASQDISN------YLN	WYQQKP DGTV KLLIY	YTSRLHS	GVPSRFSGSGSGTDYSLTISNLEHED IATYFC	QQGST--LPRT	FGGGT KLEIKRA
1mcw	-SALTQPASVSG-SPGQSITVSC	AGHTSDVA---DSNSIS	WFQQHP DKAP KLLIY	AVTFRPS	GIPLRFSGSKSGNTASLTISGLLPDD EADYFC	MSYLS-DASFV	FGSGT KVTVLRQ
3mcg	-SALTQPPSASG-SLGQSVTISC	TGTSSDVG---GYNYVS	WYQQHA GKAP KVIIY	EVNKRPS	GVPDRFSGSKSGNTASLTVSGLQAED EADYYC	SSYEGSD-NFV	FGTGT KVTVLGQ
1mam	DIQMTQTTSSLSASLGDRVTISC	RASQDIYN------YLN	WYQQKP DGTV KLLIY	YTSRLHS	GVPSRFSGSGSGTDYSLTISNLNQED MATYIC	QQGNT--LPFT	FGSGT KLEIKRA
1baf	QIVLTQSPAIMSASPGEKVTMTC	SASSSVY-------YMY	WYQQKP GSSP RLLIY	DTSNLAS	GVPVRFSGSGSGTSYSLTISRMEAED AATYYC	QQWSSYP-PIT	FGVGT KLELKRA
1hil	DIVMTQSPSSLTVTAGEKVTMSC	TSSQSLFNSGKQKNYLT	WYQQKP GQPP KVLIY	WASTRES	GVPDRFTGSGSGTDFTLTISSVQAED LAVYYC	QNDYS--NPLT	FGGGT KLELKRA
8fab	--ELTQPPSVSV-SPGQTARITC	SANALPNQ------YAY	WYQQKP GRAP VMVIY	KDTQRPS	GIPQRFSSSTSGTTVTLTISGVQAED EADYYC	QAWDN--SASI	FGGGT KLTVLGQ
1dfb	DIQMTQSPSTLSASVGDRVTITC	RASQSISR------WLA	WYQQKP GKVP KLLIY	KASSLES	GVPSRFSGSGSGTEFTLTISSLQPDD FATYYC	QQYNS----YS	FGPGT KVDIKRT
1fdl	DIQMTQSPASLSASVGETVTITC	RASGNIHN------YLA	WYQQKQ GKSP QLLVY	YTTTLAD	GVPSRFSGSGSGTQYSLKINSLQPED FGSYYC	QHFWS--TPRT	FGGGT KLEIKRA
1igf	DVLMTQTPLSLPVSLGDQASISC	RSNQTILLS-DGDTYLE	WYLQKP GQSP KLLIY	KVSNRFS	GVPDRFSGSGSGTDFTLKISRVEAED LGVYYC	FQGSH--VPPT	FGGGT KLEIKRA
6fab	DIQMTQIPSSLSASLGDRVSISC	RASQDINN------FLN	WYQQKP DGTI KLLIY	FTSRSQS	GVPSRFSGSGSGTDYSLTISNLEQED IATYFC	QQGNA--LPRT	FGGGT KLEIKRA
7fab	ASVLTQPPSVSG-APGQRVTISC	TGSSSNIG---AGHNVK	WYQQLP GTAP KLLIF	HNNA---	----RFSVSKSGTSATLAITGLQAED EADYYC	QSYDR--SLRV	FGGGT KLTVLRQ
1igm	DIQMTQSPSSLSASVGDRVTITC	QASQDISN------YLA	WYQQKP GKAP ELRIY	DASNLET	GVPSRFSGSGSGTDFTFTISSLQPED IATYYC	QQYQN--LPLT	FGPGT KVDIKRT
1bbj	DIQMTQSPASLSVSVGETVTITC	RASENIYS------NLA	WYQQKQ GKSP QLLVY	AATNLAD	GVPSRFSGSGSGTQYSLKINSLQSED FGSYYC	QHFWG--TPYT	FGGGT RLEIKRA
1bbd	DIVMTQSPSSLTVTTGEKVTMTC	KSSQSLLNSRTQKNYLT	WYQQKP GQSP KLLIY	WASTRES	GVPDRFTGSGSGTDFTLSISGVQAED LAVYYC	QNNYN--YPLT	FGAGT KLELKRA
1ncd	DIVMTQSPKFMSTSVGDRVTITC	KASQDVST------AVV	WYQQKP GQSP KLLIY	WASTRHI	GVPDRFAGSGSGTDYTLTISSVQAED LALYYC	QQHYS--PPWT	FGGGT KLEIKRA
1fvc	DIQMTQSPSSLSASVGDRVTITC	RASQDVNT------AVA	WYQQKP GKAP KLLIY	SASFLYS	GVPSRFSGSRSGTDFTLTISSLQPED FATYYC	QQHYT--TPPT	FGQGT KVEIKAT
1igi	DVVMTQTPLSLPVSLGDQASISC	RSSQSLVHS-NGNTYLN	WYLQKA GQSP KLLIY	KVSNRFS	GVPDRFSGSGSGTDFTLKISRVEAED LGIYFC	SQTTH--VPPT	FGGGT KLEIKRA
1ggi	DIVLTQSPGSLAVSLGQRATISC	RASESVDD--DGNSFLH	WYQQKP GQPP KLLIY	RSSNLIS	GIPDRFSGSGSRTDFTLTINPVEADD VATYYC	QQSNE--DPLT	FGAGT KLEIKRA
	12345678901234567890123	45678901234567890	123456 7890 12345	6789012	34567890123456789012345678 901234	56789012345	67890 1234567

Structure	HFR1	H1	HFR2	H2	HFR3	H3	HRF4
glb2	QVQLQQSGTELARPGASVRLSCKASGYTFT	T--FGIT	WVKQ RTGQ GLEWIG	EIFPGNS--KTY	YAERFKGKATLTADKSSTTAYMGLSSLTSEDSAV YFCAR	E------------IRY	WG QGTTLTVS
2hfl	-VQLQQSGAELMKPGASVKISCKASGYTFS	D--YWIE	WVKQ RPGH GLEWIG	EILPGSG--STN	YHERFKGKATFTADTSSSTAYMQLNSLTSEDSGV YYCLH	GNYD---------FDG	WG QGTTLTVS
3hfm	DVQLQESGPSLVKPSQTLSLTCSVTGDSIT	S--DYWS	WIRK FPGN RLEYMG	YVSYSG---STY	YNPSLKSRISITRDTSKNQYYLDLNSVTTEDTAT YYCAN	WD-----------GDY	WG QGTLVTVS
2fbj	EVKLLESGGGLVQPGGSLKLSCAASGFDFS	K--YWMS	WVRQ APGK GLEWIG	EIHPDSG--TIN	YTPSLKDKFIISRDNAKNSLYLQMSKVRSEDTAL YYCAR	LHYYGY--------NAY	WG QGTLVTVS
2fb4	EVQLVQSGGGVVQPGRSLRLSCSSSGFIFS	S--YAMY	WVRQ APGK GLEWVA	IIWDDGS--DQH	YADSVKGRFTISRNDSKNTLFLQMDSLRPEDTGV YFCAR	DGGHGFCSSASCFGPDY	WG QGTPVTVS
2mcp	EVKLVESGGGLVQPGGSLRLSCATSGFTFS	D--FYME	WVRQ PPGK RLEWIA	ASRNKGNKYTTE	YSASVKGRFIVSRDTSQSILYLQMNALRAEDTAI YYCAR	NYYGSTWY------FDV	WG AGTTVTVS
4fab	EVKLDETGGGLVQPGRPMKLSCVASGFTFS	D--YWMN	WVRQ SPEK GLEWVA	QIRNKPYNYETY	YSDSVKGRFTISRDDSKSSVYLQMNNLRVEDMGI YYCTG	SYYG---------MDY	WG QGTSVTVS
2f19	QVQLQQSGAELVRAGSSVKMSCKASGYTFT	S--YGVN	WVKQ RPGQ GLEWIG	YINPGKG--YLS	YNEKFKGKTTLTVDRSSSTAYMQLRSLTSEDAAV YFCAR	SFYGGSDLA--VYYFDS	WG QGTTLTVS
1mam	EVKLVESGGGLVQPGGSLRLSCATSGFTFT	D--YYMS	WVRQ PPGK ALEWLG	FIRNKADGYTTE	YSASVKGRFTISRDNSQSILYLQMNTLRAEDSAT YYCTR	DPYGP--------AAY	WG QGTLVTVS
1baf	DVQLQESGPGLVKPSQSQSLTCTVTGYSIT	S-DYAWN	WIRQ FPGN KLEWMG	YMSYSG---STR	YNPSLRSRISITRDTSKNQFFLQLKSVTTEDTAT YFCAR	GWP----------LAY	WG QGTQVSVS
1hil	EVQLVESGGDLVKPGGSLKLSCAASGFSFS	S--YGMS	WVRQ TPDK RLEWVA	TISNGGG--YTY	YPDSVKGRFTISRDNAKNTLYLQMSSLKSEDSAM YYCAR	RERYDENG------FAY	WG QGTLVTVS
8fab	AVKLVQAGGGVVQPGRSLRLSCIASGFTFS	N--YGMH	WVRQ APGK GLEWVA	VIWYNGS--RTY	YGDSVKGRFTISRDNSKRTLYMQMNSLRTEDTAV YYCAR	DPDILTAFS-----FDY	WG QGVLVTVS
1dfb	EVQLVESGGGLVQPGRSLRLSCAASGFTFN	D--YAMH	WVRQ APGK GLEWVS	GISWDSS--SIG	YADSVKGRFTISRDNAKNSLYLQMNSLRAEDMAL YYCVK	GRDYYDSGGYFTVAFDI	WG QGTMVTVS
1fdl	QVQLKESGPGLVAPSQSLSITCTVSGFSLT	G--YGVN	WVRQ PPGK GLEWLG	MIWGDGN---TD	YNSALKSRLSISKDNSKSQVFLKMNSLHTDDTAR YYCAR	ERDYR---------LDY	WG QGTTLTVS
1igf	EVQLVESGGDLVKPGGSLKLSCAASGFTFS	R--CAMS	WVRQ TPEK RLEWVA	GISSGGS--YTF	YPDTVKGRFIISRNNARNTLSLQMSSLRSEDTAI YYCTR	YSSDPFY-------FDY	WG QGTTLTVS
6fab	EVQLQQSGVELVRAGSSVKMSCKASGYTFT	S--NGIN	WVKQ RPGQ GLEWIG	YNNPGNG--YIA	YNEKFKGKTTLTVDKSSSTAYMQLRSLTSEDSAV YFCAR	SEYYGGSYK-----FDY	WG QGTTLTVS
7fab	AVQLEQSGPGLVRPSQTLSLTCTVSGTSFD	D--YYWT	WVRQ PPGR GLEWIG	YVFYTG---TTL	LDPSLRGRVTMLVNTSKNQFSLRLSSVTAADTAV YYCAR	NLIAGG--------IDV	WG QGSLVTVS
1igm	EVHLLESGGNLVQPGGSLRLSCAASGFTFN	I--FVMS	WVRQ APGK GLEWVS	GVFGSGG--NTD	YADAVKGRFTITRDNSKNTLYLQMNSLRAEDTAI YYCAK	HRVSYVLTG-----FDS	WG QGTLVTVS
1bbj	-VQLQQSDAELVKPGASVKISCKASGYTFT	D--HAIH	WAKQ KPEQ GLEWIG	YISPGND--DIK	YNEKFKGKATLTADKSSSTAYMQLNSLTSEDSAV YFCKR	SY-----------YGH	WG QGTTLTVS
1bbd	EVQLQQSGAELVRPGASVKLSCTTSGFNIK	D--IYIH	WVKQ RPEQ GLEWIG	RLDPANG--YTK	YDPKFQGKATITVDTSSNTAYLHLSSLTSEDTAV YYCDG	YYSYYD--------MDY	WG PGTSVTVS
1ncd	QIQLVQSGPELKKPGETVKISCKASGYTFT	N--YGMN	WVKQ APGK GLKWMG	WINTNTG--EPT	YGEEFKGRFAFSLETSASTANLQINNLKNEDTAT FFCAR	GEDNFGSL------SDY	WG QGTTVTVS
1fvc	EVQLVESGGGLVQPGGSLRLSCAASGFNIK	D--TYIH	WVRQ APGK GLEWVA	RIYPTNG--YTR	YADSVKGRFTISADTSKNTAYLQMNSLRAEDTAV YYCSR	WGGDGFYA------MDY	WG QGTLVTVS
1igi	-VQLQQSGPELVKPGASVRMSCKSSGYIFT	D--FYMN	WVRQ SHGK SLDYIG	YISPYSG--VTG	YNQKFKGKATLTVDKSSSTAYMELRSLTSEDSAV YYCAG	SSGNKWA-------MDY	WG HGASVTVS
1ggi	QVQLKESGPGILQPSQTLSLTCSFSGFSLS	TYGMGVS	WIRQ PSGK GLEWLA	HIFWDG---DKR	YNPSLKSRLKISKDTSNNQVFLKITSVDTADTAT YYCVQ	EG-----------YIY	WG QGTSVTVS
	123456789012345678901234567890	1234567	8901 2345 678901	234567890123	4567890123456789012345678901234567 89012	34567890123456789	01 23456789

These structures are used for framework construction, canonical structures, and for calculation of C^{α} database constraints. Those tracts of sequence in the framework regions used for fitting structures during rmsd calculations are highlighted in grey and are the most highly conserved.

Table 3. Distribution of residues on the interfacial surfaces of V_L and V_H domains of mouse and human Fv regions

Light chain position	**Human**	**Mouse**
41	W99	W98
42	Y88, F8	Y74, F12
43	Q93, L4	Q74, L22
44	Q98	Q88, E5
45	K58, H13, L10	K81, R13
46	P89, A5	P80, S13
51	K52, R27	K73, Q12
52	L72, V10	L70, R9
53	L75, I8, V11	L81, W15
54	L92	I91, V4
55	V86, F6	83
89	E42, F37, V9	A21, I13, L26, F9, V13
90	A90, G8	A67, E28
91	D38, T22, V28	D7, I11, T36, V30
92	Y99	Y99
93	Y90, F9	Y67, F30
94	C99	C99
106	F93, Y5	F91
107	G94	G92
108	G44, Q34, T10	G56, A21, S12
109	G95	G95
110	T95	T92
Heavy chain position		
155	W96	W98
156	V70, I23	V86, I10
157	R90	R53, K44
158	Q93	Q90, K5
164	G83, A6, S6	G58, R20, K8, S7
165	L95	L98
166	E88	E96
167	W98	W92
168	V46, I22, M17, L13	I61, V18, L11, M9
169	G58, A22, S15	G68, A29
215	Y98	Y96
216	V90, F9	Y80, F18
217	C97	C98
218	A82, T10	A80
219	R66, K12, P6	R83
237	W91	W95
238	G94	G96

The numbered positions are according to Pedersen *et al.* (50). The numbers following the one letter code names represent percentage occurrences at each position in the aligned heavy and light chain sequences taken from the Kabat database (65). Only frequencies greater than or equal to 5% are included.

with domain association, thus limiting the repertoire of canonical pairings. Any preferences for particular CDR combinations have not yet been described, presumably because of the limited number of structures from which to generalize, but their existence may limit the effective repertoire size of V_H–V_L gene (or phage) libraries.

2.5 The special problem of CDR H3

However comprehensive the canonical hypothesis turns out to be, it cannot at present be used to define the conformations of CDR H3 loops. This CDR occupies a central position in the combining site and has a critical effect on the combining site topography. One of the factors determining CDR conformation is the loop take-off angle, which varies substantially from one H3 to another. This positional variability may be important for binding of antigen when induced fit is required (see later). However, in the unbound state each CDR would be expected to exhibit a preference for one or a small number of energetically favourable conformers.

We have analysed the take-off angles of all six CDRs (2,28). For CDR H3, this can vary by up to 90° between different antibodies. This analysis suggested the existence of multiple structural classes of H3 based on take-off angle alone. In addition, specific sequence-dependent features may be present. For example, where a conserved Arg or Lys is present at position 94, it will normally interact with Asp, Gly, or Ala at position 101 (Kabat numbering), fixing the conformation of this region of the CDR. On the basis of the foregoing, we have defined seven structural classes of H3 loop, illustrated in *Figure 5*:

- H3a—loops of 6 residues or less
- H3b—loops of 7 residues
- H3c–f—loops of 8, 9/10, 11, or 12 residues, in which a conserved Arg or Lys at position 94 interacts with Asp, Gly, or Ala at position 101

Loops of 13 residues or more do not easily fit into this classification and require further analysis. This H3 classification is now a part of the standard antibody modelling protocol, AbM (5), the use of which will be described later.

2.6 Conservation of variable domain surfaces

The classification 'variable domain' when applied to antibodies suggests the existence of major differences between antibody molecules. As has been shown, the β-sheet core is well conserved and even in the CDRs, where the greatest variability occurs, patterns of conservation are also seen. There is one further respect in which variable domains show structural conservation and that is in the positions and types of the surface exposed residues. When

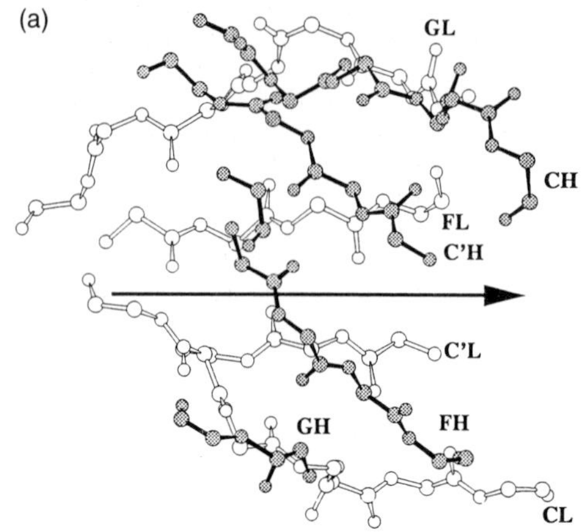
(a)
GL
CH
FL
C'H
C'L
GH
FH
CL

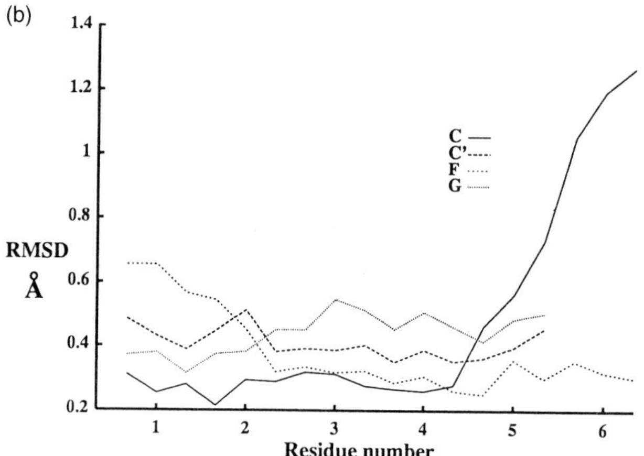
(b)
1.4
1.2
1
0.8
0.6
0.4
0.2
RMSD
Å
C
C'
F
G
1
2
3
4
5
6
Residue number

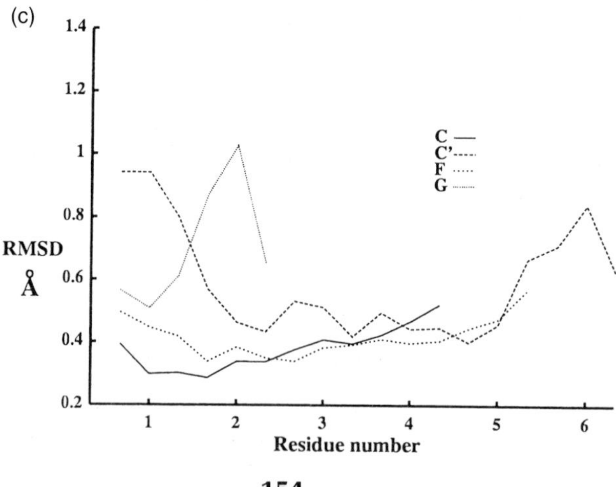
(c)
1.4
1.2
1
0.8
0.6
0.4
0.2
RMSD
Å
C
C'
F
G
1
2
3
4
5
6
Residue number

Figure 4. (a) Structure of the 'mean barrel' showing the average positions of the conserved segments of the β-barrel strands in a V region, generated according to the multiple fitting procedure described in the text. The conjugate axis of the β-barrel is shown by the *arrow* which points in the direction of the CDRs. The strands participating in the barrel are indicated by their normal letter names, as given in *Figure 2*. The light chain strands are in white and the heavy chain in black. (b) rms deviations from the mean barrel of the light chain strands that participate in the V region β-barrel. The rms deviation is calculated for N, C^{α}, C atoms at each residue position along the strands. Residue number of the strand segments is according to Kabat *et al.* (65). Light chain: C strand 35–40, C′ strand 45–49, F strand 83–88, G strand 98–102. (c) As (b) but for the heavy strands. Residue numbering is as follows. Heavy chain: C strand 36–39, C′ strand 44–49, F strand 90–94, G strand 103–104.

---

the solvent accessibility is calculated for all framework residues (that is, without CDR residues) in the Fv region, using a set of high resolution crystal structures of Fab fragments, the alignment positions of those residues defined as 'surface' (29) is conserved with 98% fidelity. *Table 5* lists the residue positions and variability seen in human and mouse sequences.

One of the most striking features of this analysis has been the fact that the identical V_L and V_H sequence families, previously classified on the basis of contiguous sequences, can be generated using the surface residue profiles alone. This has suggested that the surface patterns of Fv regions may be more conserved than was previously thought. The main conclusions of the analysis are:

- V_L and V_H framework surface residue positions are conserved
- within each species there appears to be a preference for particular residues at certain positions
- V_L and V_H sequences can be classified into families according to their framework surface residue profiles

This study allowed the development of a novel method of humanization in which the surface pattern of a murine Fv region is replaced by it nearest human Fv surface, derived from appropriately edited databases (30). One further aspect of this conservation may be important. Where antigens are flexible and with extended epitopes (e.g. carbohydrates), they may engage not only the CDR regions but also parts of the framework surface. There is some evidence that 'superantigens' exhibit this requirement (31). Clearly, if the framework–CDR surface occasionally works in concert to provide an antigen binding region, this adds a new and complicated dimension to combining site prediction.

3. Variable region modelling

3.1 Database and knowledge-based methods

As the database of antibody structures increases, the use of knowledge-based methods in determining loop conformations has gained in importance (4,32).

Table 4. The antibody canonical CDRs L1, L2, L3, H1, and H2 are defined on the basis of length and the occurrence of 'key residues' at the positions indicated† AbM numbering (5)

CDR	Loop	Class	Loop Length	Key Residues	Pattern	Structures
L1	24-40†	1	10	2,23,25,29,39,41,77	I-x(20)-C-x(1)-A-x(3)-V-x(9)-[LM]-x(1)-W-x(35)-Y	2hfl,2fbj,1baf
	24-34¶	2	11	2,23,25,29,39,41,77	I-x(20)-C-x(1)-A-x(3)-[IV]-x(9)-L-x(1)-W-x(35)-[YF]	glb2,3hfm,1rei,1igm,2f19,1mam,1fdl,6fab,1dfb
		3	17	2,23,25,29,39,41,77	I-x(20)-C-x(1)-S-x(3)-L-x(9)-L-x(1)-W-x(35)-[YF]	2mcp,1hil,1bbd,1bbj
		4	16	2,23,25,29,39,41,77	V-x(20)-C-x(1)-S-x(3)-L-x(9)-L-x(1)-W-x(35)-[YF]	1igi,4fab
		5	13	2,23,25,30,39,41,77	S-x(20)-C-x(1)-G-x(4)-I-x(8)-V-x(1)-W-x(35)-[A]	2fb4,1fb4,2rhe
		6	11	3,23,25,29,39,41,77	E-x(19)-C-x(1)-[SA]-x(3)-P-x(9)-A-x(1)-W-x(35)-[VSA]	8fab
L2	56-62† 50-56¶	1	7	54,70	[IV]-x(15)-G	2hfl,2fbj,1baf,glb2,3hfm,1rei,1igm,2f19,1mam, 6fab,1ncd,1fvc,2mcp,1hil,1bbd,4fab,1igf,1igi, 2fb4,1ggi, 1bbj, 1dfb, 1fb4, 1fdl, 1mcw, 3mcg
L3	95-105† 89-97¶	1	9	94,96,103	C-x(1)-[QNH]-x(6)-P	glb2,3hfm,1rei,1igm,2f19,1mam,6fab,1fvc, 2mcp,1hil,1bbd,4fab,1igf,1igi,1ggi, 1bbj, 1ncd 1fdl, 1rei
		2	9	94,96,102,103	C-x(1)-Q-x(5)-P-{P}	2fbj
		3	8	94,96,103	C-x(1)-Q-x(6)-P	2hfl
H1	26-37† 31-35B¶	1	10	26,27,29,36,100,102	G-[GFY]-x(1)-[IFLV]-x(6)-[MIVLT]-x(63)-C-x(1)-[RGHTKE]	2f19,1fdl,1igf,1mam,2fb4,2fbj,2hfl,2mcp,4fab, 6fab,8fab,glb2,1igm,1bbd,1ncd,1igi,1fvc,1hil 1bbj, 1dfb
		2	10	26,27,29,36,100,102	G-[TD]-x(1)-[FI]-x(6)-[WS]-x(63)-C-x(1)-[RN]	3hfm, 7fab
		3	11	26,27,29,36,100,102	G-[FGY]-x(1)-[LI]-x(6)-[WV]-x(63)-C-x(1)-[RH]	1baf
H2	52-63†	1	9	57,76	[DG]-x(18)-[KRI]	1fdl,1baf,1ggi,3hfm
	50-65¶	2	10	55,58,76	[PTA]-x(2)-[GS]-x(17)-[ALT]	2hfl,glb2,1fvc,1ncd
		3	10	57,76	[GSND]-x(18)-R	1igf,2fb4,1igm,1hil,2fbj, 1dfb, 8fab
		4	12	59,60,76	[GKN]-Y-x(15)-R	2mcp,4fab, 1mam

¶ Kabat numbering (65)
The first residue of the canonical pattern corresponds to the first canonical in the list of key residues. Residues in square brackets indicate sites where more than one residue type may be present, and n spaces in the sequence is denoted by x(n).

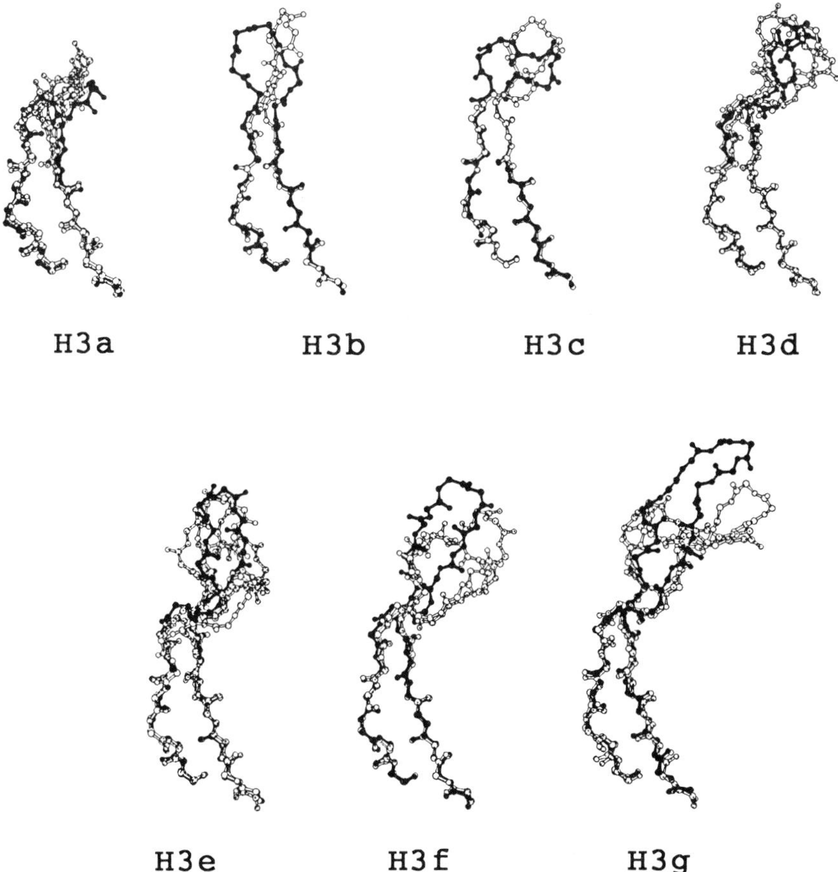

Figure 5. The effect of the difference in take-off angle on the conformations of the seven different H3 CDR classes H3a–g. The backbone positions of the framework β-strands are shown for each CDR to illustrate the identity of these positions, seen as overlapping regions at the base of each of the loop clusters. Within classes similar take-off angles are seen, whereas between classes the differences in take-off angles leads to markedly different CDR conformations. Each group contains H3 loops taken from antibody (or fragment) structures clustered so that they differ from each other by no more than 35°. The angular difference is calculated as the angle between the planes defined by the N-terminal C^{α} atom, the centre of geometry, and the C-terminal C^{α} atom, for each pair of structures. Residue numbering is as defined in ref. 50.

Their advantages when compared to *ab initio* methods are:

- the starting structures are known to exist
- they are computationally more efficient at saturating conformational space

The method of Jones and Thirup (33) for example, identifies useful database loops by employing a set of α-carbon distance constraints during the search.

A variation on this method is implemented in AbM and will be described later. Other methods include that of Sutcliffe *et al.* (34) which uses a high resolution database to identify structurally conserved regions, and that of Stanford and Wu (35) which generates backbone structures from an analysis of tripeptides in β-sheet proteins.

3.2 *Ab initio* methods

With these methods the generation of all possible loops is achieved by conformational search methods. In one such method used in AbM (CONGEN) (36), conformational space is searched by rotation about the backbone ϕ and ψ dihedral angles, generating large numbers of loop conformations. The problem with this approach is that the number of conformations increases in size exponentially as the numbers of degrees of freedom (i.e. length of loop) increases. It is therefore only suitable for short loops. Conformational searching, however, has two advantages over random methods such as Monte Carlo simulated annealing (37,38) or molecular dynamics (39,40) methods. First, it is carried out on a regular grid with discrete search steps, unlike dynamics or Monte Carlo methods that sequentially perturb one conformation into another by small increments and hence may sample the same space many times. Secondly, conformational searching does not entail the cost of determining energy derivatives, as in molecular dynamics, and permits the examination of non-continuous energy surfaces (41).

In addition, the *ab initio* methods generate multiple conformations that must be evaluated at some stage by an objective function, usually consisting of an energy term. This evaluation is difficult since many low energy conformations are produced that have only slight differences from one another. This problem is being addressed by the development of more sophisticated force fields and the inclusion of solvent models for more accurate free energy calculations (42–45).

3.3 A combined algorithm

A combined approach to the construction of antibody loops was proposed by us (3,4) in an attempt to overcome some of the limitations of the above methods. Where conformational search or database methods alone are ineffectual, such as in longer loops, database methods are used to define the loop 'stem' position, while conformational searching is applied to the central region of the loop, which is likely to be more flexible. A solvent modified energy term is then used to screen conformers, followed by a structural filter derived from the algorithm described by Sutcliffe *et al.* (34).

The algorithm we have developed, CAMAL, is shown in outline in *Figure 6* and forms the core of the program AbM (5). The essential steps are outlined in the worked example below. In the example used to illustrate the protocol, we have selected the antigen-free structure of D1.3, the X-ray

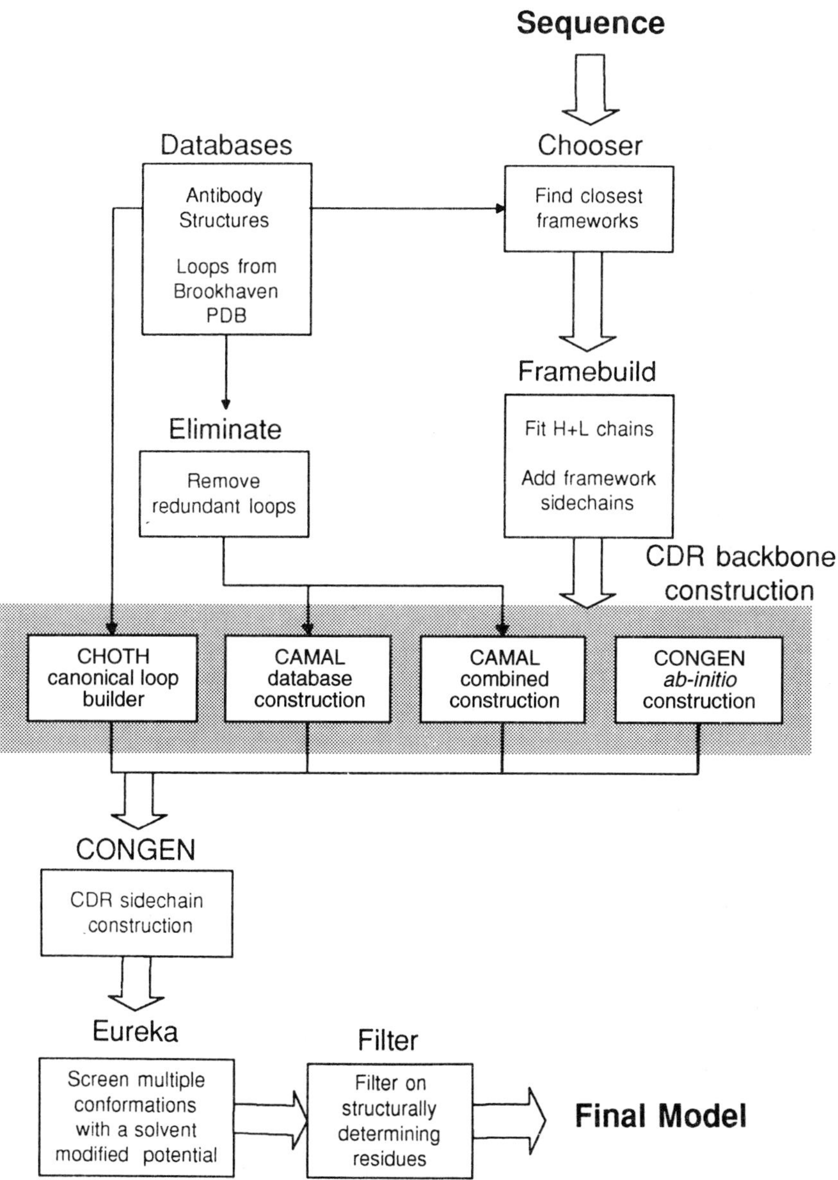

Figure 6. Flow chart of the antibody modelling algorithm AbM (5). The shaded area contains those parts of the program that derive the CDR conformations.

coordinates of which were supplied to us (by R. Poljak) only after the prediction was complete.

3.4 Modelling of antibody D1.3 according to the AbM protocol

1. Sequence entry. The V domain sequences of D1.3 are entered and aligned against a database of antibodies with known crystal structures. The database is non-redundant with a total of 49 light chains (taking only one of each LC dimer) and 41 heavy chains at the time of writing, most of which are derived from the Protein Data Bank. Alignment is based on sequence identity, with gaps corresponding to regions of highest structural variation within the CDRs.

2. Framework construction. Due to their conserved nature, frameworks for the light and heavy chain domains are selected by similarity alone. For each database antibody, each residue is compared with the corresponding residue in the aligned D1.3 sequence. Exact identity adds one to the similarity score, while alignment gaps in the conserved framework away from chain termini introduces a penalty of -200 for the database antibody being assessed. The antibodies REI and ANO2 (*Table 1*) supplied the most similar light and heavy frameworks for D1.3 with 67% and 59% similarity respectively. The REI light chain and ANO2 heavy chain frameworks were paired by fitting on to the average barrel described earlier. Side chains were then replaced using the iterative algorithm in CONGEN to match those of the D1.3 sequence.

3. CDR modelling—L1–3, H1, and H2 loops. The D1.3 sequence has five CDRs with canonical structures: L1, L2, L3, H1, and H2. The canonical classes are defined in *Table 3* and the selection summarized below. The most homologous candidate within each class is identified for each CDR, by reference to the Dayhoff mutation matrix.

CDR loop	**Canonical class**	**CDR template identified**
L1	2	REI, PDB code: 1rei
L2	1	REI
L3	1	REI
H1	1	3D6, PBD code: 1dfb
H2	1	ANO2, PBD code: 1baf
H3		3D6 (take-off angle class)

Loops within a canonical class have identical length so that only the side chains require replacement by those of D1.3. Side chains that are within a van der Waals interaction energy threshold, with respect to the surrounding framework and CDR residues, are added on a first found basis. This is carried out using CONGEN after the framework has been fully constructed.

When database searching is employed, it is normally conducted on a non-redundant database of Protein Data Bank entries to better than 3 Å resolution and typically identifies between a few (sometimes none) and several hundred thousand loops. A large number of hits is usually seen with short to medium length CDRs that have only a few α-carbon distance criteria. However, many of these will have similar conformations and, to avoid undue redundancy, the loops are clustered into groups so that the members of each group have backbone torsion angles that differ from other groups by at least 30°.

By contrast, a low hit rate is seen when there are stringent α-carbon constraints. This is often the case for CDR L1, where few non-antibody proteins will contain loops that are both long enough and satisfy the constraints applied. Long L1s accommodate the length by incorporating a bulge in the N-terminal region of the arch, formed without disrupting the interaction between the conserved residues, 29 and 32 (25,26). When the database search method only identifies a small number of loops, conformational space will be inadequately saturated. Amplification is then achieved by regenerating a stretch of the loop with the conformational search procedure embodied in CONGEN. The region subjected to this search is normally the central part of the loop, since this is likely to be the most flexible. In addition, if the structural residues glycine and proline are present in a sequence, these may also be included.

4. CDR modelling—H3. For CDR H3, the most similar loop within its class of take-off angle (*Figure 5*) is identified. This defines the global orientation of the loop stem. The backbone is then constructed by the combined algorithm in AbM. CDR H3 of D1.3 is eight residues long. The construction of its central region, after selection of the appropriate H3 class, used a combination of database and conformational search (CAMAL). The database search identified 920 hits that clustered to 342 when the 30° torsional cut-off was applied. Subsequent conformational search of the central part of the loop gave 6672 conformations, complete with side chain atoms generated by the iterative method in CONGEN immediately following the conformational search. These 6672 conformations were then subjected to energy screening.

5. Screening of database loops. Each candidate loop, generated either from the PDB database or CONGEN search, is grafted on to the D1.3 framework model and its energy evaluated using a solvent modified Eureka potential (46). High energy conformers are discarded to leave a small number of low energy loops, ranked using a structurally determining residue (SDR) algorithm which scores the resemblance of each loop residue conformation to known loop structures in the protein database. The final model is energy minimized to relax any poor contacts and torsion angles.

6. Results. The D1.3 model had a global N, C^{α}, C, 0 rms deviation of 0.91 Å from the crystal coordinates (PDB entry: 1fdl) when modelled by the

above procedure, which includes the new take-off angle rules for H3 (rmsd 1.76 Å). When these H3 rules are not included the rmsd is somewhat worse (2.19 Å). See Appendix 1 for details.

3.5 Modelling of other antibodies

The performance of AbM on other antibody structures is illustrated in Appendices 1 and 2. In Appendix 1, the CDRs from seven high resolution antibody structures (including D1.3 above) from the PDB have been modelled using AbM.

In Appendix 2 we show the results of H3 prediction only. In these example, one antibody from each of the H3 length classes previously described has been modelled in the following way: the V region framework and CDRs L1–3, H1, and H2 were taken directly from the X-ray structures; H3 was then modelled by the various methods used in AbM—database only, conformational search, or a combination of both. This choice is made within AbM. These results highlight a number of problems with the present algorithm. In three instances (1cgs, 1mam, and 1hil) poor global rms values are obtained when the structural filter is used to screen the five lowest energy conformations. However, when the energy alone is used to screen dramatic improvements are seen for 1cgs and 1mam while the 1hil the best conformation is the third lowest energy conformer. We have identified the origin of this problem, which resides within the conformational search procedure CONGEN, and are hopeful that the further improvements being implemented to AbM will result in a CDR modelling procedure which predicts the conformations of H3 loops up to 14 residues with high accuracy.

Of course, where CDRs are innately flexible so that in solution they are moving between different, energetically similar conformations, selection of a single conformation may be inappropriate. The conformational change seen in a number of H3 loops on antigen binding (47–49) suggests that prediction of a family of closely related conformers likely to be accessible to the loop may be more meaningful.

4. Accuracy, limitations, and future developments

Modelling of the antibody combining site has made great advances since the earliest attempts. While for many antibodies the models will have an overall accuracy approaching that of medium resolution X-ray structure, some parts of the variable region will be more accurately defined than others. For example, the framework regions generated by the docking of two previously unpaired V_L and V_H domains will normally have rms derivations in the range 0.5–0.8 Å (51). For CDRs conforming exactly to a canonical structure, the local backbone rms deviations have been found to be in the range 0.7–1.2 Å while global rms deviations are in the range 1–3 Å (51). The larger global deviations are due to errors in take-off angles of the CDRs with respect to the framework. These can be reduced when the take-off angle preferences, embodied in AbM, are included.

Where CDRs do not conform to canonical classes they can be modelled by the CAMAL method. The four CDRs in the antibodies of Appendices 1 and 2 not conforming to canonical structures (e.g. L3 in 2hfl; L1 in 4fab; H2 in 1mam; L1 in bi3i, see *Tables 1* and *4*), have global (N, C^{α}, C, O) rms deviations of 2.52 Å, 2.47 Å, 2.98 Å, and 2.67 Å respectively. These rms figures account for both local backbone and take-off angle orientation and are thus comparable to those obtained for exact canonical structures.

For CDR H3, where no canonical structures exist other than the classes we have described, rms deviations will vary. The accuracy obtained will largely depend on the length of H3. The modelling protocol embodied in AbM is satisfactory for H3 loops of 14 residues or less (see Appendices 1 and 2). Above that, the increased flexibility of a 'long' loop will makes its prediction somewhat more difficult. This is no great disadvantage since the majority of H3 loops in the Kabat database are between 4 and 12 residues long (52). However, it is a limitation that requires further development.

The correct prediction of side chain conformation is essential where the identity of possible antigen contact residues is to be made. For canonical structures, side chains are taken directly from the CDR selected as the canonical template or, where a residue differs from this template, from another CDR which is in the same canonical class (51). This clearly has limitations where no canonical CDR can be found in the structural database. AbM use several different approaches depending on the method selected to model a particular CDR;

(a) If CONGEN is used alone, the side chains are generated by an energy screening method.

(b) If CAMAL is used, the conformations present in the selected database loop are retained while for the deleted and reconstructed region method (a) is used.

(c) Where a database loop is used with no reconstruction, side chain conformations are retained where possible and where different residues are found, a maximum overlap procedure or a Monte Carlo simulated annealing protocol is used.

The details of these procedures can be found in AbM (5) and Pedersen *et al.* (50).

5. Applications

5.1 Site-directed mutations

When the first *in vitro* mutagenesis of the antibody combining site was carried out to improve antigen affinity, it was guided by crude models. Even so, the affinity of an anti-lysosyme antibody could be improved by a factor of ten or so (53). Since these early experiments, modelling has improved to the point where the effects of single point mutations can be predicted with better reliability. Of course, if an X-ray structure is available then the positions of CDR residues at or nearer the putative antigen binding region can be more accurately specified. Some examples of mutations to modify specificity or affinity can be found in refs 54–56.

In general, a few simple rules should be observed when designing mutations, as follows:

(a) Where possible make a series of changes, from minimal to large effect;
  - lys → arg — maintains charge
  - lys → gln — loss of charge and some volume but retains the H-bonding potential
  - lys → ala — loss of charge and H-bonding potential and large change in volume.

(b) Avoid residues that may have a drastic structural effect (e.g. key residues defining a canonical class).

(c) Target solvent accessible residues since these are more likely to be involved in antigen contact.

(d) Avoid CDR residues that are close to framework residues. Certain of these pairwise interactions may be important for maintaining CDR shape.

(e) Make some changes in combination. Two residues mutated individually may have a minimal effect, but their combined modification may allow an improved orientation of the antigen (53,54).

5.2 Introduction of new properties

In some circumstances it may be desirable to modify a combining site so that it is capable of binding a co-factor such as a metal ion. This can be approached in two ways. Existing residues in the combining site, spatially conforming to a template appropriate for a set of coordinating ligands, but sequentially discontinuous, may be changed to histidine, cysteine, or glutamate residues, or a combination of these depending on the metal preferences. An example of this template-based approach can be found in Roberts *et al.* (57,58). An alternative method, which uses a computer program to search for sites within a protein that have a high potential for metal binding, can be used (59). In one experimental example of this latter method, an anti-lysozyme antibody was modified within CDR L1 by introduction of three Zn^{2+} liganding residues. A fourth ligand was introduced at the N-terminus of the light chain. As predicted, the apo-antibody had the identical affinity for its antigen as the wild-type whereas when various metals were bound ($K_a \sim 10^8$–10^9 M^{-1}), antigen affinity could be modulated by up to 12-fold (D. Staunton and A. R. Rees, unpublished).

5.3 Humanization of antibodies

Humanization or 'reshaping' or murine antibodies is an attempt to transfer the full antigen specificity and binding ability of the murine combining site to a human framework region by CDR grafting (60). However, to preserve the binding affinity, the majority of CDR grafted antibodies require additional amino acid changes in the framework region, because such residues are either conformationally important or are in direct contact with the antigen (61,62). Such necessary framework changes may introduce new antigenic epitopes and, if many changes are needed, the advantage of CDR grafting over chimeric antibody constructions (63) will be lost. Minimization of back mutations is therefore desirable. This can only be carried out in a rational manner using reliable methods to model the entire variable regions, requiring correct docking of V_L and V_H domains and accurate modelling of all six CDRs.

5.4 Humanization by resurfacing

In this method of humanization the surface residues only on the murine Fv are replaced. As previously described, the framework surface residue positions are highly conserved (*Table* 5). First, the V_L and V_H surface sets are searched against a human Fv sequence database and the nearest human V_L–V_H pair selected (30). Surprisingly, when this set is superimposed or the murine V_L–V_H set very few changes are required to convert one to the other. In antibody N901 (64) only three murine residues in V_L and seven residues in V_H required change to convert the murine surface to a human surface. The final 'resurfaced' antibody was modelled by AbM and compared with a model of murine N901. Any residue in the framework region within 5 Å of a CDR residue atom was inspected. In one instance residue Leu 3 of the murine framework had an interaction with residues Arg 24, Ser 25, and Ser 26 (numbering as in ref. 50). In the human sequence selected for its surface identity to N901, position three is a valine. The two models indicated a major difference in the conformation of Arg 24 when valine was present. On conversion back to leucine, the murine conformation was reproduced. Synthesis and expression of this resurfaced N901, back-mutated only at this one position, led to a humanized antibody with the identical affinity to the murine parent.

This example demonstrates two points:

- humanization can be carried out successfully either by CDR grafting or by resurfacing
- models of combining sites during either process are essential

Table 5. Distribution of accessible residues in human and mouse light chain and heavy chain variable framework regions

Light chain position	Human	Mouse
1	D51, V34, A5, S5	D76, Q9, E6
3	V38, Q24, S24, Y6	V63, Q22, L5
5	T61, L37	T87
9	S26, P26, G17, A14, L7	S36, A29, L17, P5
15	P62, V25, L12	L47, P30, V8, A7
18	R57, S18, T13, P6	R38, K22, S13, Q12, T9
46	P94	P82, S9
47	G89	G71, D18
51	K43, R31	K70, Q13, R8, T5
63	G91	G98
66	D43, S25, A9	D38, S26, A26
73	S96	S90, I5
76	D43, S16, T18, E15	D67, S15, A5, K5
86	P44, A27, S17, T8	A50, P11, T8, E7, Q6
87	E71, D11, G7	E91, D6
111	K74, R12, N6	K93
115	K54, L40	K87, L5
116	R60, G33, S5	R89, G9
117	Q50, T37, E6, P6	A74, Q14, P5, R5
118	E47, Q46	E59, Q29, D10
120	Q83, T7	Q68, K26
122	V59, Q13, L15	Q57, V27, L5, K5
126	G54, A23, P18	G36, P30, A29
127	G53, E22, A14, D7	E45, G43, S6
128	L61, V31, F7	L96
130	K46, Q41, E5	K52, Q27, R17
131	P95	P91, A5
132	G74, S16, T7	G82, S17
136	R53, K23, S17, T7	K66, S17, R13
143	G96	G98
145	T46, S32, N9, I7	T63, S19, N7, A5, D5
160	P84, S10	P89, H7
161	G93	G71, E24
162	K76, Q10, R8	K50, Q30, N10, H5
183	D26, P25, A17, Q10, T7	E31, P22, D17, Q11, A12
184	S70, K9, P8	K42, S37, T6
186	K53, Q22, R7, N7	K83, Q7
187	G66, S21, T5	G62, S18, D10
195	T30, D26, N19, K7	T36, K30, N26, D6
196	S91	S76, A16
197	K65, T8, I8, R5	S46, K34, Q11
208	R46, T18, K17, D6	T55, R26, K8
209	A50, P21, S13, T8	S67, A14, T11
210	E46, A18, D13, S9, V5, Z8	E88, D7
212	T91	T53, S43

CDR residues are excluded. Only the most commonly observed residues (greater than or equal to 5% frequency) are given for each position. The numbers after the one letter code names indicate the percentage frequency of occurrence in the aligned light and heav y chain sequences, taken from the Kabat database (65) and from Tomlinson *et al.* (66). The definition of surface accessible is as given in Pedersen *et al.* (29). All residues are given in the single letter code; Z = Glx.

6. Future prospects

The antibody is unique in the repertoire of protein molecules. The construction of the variable region allows for both rigid antibody–antigen docking and for induced fit mediated either by V_L–V_H domain rotational or sliding motions or by CDR conformational change. The V_L–V_H association is stable ($K_A \approx 10^{10}\ M^{-1}$) and can be engineered by cross-linking to be even more kinetically stable. The apparently limitless variation in CDR length and sequence combinations provides a goldmine of binding molecules. Indeed, the antibody has become the 'gold standard' in discussions of molecular recognition. All of these characteristics can be harnessed by tapping the natural repertoire. Why then should there be a role for structure prediction and antibody engineering?

There are three reasons:

(a) Antibodies are a recent evolutionary product. The genes encoding variable domains are governed by strict laws of natural selection and a somewhat slow evolutionary rate. The engineer is not restricted and, either by molecular library approaches or rational replacement of residues or whole CDRs, can create combining site shapes and chemical constitutions not yet discovered by nature.

(b) Antibodies are large and contain constant regions without which the binding function would be pointless. These regions can be replaced with extraordinary ease to produce multifunctional proteins, or 'magic bullets'.

(c) The rules that govern the interaction between antibodies and the enormous variety of antigenic shapes, sizes, and chemical character will only be deduced by a combination of X-ray crystallographic, molecular modelling, and protein engineering studies. Only when such rules of recognition arrive will it be possible to generate antibodies with predetermined specificity and affinity by design.

Acknowledgements

We would like to thank the SERC Biotechnology Directorate (now part of BBSRC), British Biotechnology plc, Wellcome Research plc, Amersham International, and Oxford Molecular plc for financial support. SJS and AHH thank BBSRC for studentships.

References

1. Bernstein, F. C., Koetzle, T. F., Williams, G. J. B., Meyer, E. F., Brice, M. D., Rodgers, J. R., *et al.* (1977). *J. Mol. Biol.*, **112**, 535.
2. Rees, A. R., Staunton, D., Webster, S. D. M., Searle, S. M. J., Henry, A. H., and Pedersen, J. T. (1994). *Trends Biotechnol.*, **12**, 199.

3. Martin, A. C. R., Cheetham, J. C., and Rees, A. R. (1989). *Proc. Natl. Acad. Sci. USA*, **86**, 9268.
4. Martin, A. C. R., Cheetham, J. C., and Rees, A. R. (1991). *Methods in enzymology* (ed. John J. Langone), Vol. 203, pp. 121–52. Academic Press.
5. Rees, A. R., Martin, A. C. R., Pedersen, J. T., and Searle, S. M. J. (1992). ABM™, *a computer program for modelling variable regions of antibodies.* Oxford Molecular Ltd., Oxford, UK.
6. Porter, R. (1962). *Nature*, **182**, 670.
7. Edelman, G. M. and Poulik, M. D. (1961). *J. Exp. Med.*, **113**, 867.
8. Poljak, R. J., Amzel, L. M., Avery, H., Chen, B. L., Phizackerley, R. P., and Saul, F. (1973). *Proc. Natl. Acad. Sci. USA*, **70**, 3305.
9. Deisenhofer, J., Colman, P. M., and Huber, R. (1976). *Hoppe Seyler Z. Physiol. Chem.*, **357**, 435.
10. Silverton, T. W., Naria, M. A., and Davies, D. R. (1977). *Proc. Natl. Acad. Sci. USA*, **74**, 5140.
11. Rajan, S. S., Ely, K. R., Abola, E. E., Wood, M. K., Colman, P. M., Athay, R. J., *et al.* (1983). *Mol. Immunol.*, **20**, 797.
12. Harris, L. J., Larson, S. B., Hasel, K. W., Day. J., Greenwood, A., and McPherson, A. (1992). *Nature*, **360**, 369.
13. Barclay, A. N., Birkeland, M. L., Brown, M. H., Beyers, A. D., Davis, S. J., Somoza, C., *et al.* (1993). *The leucocyte antigen facts book.* Academic Press, London. (ISBN 0-12-M8180-1).
14. Lesk, A. M. and Chothia, C. (1982). *J. Mol. Biol*, **160**, 325.
15. Chothia, C. and Janin, J. (1981). *Proc. Natl. Acad. Sci. USA*, **78**, 4146.
16. Colcher, D., Milenic, D., Roselli, M., Raubitschek, A., Yarranton, D., King, D., *et al.* (1989). *Cancer Res.*, **49**, 1738.
17. Chothia, C., Novotny, J., Bruccoleri, R. E., and Karplus, M. (1985). *J. Mol. Biol.*, **186**, 617.
18. Novotny, J., Bruccoleri, R., Newell, J., Murphy, D., Huber, E., and Karplus, M. (1983). *J. Biol. Chem.*, **258**, 14433.
19. Novotny, J., Bruccoleri, R., and Newell, J. (1984). *J. Mol. Biol*, **177**, 567.
20. Kabat, E. and Wu, T. (1971). *Ann. N.Y. Acad. Sci.*, **190**, 382.
21. Darsley, M. J., Phillips, D. C., Rees, A. R., Sutton, B. J., and de la Paz, P. (1985). In *Investigation and exploitation of antibody combining sites* (ed. G. Reid, G. M. W. Cook, D. J. Morré), pp. 63–8. Plenum Press.
22. de la Paz, P., Sutton, B. J., Darsley, M., and Rees, A. R. (1986). *EMBO J.*, **5**, 415.
23. Wilmot, C. M. and Thornton, J. M. (1988). *J. Mol. Biol.*, **203**, 221.
24. Rose, G. D., Gierasch, L. M., and Smith, J. A. (1985). *Adv. Protein Chem.*, **37**, 1.
25. Chothia, C. and Lesk, A. M. (1987). *J. Mol. Biol.*, **196**, 901.
26. Chothia, C., Lesk, A. M., Tramontano, A., Levitt, M., Smith-Gill, S. J., Air, G., *et al.* (1989). *Nature*, **342**, 877.
27. Chothia, C., Lesk, A. M., Gerardi, E., Tomlinson, I. M., Walter, G., Monks, J. D., *et al.* (1992). *J. Mol. Biol.*, **227**, 799.
28. Rees, A. R., Pedersen, J., Searle, S. J., Henry, A. H., and Webster, D. M. (1996). In *Antibody engineering: a manual* (ed. K. Borrebaeck) pp. 3–44. Oxford University Press.
29. Pedersen, J., Henry, A. H., Searle, S. J., Guild, B. C., Roguska, M., and Rees, A. R. (1994). *J. Mol. Biol.*, **235**, 959.

30. Roguska, M. A., Pedersen, J. T., Keddy, C. A., Henry, A. H., Searle, S. M. J., Lambert, J. M., *et al.* (1994). *Proc. Natl. Acad. Sci. USA*, **91**, 969.
31. Potter, K., Li, Y. C., and Capra, J. D. (1994). *Scand. J. Immunol.*, **40**, 43.
32. Bajoreth, J., Stenkemp, R., and Aruffo, A. (1993). *Protein Sci.*, **2**, 1798.
33. Jones, T. A. and Thirup, S. (1986). *EMBO J.*, **5**, 819.
34. Sutcliffe, M. J., Haneef, I., Carney, D. and Blundell, T. L. (1987). *Protein Eng.*, **1**, 377.
35. Stanford, J. M. and Wu, T. T. (1981). *J. Theor. Biol.*, **88**, 421.
36. Bruccoleri, R. E. and Karplus, M. (1987). *Biopolymers*, **26**, 137.
37. Higo, J., Collura, V., and Garnier, J. (1992). *Biopolymers*, **32**, 33.
38. Collura, V., Higo, J. and Garnier, J. (1993). *Protein Sci.*, **2**, 1502.
39. Fine, R. M., Wang, H., Shenkin, P. S., Yarmush, D. L., and Levinthal, C. (1986). *Proteins: Struct. Funct. Genet.*, *1*, 342.
40. van Gelder, C. W. G., Leusen, F. J. J., Leunissen, J. A. M., and Noordik, J. H. (1994). *Proteins: Struct. Funct. Genet.*, **18**, 174.
41. Bruccoleri, R. E. (1993). *Mol. Simulation*, **10**, 151.
42. Vila, J., Williams, R. L., Vasquez, M., and Scheraga, H. A. (1991). *Proteins: Struct. Funct. Genet.*, **10**, 199.
43. Schiffer, C. A., Caldwell, J. W., Kollman, P. A., and Stroud, R. M. (1993). *Mol. Simulation*, **102**, 121.
44. Smith, K. C. and Honig, B. (1994). *Proteins: Struct. Funct. Genet.*, **18**, 119.
45. Jackson, R. M. and Sternberg, M. J. E. (1994). *Protein Eng.*, **7**, 371.
46. Dauber-Osguthorpe, P., Roberts, V. A., Osguthorpe, D. J., Wolff, J., Genest, M., and Hagler, A. T. (1988). *Proteins: Struct. Funct. Genet.*, **4**, 31.
47. Rini, J. M., Schulze-Gahmen, U., and Wilson, I. A. (1992). *Science*, **255**, 959.
48. Rini, J. M., Stanfield, R. L., Stura, E. A., Salinas, P. A., Profy, A. T., and Wilson, I. A. (1993). *Proc. Natl. Acad. Sci. USA*, **90**, 6325.
49. Webster, D. M., Henry, A. H., and Rees, A. R. (1994). *Curr. Opin. Struct. Biol.*, **4**, 123.
50. Pedersen, J. T., Searle, S. J., Henry, A. H., and Rees, A. R. (1992). *Immunomethods*, **1**, 126.
51. Lesk, A. M. and Tramontano, A. (1991). In *Antibody engineering manual* (1st edn), pp. 1–38. W. H. Freeman, NY.
52. Wu, T. T., Johnson, G., and Kabat, E. A. (1993). *Proteins: Struct. Funct. Genet.*, **16**, 1.
53. Roberts, S., Cheetham, J. C., and Rees, A. R. (1987). *Nature*, **328**, 731.
54. Sharon, D. (1990). *Proc. Natl. Acad. Sci. USA*, **87**, 4814.
55. Denzin, L. K., Whitlow, M., and Voss, G. W. Jr. (1991). *J. Biol. Chem.*, **266**, 14095.
56. Near, R. I., Mudgett-Hunter, M., Novotny, J., Bruccoleri, R., and Chung Ng, S. (1993). *Mol. Immunol.*, **30**, 369.
57. Roberts, V. A., Iversen, B. L., Iversen, S. A., Benkovic, S. J., Lerner, R. A., Getzoff, E. D., *et al.* (1990). *Proc. Natl. Acad. Sci. USA*, **87**, 6654.
58. Wade, W. S., Loh, J. S., Han, N., Hocstan, D. M., and Lerner, R. A. (1993). *J. Am. Chem. Soc.*, **115**, 4449.
59. Gregory, D. S., Martin, A. C. R., Cheetham, J. C., and Rees, A. R. (1993). *Protein Eng.*, **6**, 29.
60. Riechman, L., Clark, M., Waldmann, H., and Winter, G. (1988). *Nature*, **332**, 323.

61. Kettleborough, C. A., Saldanha, J., Heath, V. J., Morrison, C. J., and Bendig, M. M. (1991). *Protein Eng.*, **4**, 773.
62. Foote, J. and Winter, C. T. (1992). *J. Mol. Biol.*, **224**, 487.
63. Morrison, S., Johnson, M. J., Hersenberg, S. A., and Oi, V. T. (1984). *Proc. Natl. Acad. Sci. USA*, **81**, 6851.
64. Griffin, J. D., Hercend, T., Beveridge, R., and Schlossman, S. F. (1983). *J. Immunol.*, **130**, 2947.
65. Kabat, E. A., Wu, T. T., Reid-Miller, M., Perry, H. M., and Gottesman, K. S. (1992). *Sequences of proteins of immunological interest* (5th edn). US Department of Health and Human Services, USA.
66. Tomlinson, I., Walter, G., Marks, J., Llewelyn, M., and Winter, G. (1992). *J. Mol. Biol.*, **227**, 776.

Appendix 1

Results of modelling seven crystal structures of antibodies using the modelling program AbM (5). The rmsd figures are in ångstroms (Å) and are obtained after global least squares fitting of the entire Fv structure of the crystal structure and the model.

† This structure not yet in the PDB.

(a) Results obtained for six antibodies without the inclusion of the improved take-off angle analysis described in the text.

Structure	CDR	CDR length	Global rmsd (N, C^a, C, O)	Canonical structure known
glb2†	L1	11	1.161	+
	L2	7	0.647	+
	L3	9	1.031	+
	H1	5	1.785	+
	H2	10	1.609	+
	H3	4	1.273	−
	Total		1.251	
2.hfl	L1	10	1.150	+
	L2	7	0.712	+
	L3	8	2.524	−
	H1	5	1.261	+
	H2	10	2.155	+
	H3	7	2.310	−
	Total		1.685	
2mcp	L1	17	0.784	+
	L2	7	0.538	+
	L3	9	0.739	+
	H1	5	1.004	+

Structure	CDR	CDR length	Global rmsd (N, C^a, C, O)	Canonical structure known
	H2	12	2.014	+
	H3	11	2.306	−
	Total		1.231	
4fab	L1	16	2.470	−
	L2	7	0.792	+
	L3	9	1.255	+
	H1	5	0.721	+
	H2	12	2.028	+
	H3	7	2.132	−
	Total		1.566	
3hfm	L1	11	0.775	+
	L2	7	1.021	+
	L3	9	0.394	+
	H1	5	2.012	+
	H2	9	0.942	+
	H3	5	1.683	−
	Total		1.302	
bi3i	L1	16	2.667	−
	L2	7	0.763	+
	L3	9	0.877	+
	H1	5	1.310	+
	H2	10	1.202	+
	H3	10	2.97	−
	Total		1.632	

(b) Modelling of an additional antibody (d1.3) in the most recent version of AbM which incorporates new antibody structure data, allowing improved specification of CDR constraints, and either with (new) or without (old) implementation of the H3 take-off angle algorithm. Structure names are as found in *Table 1*.

Structure	CDR	CDR length	Global rmsd (N, C^a, C, O)	Canonical structure known
d1.3 (old)	L1	11	0.799	+
	L2	7	0.928	+
	L3	9	1.138	+
	H1	5	0.846	+
	H2	9	1.413	+
	H3	8	2.188	−
	Total		1.219	

Structure	CDR	CDR length	Global rmsd (N, C^a, C, O)	Canonical structure known
d1.3 (new)	L1	11	0.75	+
	L2	7	0.60	+
	L3	9	0.56	+
	H1	5	1.22	+
	H2	9	0.55	+
	H3	8	1.76	−
	Total		0.91	

Appendix 2

Modelling of H3 loops in the presence of the crystal structure conformations of the framework and L1–3, H1, and H2 CDRs. Loops are constructed either by database searching or by the algorithm CAMAL in which database searching and CONGEN are combined. Final screening of loops is carried out either by a structural filter or by energy. In both cases the five lowest energy loops are scored. LE = the lowest energy.

PDB entry	H3 length	Local rmsd (Å)	Global rmsd (Å)	Construction and screening method
1bbj	5	0.856	1.207	Database
icgs	7	2.436	3.498	Database+filter
		1.611	2.205	Database+LE
1mam	8	2.190	4.675	CAMAL+filter
		1.079	1.251	CAMAL+LE
1fbj	9	1.110	1.819	CAMAL+LE
1for	10	1.282	1.549	CAMAL+LE
1igf	10	1.309	1.933	CAMAL+filter
1hil	11	2.309	3.925	CAMAL+filter
		1.448	1.480	CAMAL+3rd LE
1igm	12	1.841	2.378	CAMAL+filter
1gig	14	0.668	1.047	CAMAL+filter

8

Protein folds and their recognition from sequence

DAVID T. JONES, CHRISTINE A. ORENGO, and JANET M. THORNTON

1. Introduction

Until relatively recently there had been little progress in methods for protein structure prediction (1). Secondary structure prediction had hardly changed since the early 1970s with little improvement in accuracy despite much work and an increase in the size of the database used to derive the parameters (2). Methods for tertiary structure prediction were hampered by the relatively low accuracy of secondary structure prediction. The latter (3–7) were predominantly knowledge-based, in that most methods relied on secondary structure propensities derived from proteins of known structure. As the number of structures rapidly increased during the 1980s, it became apparent that there are also rules which govern the arrangement of secondary structures into their globular fold. For example the connections in βαβ units are always right-handed (8,9), and helices and sheets tend to pack in preferred orientations (10–12). The challenge at the end of the 1980s was how to use this information to improve tertiary structure prediction.

In this chapter we describe progress in methods for tertiary structure prediction, based on fold recognition from amino acid sequence, see refs 13 and 14 for recent reviews. These methods have arisen from the observation that two structures may have very similar folds despite lacking any statistically significant sequence similarity. Although some of these protein pairs have similar functions, many show no functional relationship. There are now many published examples, some of which are listed in *Table 1*. This has led to the suggestion that there may be a limited number of possible topologies or folds (15,16), and therefore a sensible approach to predicting a structure is to ask if the sequence could adopt one of the currently known set of protein folds.

At this stage it is important to distinguish two distinct though related problems. The first, which is the usual question asked by experimentalists, is 'Given a novel sequence, does it adopt a known fold?' Methods (17–19) to

Table 1. Table listing some well populated fold families together with proteins belonging to them which have insignificant sequence similarity (< 25%) and different functions

Fold	Structures (PDB code)
Globin	Hemoglobin (1thb), Colicin A (1col), Phycocyanin (1cpc)
Alpha Up-Down	Hemerythrin (1hmz), Cytochrome b562 (256bA), Apolipoprotein-E3 (1le2), H-Ferritin (1fha), Ligand-Binding Domain (1lig), Interleukin-4 (1rcb), Granulocyte-Macrophage CS Factor (1gmfA), Tobacco Mosaic Virus (2tmvP)
Greek Key - Ig	Immunoglobulin (2rhe), Macromomycin (2mcm), CD4 (2cd4), Telokin (1tlk), Prealbumin (Human Plasma) (2pab), Superoxide Dismutase (2sodB), Human Tenascin (1ten), Human Class I Histocompatability (3hla),
Trefoil	Interleukin 1-beta (1i1b), Erythrina Trypsin Inhibitor (1tie), Fibroblast Growthfactor (3fgf), Ricin (1rtc)
Jelly Roll	Foot and Mouth Disease Virus (1bbt), Tumour Necrosis Factor (1tnf)
TIM Barrel	Triosephosphate Isomerase (2tim), Aldolase (1ald), Flavocytochrome B2 (1fcb), Anthranilate Isomerase (1pii), Tryptophan Synthase (1wsyA), Xylose Isomerase (6xia), Rubisco (5rub) Taka-Amylase A (2taa), Enolase (4enl), Glycolate Oxidase (1gox)
Doubly Wound	Ras p21 protein (5p21), Flavodoxin (4fxn), Carboxypeptidase A (5cpa) Elongation factor tu (1etu), Subtilisin Carlsberg (1cse)
Split Alpha-Beta Sandwich	Scorpion Neurotoxin (1sn3), Gamma-1-P Thionin (1gps), Ferredoxin (2fxb), Crambin (1crn), L7/12 Ribosomal Protein (1ctf), Acylphosphatase (1aps), His-Containing Phosphocarrier (2hpr), Procarboxypeptidase (1pba), Bovine Papillomavirus (1bop), N-Term. Fragment Ribonucleoprot (1nrcA) Ribulose Carboxylase/Oxygenase (3rubS)
UB Alpha-Beta Roll	Ubiquitin (1ubq), Ferredoxin I (1fxiA)
OB Alpha-Beta Roll	Verotoxin (1bovA), Heat-Labile Enterotoxin (1ltsA), Staphylococcal Nuclease (1sns)

answer this question match a single sequence against a library of folds, and the best match is somehow identified.

A different question is ‘Given a structure, can we identify all the sequences in the sequence data banks which will adopt the same fold?’ This problem is very similar to the ‘inverse folding problem in which all *possible* sequences, which will fold into a given structure, are sought (20,21). This is thought to be easier to solve than the conventional folding problem, since many sequences appear to adopt the same fold and therefore the constraints on sequence are weaker and will be more forgiving of errors (22). Methods to identify sequences which match a fold, search the protein sequence data banks, and calculate a score which evaluates the fit of each sequence to the given structure, generally utilizing some sort of one-dimensional (1D) profile. These scores are then compared and exceptionally significant scores identified as outliers from the pack. Such methods work in 1D since the sequence data banks are so large that they can only be searched using very rapid methods of alignment such as dynamic programming. In practice, although fold recognition and inverse folding are related, current methods often cannot be applied to both problems because of normalization effects.

Algorithms which work in 1D (i.e. a structure is defined as a series of properties associated with each residue and no specific pair information is included) can be broadly described as profile-fitting ('A sequence is matched against a profile in 1D'). Algorithms which work in three-dimensions (so that specific pair interactions can be explicitly included) can be described as threading ('A sequence is threaded on to a structure in 3D').

1.1 Practical procedure

Given a novel sequence it is most important to explore all the conventional sequence comparison approaches before attempting to recognize distant relationships by threading or profiling. The recommended approach is shown schematically in the flow diagram in *Figure 1*. The first step is to search the protein sequence data banks for a matching sequence. If a sequence or a family of sequences is found, then the novel sequence should be aligned with all the hits and from this a pattern of conservation may be detected.

If a structure is known for one of the sequences in the family, with a sequence identity greater than about 30%, then it is now relatively easy to

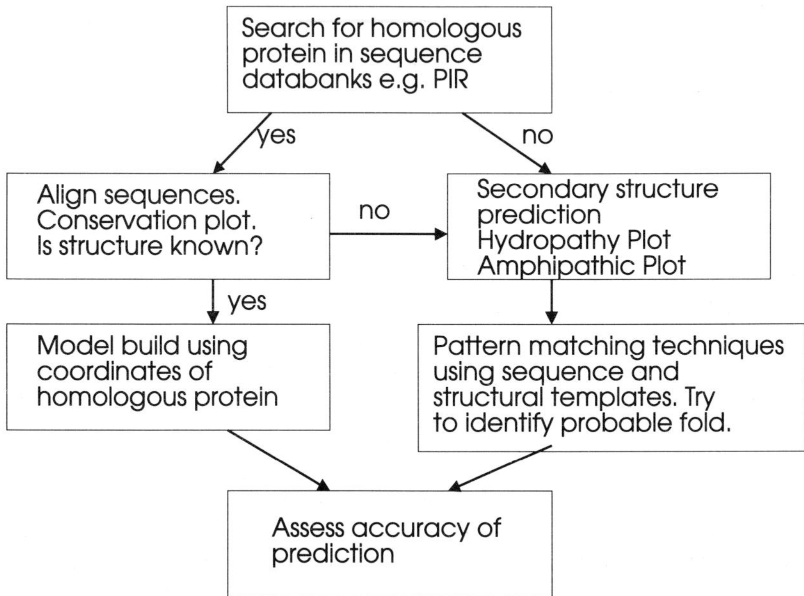

Figure 1. Flow diagram showing the different stages in predicting the tertiary structure of a protein from the amino acid sequence.

build a three-dimensional model using homology modelling techniques (23,24 and Chapters 6 and 7). Extensive data show that at this level of sequence identity, the topology of the protein will be conserved, with the core regions maintained. Such techniques, of varying sophistication, are increasingly available in the commercial modelling packages. The accuracy of a model derived in this way is variable across the model, but a broad level of expected accuracy can be derived from the relationship between per cent sequence identity of a pair of sequences and the rms difference in ångstroms between their structures, as derived by Chothia and Lesk (25) and Hubbard and Blundell (26). Thus if a sequence clearly matches a protein with a known structure, then this is the best route to derive a reasonably accurate model.

However it is often the case that either no matching sequence is located or no matching structure is known for any of the matching sequences. Until recently the only possible route forward was to make a secondary structure prediction and analyse the patterns of hydrophobic and hydrophilic residues. Recently, some remarkably accurate predictions of secondary structure from large families of protein sequences have been obtained prior to structure determination (27–29 and Chapter 4). However attempts to extend such predictions into three-dimensions are still in their infancy. Even with recent advances and the use of multiply aligned sequences, prediction scores hover around an average of 70%, with a range stretching between 40–90% for different specific examples. Except in favourable cases, this is still rather low as a starting point for tertiary structure prediction. Furthermore our inability to predict secondary structure packing using conventional force fields and simulation techniques, renders such approaches difficult, if not impossible.

1.2 Fold recognition

Rather than model a tertiary structure from scratch, the alternative is to use known folds as possible models. What are the stages involved in recognizing a fold from the sequence? Essentially there are three basic components:

(a) A library of all non-duplicate known folds or topologies must be derived from the structural database against which to match the sequence.

(b) A set of potentials or scores is required to evaluate the match of the sequence on to a fold.

(c) An algorithm to find the optimal alignment of a sequence on to a structure.

In this chapter we shall give a brief description of the available methods to establish a library of known folds, although in practice these will best be obtained from a centralized library of protein folds over the network. Automated structure comparison is fraught with difficulties, and several groups have spent much time comparing structures and deriving suitable libraries (30–36). We shall describe the results of our own comparisons and what they tell us about the world of protein structures and structural families.

Several different types of potentials and 'propensities', which are all derived from proteins of known structure, have been used in profile matching and in threading. Herein we shall describe their application to the problem of evaluating structural models.

Similarly several methods have been used to find the optimal alignment of sequence on a structure which include simulations, stochastic searches, and single and double dynamic programming techniques. In three-dimensions, as required for optimal sequence threading, this is without doubt the computationally difficult part of the process, since many threadings are possible and it is an extraordinarily difficult problem to find the best one.

2. Fold libraries

Although the protein structure data bank contains more than 3000 structures to date, many of these are very similar. For example, mutation studies of the protein lysozyme have resulted in the deposition of several hundreds of nearly identical lysozyme structures. Many other similarities are due to the fact that the structures of proteins are more highly conserved than their sequences. Insertions and deletions occurring during evolution are confined mainly to loop regions between secondary structures and do not alter the fold. This gives rise to families of folds having related structures but varying sequence identities.

Chothia and Lesk (25) first showed that for proteins with more than 50% sequence identity 85% of the residues would adopt the same conformation. Other groups subsequently extended these studies and have showed that the fold remains the same even if the sequence identity falls as low as 30% (2,26,37,38). Protein structures which are presumed to have diverged from a common ancestor in this way are described as homologous.

More recently, (see ref. 39 for a review) several examples have been found of proteins having very little sequence identity (less than 5%) but still adopting the same fold. In these cases the possibility of a common ancestor is more remote, especially where no common function can be found. Similarity may be due to the fact that the total number of folds is limited by various physical constraints (15). Therefore, the relationship between the two structures is coincidental and the folds are often described as analogous.

Most procedures for generating fold libraries first compare the sequences of all proteins in the Brookhaven data bank. Aligning linear sequences is far less computationally expensive than comparing three-dimensional structures and proteins having more than 30% of their sequences in common can now be assumed to adopt the same folds. However, the discovery of protein pairs having no sequence identity but the same fold, means that at some stage structures must be compared directly. Over the last few years, fast and flexible algorithms have been developed which make this task possible. Below we describe the most robust procedures currently available for generating fold libraries.

2.1 Comparing protein sequences

Before comparing proteins, any poorly resolved or model structures are usually excluded. For the April 1993 release of the Brookhaven Protein Structure Data Bank (40), this gave a set of 2500 structures. Most approaches then use conventional dynamic programming algorithms (41) to align protein sequences. Residue identities are compared rather than physico-chemical properties. Hobohm *et al.* (42) describe two automatic algorithms for clustering related sequences both based on a cut-off of 30% sequence identity. We use a slightly more cautious threshold of 35% on the premise that any similarities missed are subsequently recognized when the structures are compared directly (31).

Performing pairwise comparisons between all 2500 protein sequences takes approximately two CPU days on a SUN4. Single linkage cluster analysis can then be used to join proteins with more than 35% sequence similarity. This gives 323 sequence-based (35Seq) families. A single representative can be selected from each family to satisfy any chosen criteria. We use the best resolved structure, though for some purposes it may be more appropriate to choose a structure which has the largest number of close neighbours within the family.

2.2 Including structural information

Representatives from each sequence-based family are now structurally compared. Many different algorithms are available, exhibiting a variety of speeds and sensitivities. Generally the most robust algorithms compare intramolecular relationships between residues. For example Holm and Sander (36) compare distance maps of hexapeptide fragments. Sali and Blundell (33) compare hydrogen bonding relationships and distances between C^{α} atoms. The algorithm we adopt SSAP developed by Taylor and Orengo (30), compares residue structural environments. These are defined as the set of vectors from the C^{β} atom of a given residue to the C^{β} atoms of all other residues in the protein.

Comparing relationships improves the alignment of distantly related structures. Although secondary structure shifts reduce the number of superposable residues between distant proteins, the structural relationships defining fold shape are more easily recognized. Often relationship data is supplemented by information on residue features (e.g. accessibility or torsional angles) (33,43). Relationships are harder to compare than features and several algorithms use optimization strategies, such as simulated annealing (33) or Monte Carlo optimization (36). SSAP employs dynamic programming at two levels. Initially to compare residue environments between pairs of residues and finally to obtain an alignment from accumulated data on residue pairs. This algorithm has the added advantage of coping in a robust way with insertions and deletions between proteins.

2.3 Protein families

Few procedures for automatic data bank classification have yielded satisfactory results. Holm and Sander (36) adopt a procedure using three algorithms of varying sensitivity. Their approach requires one months CPU time on a SUN4 to compare 200 structures, all against all. We employ a fast version of SSAP (44) which first checks whether the arrangements of secondary structures are similar and then proceeds to a full residue comparison. Five CPU days on a SUN4 are required to compare 323 structures, all against all.

Families whose representatives match with a high SSAP score are combined, provided that at least 70% of the smaller fold is equivalenced. This prevents folds being merged on the basis of common motifs. Two cut-offs on the SSAP score are used. Values above 80 indicate highly similar folds. Often these have related functions suggesting divergence from a common ancestor (homologous folds). Structures yielding scores between 70 and 80 whilst similar show greater variation in loops and orientations of secondary structures (analogous folds). Both cut-offs were established through empirical trials (44). Applying single linkage clustering with a cut-off of 80 on the SSAP score gives 200 (80Str) families. By softening the cut-off to 70, the number of fold families is reduced to 150 (70Str) families (see *Table 2* and *Figure 2*). This represents a nearly 20-fold reduction in the size of the data bank.

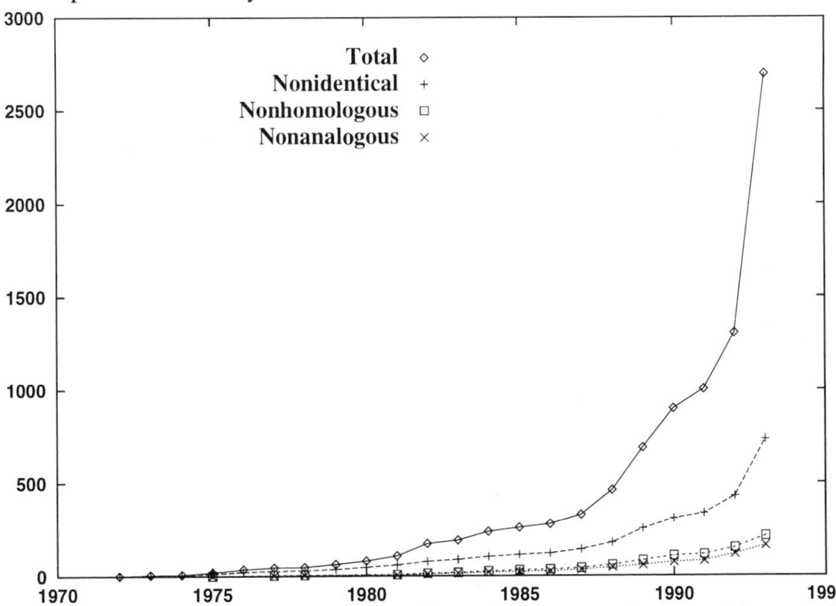

Figure 2. Annual increase in the numbers of protein structures determined by crystallographic and NMR techniques. The top line gives the total number of protein chains in the PDB (◇), the lines below show the increase in non-identical chains (< 98% sequence identity (+)), non-homologous folds (< 35% sequence identity and SSAP score < 80 (□)) and non-analogous folds (< 35% sequence identity and SSAP score < 70 (x)).

Table 2. Table of sequence families (35Seq), non-homologous families (80Str), and non-analogous families (70Str) identified by clustering the Brookhaven PDB data bank using both sequence and structure alignment methods (3)

	35Seq Rep	80Str Rep	70Str Rep	Title	Res Å	Length
ALPHA: GLOBIN						
13	1mbc	1mbc*	1mbc+	Myoglobin (*sperm whale*)	1.5	153
5	1mba			Myoglobin (*Sea Hare*)	1.6	146
4	1ecd			Erythrocruorin (*Chironomous Thummi Thummi*)	1.4	136
52	1thbA			Hemoglobin (*Human*)	1.5	141
14	2lh3			Leghemoglobin (*Yellow Lupin*)	2.0	153
4	2sdhA			Hemoglobin (*Clam*)	2.4	146
2	1ithA			Hemoglobin (*Innkeeper Worm*)	2.5	141
1	2lhb	↓		Hemoglobin (*Sea Lamprey*)	2.0	149
2	1colA	1colA	↓	Colicin A (*Escherichia Coli*)	2.4	197
ALPHA: ORTHOGONAL						
2	1lmbA	1lmbA*	1lmbA*	Lambda Repressor Operator Complex (*Bacteriophage Lambda*)	1.8	87
1	1r69	↓	↓	434 Repressor (*Phage 434*)	2.0	63
4	1utg	1utg	1utg+	Uteroglobin (*Rabbit*)	1.3	70
5	1fiaA	1fiaA		Factor for Inversion Stimulation (*Escherichia Coli*)	2.0	79
3	2wrpR	2wrpR*		DNA Binding Regulatory Protein (*Escherichia Coli*)	1.6	104
4	1hddC	↓	↓	Engrailed Homeodomain Complex (*Fruit Fly*)	2.8	57
2	3sdpA	3sdpA	3sdpA	Iron superoxide Dismutase (*Pseudomonas Ovalis*)	2.1	186
ALPHA: EFHAND						
16	4cpv	4cpv*	4cpv+	Calcium Binding Parvalbumin (*Carp*)	1.5	108
2	2scpA	↓		Sarcoplasmic Calcium Binding Protein (*Sandworm*)	2.0	174
3	4icb	4icb	↓	Bovine Calbindin D9K (*Bovine*)	1.6	76
ALPHA: UP/DOWN						
2	256bA	256bA*	256bA*	Cytochrome b562 (*Escherichia Coli*)	1.4	106
2	2ccyA			Cytochrome C' (*Rhodospirillum Molischianum*)	1.7	127
17	2hmzA			Hemerythrin (*Sipunculid worm*)	1.7	114
1	1le2			Apolipoprotein E2 (*Human*)	3.0	144
1	1ropA	↓	↓	Rop: Col E1 repressor (*Escherichia Coli*)	1.7	56
1	2tmvP	2tmvP	2tmvP	Tobacco Mosaic Virus (*Tobacco Mosaic Virus*)	2.9	154
2	1gmfA	1gmfA	1gmfA	Granulocyte-Macrophage Colony-Stimulating Factor (*Human*)	2.4	119
ALPHA: COMPLEX UP/DOWN						
2	1prcL	1prcL*	1prcL*	Photosynthetic Reaction Centre (*Rhodopseudomonas Viridis*)	2.3	273
2	1prcM	↓	↓	Photosynthetic Reaction Centre (*Rhodopseudomonas Viridis*)	2.3	323
ALPHA: METAL RICH						
13	1ycc	1ycc*	1ycc+	Cytochrome C (reduced) (*Bakers yeast*)	1.2	108
2	451c			Cytochrome C551 (*Pseudomonas Aeruginosa*)	1.6	82
1	1cc5	↓		Cytochrome C5 (*Azobacter Vinelandii*)	2.5	83
1	1c5a	1c5a	↓	Des-Arg-74 Complement c5a (*Pig*)	nmr	65
1	1cy3	1cy3	1cy3	Cytochrome c3 (*Desulfovibro Desulfuricans*)	2.5	118
1	1prcC	1prcC	1prcC	Photosynthetic Reaction Centre (*Rhodopseudomonas Viridis*)	2.3	332

	35Seq Rep	80Str Rep	70Str Rep	Title	Res Å	Length
BETA: ORTHOGONAL BARREL						
3	1ifc	1ifc	1ifc+	Intestinal fatty Acid Binding Protein (*Rat*)	1.2	131
1	1rbp	1rbp*	│	Retinol Binding Protein (*Human*)	2.0	174
4	1bbpA	↓	↓	Bilin Binding Protein (*Cabbage Butterfly*)	2.0	173
BETA: SUPER BARREL						
1	2por	2por	2por	Porin (*Rhodobacter Capsulatus*)	1.8	301
BETA: GREEK KEY						
68	4ptp	4ptp*	4ptp+	Beta Trypsin (*Bovine*)	1.3	223
1	1sgt	│	│	Trypsin (*Streptomyces Griseus*)	1.7	223
7	2sga	│	│	Proteinase A (*Streptomyces Griseus*)	1.5	181
19	2alp	↓	│	Alpha-Lytic Protease (*Lysobacter Enzymogenes*)	1.7	198
1	1snv	1snv	↓	Sindbis Virus capsid Protein (*Sindbis Virus*)	3.0	151
38	2rhe	2rhe*	2rhe+	Immunoglobulin (*Human*)	1.6	114
1	1cd8	│	│	CD8 (*Human*)	2.6	114
2	2cd4	↓	│	CD4 (*Human*)	2.4	176
29	2fb4H	2fb4H*	│	Immunoglobulin FAB (*Human*)	1.9	229
2	3hlaB	↓	↓	Human Class I Histocompatability Antigen (*Human*)	2.6	99
4	1hoe	1hoe	1hoe+	Amylase Inhibitor (*Streptomyces Tendae*)	2.0	74
1	1acx	1acx	│	Actinoxanthin (*Actinomyces Globisporus*)	2.0	107
6	1cobA	1cobA	↓	Superoxide Dismutase (*Bovine*)	2.0	151
2	1paz	1paz*	1paz+	Pseudoazurin (*Alcaligenes Faecalis*)	1.5	120
7	1pcy	↓	│	Plastocyanin (*Poplar*)	1.6	99
3	2azaA	2azaA	↓	Azurin (*Alcaligene Denitrificans*)	1.8	129
2	2pabA	2pabA	2pabA	Prealbumin (*Human*)	1.8	114
2	1gcr	1gcr	1gcr	Gamma Crystallin (*Calf*)	1.6	174
BETA: JELLY ROLLS						
1	2stv	2stv	2stv+	Satellite Tobacco Necrosis Virus (*Tobacco Necrosis Virus*)	2.5	184
1	1bmv1	1bmv1	│	Bean Pod Mottle Virus (*Bountiful Bean*)	3.0	185
1	1bmv2	1bmv2	│	Bean Pod Mottle Virus (*Bountiful Bean*)	3.0	374
15	2plv1	2plv1	│	Poliovirus (*Human*)	2.9	287
3	1tnfA	1tnfA	│	Tumour Necrosis Factor (*Human*)	2.6	152
1	2mev1	2mev1	│	Mengo Encephalomyocarditis Virus Coat Protein (*Monkey*)	3.0	268
1	2mev2	2mev2*	│	Mengo Encephalomyocarditis Virus Coat Protein (*Monkey*)	3.0	249
15	2plv2	↓	│	Poliovirus (*Human*)	2.9	268
1	2mev3	2mev3*	│	Mengo Encephalomyocarditis Virus Coat Protein (*Monkey*)	3.0	231
15	2plv3	↓	│	Poliovirus (*Human*)	2.9	235
3	4sbvA	4sbvA*	│	Southern Bean Mosaic Virus (*Southern Bean Mosaic Virus*)	2.8	199
3	2tbvA	↓	↓	Tomato Bushy Stunt Virus (*Tomato Bushy Stunt Virus*)	2.9	283
2	2cna	2cna	2cna	Concanavalin A (*Jack Bean*)	2.0	237
2	2ltnA	2ltnA	2ltnA	Pea Lectin (*Garden Pea*)	1.7	181

	35Seq Rep	80Str Rep	70Str Rep	Title	Res Å	Length
BETA: COMPLEX SANDWICH						
23	2er7E	2er7E*	2er7E+	Endothia Aspartic Protease (*Chestnut Blight Fungus*)	1.6	330
6	1psg	|	|	Pepsinogen (*Porcine*)	1.6	365
20	5hvpA	↓	|	HIV-1 Protease (*NY5 Strain of Human Immunodeficiency Virus I*)	2.0	99
6	2rspA	2rspA	↓	Rous Sarcoma Virus Protease (*Rous Sarcoma Virus*)	2.0	115
1	1f3g	1f3g	1f3g	Phosphocarrier III (*Escherichia Coli*)	2.1	150
1	1hcc	1hcc	1hcc	16th Complement Control Protein (*Human*)	nmr	59
BETA: TREFOIL						
2	3fgf	3fgf*	3fgf*	Basic Fibroblast Growth Factor (*Human*)	1.6	124
9	8i1b	↓	↓	Interleukin 1-Beta (*Murine*)	2.4	146
BETA: PROPELLOR						
2	1nsbA	1nsbA	1nsbA	Neuraminidase Sialidase (*Influenza Virus*)	2.2	390
BETA: METAL RICH						
6	5rxn	5rxn	5rxn	Rubredoxin (*Clostridium Pasteurianum*)	1.2	54
2	2hipA	2hipA	2hipA	High Potential Iron Sulfur Protein (*Ectothiorhodospira Halophila*)	2.5	71
BETA: DISULPHIDE RICH						
2	1pi2	1pi2	1pi2	Bowman Birk Proteinase Inhibitor (*Tracy Soybean*)	2.5	61
1	1atx	1atx*	1atx*	Sea Anemone Toxin (*Sea Anemone*)	nmr	46
1	1sh1	|	|	Neurotoxin (*Sea Anemone*)	nmr	48
1	2sh1	↓	↓	Neurotoxin (*Sea Anemone*)	nmr	48
13	3ebx	3ebx	3ebx	Erabutoxin (*Sea Snake*)	1.4	62
4	1epg	1epg	1epg	Epidermal Growth Factor (*Mouse*)	nmr	53
1	4tgf	4tgf	4tgf	Transforming Growth Factor (*Human*)	nmr	50
ALPHA/BETA: TIM BARREL						
16	2timA	2timA	2timA+	Triosephosphate Isomerase (*Trypanosoma Brucei*)	1.8	249
1	1gox	1gox	|	Glycolate Oxidase (*Spinacia Oleracia*)	2.0	350
1	1ald	1ald	|	Aldolase (*Human*)	2.0	363
2	1fcbA	↓	|	Flavocytochrome B2 (*Yeast*)	2.4	494
1	1pii	1pii	|	Anthranilate Isomerase (*Escherichia Coli*)	2.0	452
1	1wsyA	1wsyA	|	Tryptophan Synthase (*Salmonella Typhimurium*)	2.5	248
21	6xia	6xia	|	Xylose Isomerase (*Streptomyces Albus*)	1.6	387
8	5rubA	5rubA	↓	Rubisco (*Rhodospirillum Rubrum*)	1.7	436
1	2taaA	2taaA	2taaA	Taka-Amylase A (*Aspergillus Oryzae*)	3.0	478
5	4enl	4enl	4enl	Enolase (*Bakers Yeast*)	1.9	436

35Seq Rep		80Str Rep	70Str Rep	Title	Res Å	Length
ALPHA/BETA: DOUBLY WOUND						
2	5p21	5p21*	5p21+	Ras P21 Protein (*Human*)	1.4	166
1	1etu	↓	|	Elongation Factor Tu (*Escherichia Coli B*)	2.9	177
2	4fxn	4fxn*	|	Flavodoxin (*Clostridium MP*)	1.8	138
5	2fx2	↓	|	Flavodoxin (*Desulfovibrio Vulgaris*)	1.9	147
1	2fcr	2fcr	|	Flavodoxin (*Chondrus Crispus*)	1.8	173
2	3chy	3chy	|	Chey Protein (*Escherichia Coli B*)	1.7	128
7	5cpa	5cpa	|	Carboxypeptidase A (*Bovine*)	1.5	307
3	2trxA	2trxA*	|	Thioredoxin (*Escherichia Coli*)	1.7	108
2	3trx	|	|	Thioredoxin (*Human*)	nmr	105
2	1ego	|	|	Glutaredoxin (*Escherichia Coli*)	nmr	85
2	1gp1A	↓	|	Glutathione Peroxidase (*Bovine*)	2.0	183
20	1cseE	1cseE	↓	Subtilisin Carlsberg (*Bacillus Subtilis*)	1.2	274
9	4dfrA	4dfrA*	4dfrA*	Dihydrofolate Reductase (*Escherichia Coli B*)	1.7	159
13	8dfr	|	|	Dihydrofolate Reductase (*Chicken*)	1.7	186
1	3dfr	↓	↓	Dihydrofolate Reductase (*Lactobacillus Casei*)	1.7	162
5	3adk	3adk	3adk+	Adenylate Kinase (*Porcine*)	2.1	194
1	1gky	1gky	↓	Guanylate Kinase (*Bakers yeast*)	2.0	186
1	3pgm	3pgm	3pgm	Phoshphoglycerate Mutase (*Dried Bakers Yeast*)	2.8	230
ALPHA/BETA: TWO DOUBLY WOUND DOMAINS (MULTIDOMAIN)						
1	1rhd	1rhd	1rhd	Rhodanese (*Bovine*)	2.5	293
7	4pfk	4pfk	4pfk	Phosphofructokinase (*Bacillus Stearothermophilus*)	2.4	319
1	3pgk	3pgk	3pgk	Phosphoglycerate Kinase (*Bakers Yeast*)	2.5	415
2	2yhx	2yhx	2yhx	Yeast Hexokinase B (*Bakers Yeast*)	2.1	457
2	2gbp	2gbp*	2gbp+	Glucose Binding Protein (*Escherichia Coli*)	1.9	309
7	8abp	↓	|	Arabinose Binding Protein (*Escherichia Coli*)	1.5	305
2	2liv	2liv	↓	Leucine/Isoleucine/Valine Binding Protein (*Escherichia Coli*)	2.4	344
ALPHA/BETA: ONE DOUBLY WOUND (MULTIDOMAIN)						
4	3grs	3grs*	3grs*	Glutathione Reductase (*Human*)	1.5	461
1	1trb	↓	↓	Thioredoxin Reductase (*Escherichia Coli*)	2.0	315
3	8catA	8catA	8catA	Catalase (*Bovine*)	2.5	498
9	6ldh	6ldh*	6ldh*	Lactate Dehydrogenase (*Dogfish*)	2.0	329
2	4mdhA	↓	↓	Malate Dehydrogenase (*Porcine*)	2.5	333
1	1ipd	1ipd*	1ipd*	Isopropylmalate Dehydrogenase (*Thermus Thermophilus*)	2.2	345
7	4icd	↓	↓	Phosphorylated Isocitrate Dehydrogenase (*Escherichia Coli*)	2.5	414
1	1pgd	1pgd	1pgd	6-Phosphogluconate Dehydrogenase (*Sheep*)	2.5	469
5	8adh	8adh	8adh	Alcohol Dehydrogenase (*Horse*)	2.4	374
16	1gd1O	1gd1O	1gd1O	Glyceraldehyde Phosphate Dehydrogenase (*Bacillus Stearothermophilus*)	1.8	334
9	7aatA	7aatA	7aatA	Aspartate Aminotransferase (*Chicken*)	1.9	401
4	2ts1	2ts1	2ts1	Tyrosyl-Transfer RNA Synthetase (*Bacillus Stearothermophilus*)	2.3	317
2	1phh	1phh	1phh	P-Hydroxybenzoate Hydroxylase (*Pseudomonas Fluorenscens*)	2.3	394

	35Seq Rep	80Str Rep	70Str Rep	Title	Res Å	Length
ALPHA + BETA: MAINLY ALPHA						
92	3lzm	3lzm	3lzm	Lysozyme (*Bacteriophage T4*)	1.7	164
26	1lz1	1lz1	↓	Lysozyme (*Human*)	1.5	130
12	4bp2	4bp2	4bp2	Prophospholipase A2 (*Bovine*)	1.6	115
ALPHA + BETA: SANDWICHES						
1	1rnh	1rnh*	1rnh*	Selenomethionyl Ribonuclease (*Escherichia Coli*)	2.0	145
2	1hrhA	↓	↓	Ribonuclease H Domain of HIV Reverse Transcriptase (*Human Immunodeficiency Virus Type I*)	2.4	125
4	1rveA	1rveA	1rveA	Eco Rv Endonuclease (*Escherichia Coli*)	2.5	244
4	2sicI	2sicI	2sicI	Subtilisin Inhibitor (*Bacillus Amyloliquefaciens*)	1.8	107
8	1cseI	1cseI*	1cseI*	Subtilisin Inhibitor (*Bacillus Subtilis*)	1.2	63
2	2ci2I	↓	↓	Chymotrypsin Inhibitor (*Barley*)	2.0	65
4	1il8A	1il8A*	1il8A*	Interleukin 8 (*Human*)	nmr	71
2	1mcaA	↓	↓	Monocyte Chemo-attractant and Activating Protein (*Human*)	nmr	68
19	5pti	5pti	5pti	Trypsin Inhibitor (*Bovine*)	1.0	58
3	1tpkA	1tpkA	1tpkA	Kringle 2 (*Human*)	2.4	88
8	9wgaA	9wgaA	9wgaA	Wheat germ Agglutinin (*Wheat*)	1.8	171
5	2ovo	2ovo*	2ovo+	Ovomucoid Third Domain (*Silver Pheasant*)	1.5	56
3	1tgsI	↓	|	Trypsinogen Inhibitor (*Bovine*)	1.8	56
4	3sgbI	3sgbI	↓	Proteinase Inhibitor (*Streptomyces Griseus*)	1.8	50
1	1sn3	1sn3	1sn3	Scorpion Neurotoxin (*Scorpion*)	1.8	65
1	1crn	1crn*	1crn+	Crambin (*Abyssinian Cabbage*)	1.5	46
1	1ctf	↓	|	L7/L12 50 S Ribosomal Protein (*Escherichia Coli*)	1.7	68
1	1aps	1aps	|	Acetylphosphatase (*Horse*)	nmr	98
1	1fxd	1fxd*	|	Ferredoxin II (*Desulfovibrio Gigas*)	1.7	58
1	2fxb	|	|	Ferredoxin (*Bacillus Thermoproteolyticus*)	2.3	81
3	4fd1	|	|	Ferredoxin (*Azotobacter Vinelandii*)	1.9	106
1	1fdx	↓	↓	Ferredoxin (*Peptococcus Aerogenes*)	2.0	54
19	8atcB	8atcB	8atcB	Aspartate Carbamoyl Transferase (*Escherichia Coli*)	2.5	146
6	2tscA	2tscA	2tscA	Thymidilate Synthase (*Escherichia Coli*)	2.0	264
4	4cla	4cla	4cla	Chloramphenicol Acetyltransferase (*Escherichia Coli*)	2.0	213
1	1pyp	1pyp	1pyp	Inorganic Pyrophosphatase (*Bakers Yeast*)	3.0	280
ALPHA + BETA: ROLLS						
5	1bovA	1bovA	1bovA	Verotoxin (*Escherichia Coli*)	2.2	69
4	1snc	↓	↓	Staphylococcal Nuclease (*Staphylococcus Aureus*)	1.6	135
1	1ubq	1ubq*	1ubq*	Ubiquitin (*Human*)	1.8	76
3	1fxiA	|	|	Ferredoxin I (*Blue Green Algae*)	2.2	96
2	2gb1	↓	↓	Protein G (*Streptomyces Griseus*)	nmr	56
27	7rsa	7rsa	7rsa	Ribonuclease A (*Bovine*)	1.3	124
10	9rnt	9rnt	9rnt+	Ribonuclease T1 (*Aspergillus Oryzae*)	1.5	104
4	2sarA	2sarA*	|	Ribonuclease SA (*Streptomyces Aurofaciens*)	1.8	96
1	1rnbA	↓	↓	Barnase (*Bacillus Amyloliquefaciens*)	1.9	109
2	1msbA	1msbA	1msbA	Mannose Binding Protein (*Rat*)	2.3	115
1	1fkf	1fkf	1fkf	FK506 Binding Protein (*Human*)	1.7	107
ALPHA + BETA: METAL RICH						
1	2cdv	2cdv	2cdv	Cytochrome c3 (*Desulfovibrio Vulgaris*)	1.8	107
1	3b5c	3b5c	3b5c	Cytochrome b5 (*Bovine*)	1.5	85

	35Seq Rep	80Str Rep	70Str Rep	Title	Res Å	Length
MULTI DOMAIN						
10	9pap	9pap	9pap	Papain (*Papaya*)	1.6	212
1	3blm	3blm	3blm	Beta Lactamase (*Staphylococcus Aureus*)	2.0	257
2	3hlaA	3hlaA	3hlaA	Human Class I Histocompatibility Antigen (*Human*)	2.6	229
11	2cpp	2cpp	2cpp	Cytochrome P450Cam (*Pseudomonas Putida*)	1.6	405
19	8atcA	8atcA	8atcA	Aspartate Carbamoyl Transferase (*Escherichia Coli*)	2.5	310
12	1csc	1csc	1csc	Citrate Synthase (*Chicken Heart Muscle*)	1.7	429
1	1ace	1ace	1ace	Acetyl Cholinesterase (*Electric Ray*)	2.8	526
1	1cox	1cox	1cox	Cholesterol Oxidase (*Brevibacterium Sterolicum*)	1.8	502
1	1cpkE	1cpkE	1cpkE	C-Amp-Dependent Protein Kinase (*Recombinant Mouse*)	2.7	336
6	1fbpA	1fbpA	1fbpA	Fructose 1,6-Bisphosphatase (*Pig*)	2.5	316
2	1fnr	1fnr	1fnr	Ferredoxin (*Spinach*)	2.2	296
2	1gstA	1gstA	1gstA	Isoenzyme 3-3 of Glutathione S-Transferase (*Rat*)	2.2	217
1	1gly	1gly	1gly	Guanylate Kinase (*Aspergillus Awamori*)	2.2	470
1	1lap	1lap	1lap	Leucine Aminopeptidase (*Bovine*)	2.7	481
4	1ovaA	1ovaA*	1ovaA*	Ovalbumin (*Hen*)	1.9	385
3	7apiA	↓	↓	Modified Alpha-1 Antitrypsin (*Human*)	3.0	339
2	1vsgA	1vsgA	1vsgA	Variant Surface Glycoprotein (*Trypanosoma Brucei*)	2.9	362
3	1lfi	1lfi	1lfi	Lactoferrin (*Human*)	2.1	688
1	1wsyB	1wsyB	1wsyB	Tryptophan Synthase (*Salmonella*)	2.5	385
5	2cyp	2cyp	2cyp	Cytochrome C Peroxidase (*Bakers Yeast*)	1.7	293
12	2glsA	2glsA	2glsA	Glutamine Synthetase (*Salmonella Typhimurium*)	3.5	468
2	2pmgA	2pmgA	2pmgA	Phosphoglucomutase (*Rabbit*)	2.7	561
1	3bcl	3bcl	3bcl	Bacteriochlorophyll A (*Prosthecochloris Aestuarii*)	1.9	344
4	8acn	8acn	8acn	Aconitase (*Bovine*)	2.0	753
1	2reb	2reb	2reb	RecA protein (*Escherichia Coli*)	2.3	303
11	6tmnE	6tmnE	6tmnE	Thermolysin (*Bacillus Thermoproteolyticus*)	1.6	316
33	1hgeA	1hgeA	1hgeA	Hemagglutinin (*Influenza Virus*)	2.6	328
33	1hgeB	1hgeB	1hgeB	Hemagglutinin (*Influenza Virus*)	2.6	175
6	2ca2	2ca2	2ca2	Carbonic Anhydrase (*Human*)	1.9	256
6	3gapA	3gapA	3gapA	Gene regulatory Protein (*Escherichia Coli*)	2.5	208
2	1prcH	1prcH	1prcH	Photosynthetic Reaction Centre (*Rhodopseudomonas Viridis*)	2.3	258

Sequence families whose representatives have related structures (SSAP score > 80 or SSAP < 70) are put into the same structural families (80Str or 70Str respectively). The structure having best resolution is chosen as the representative of the family.

Multidimensional scaling on the complete pairwise SSAP score matrix shows the division of the data bank into the three major structural classes (mainly α, mainly β, and alternating α/β) (*Figure 3*). α + β Folds were distributed throughout as they contain structural motifs from all classes. Globins and the mainly α up/down folds are most highly conserved. Structures within these families are very similar even when the sequence identity falls as low as 9%. More variation is found within families in the mainly β and α + β classes.

Schematic TOPS representations (45) can be drawn for each fold family (see *Figure 4*) and reveal further relationships between folds. For example within the alternating α/β class there are similarities between three large families (8dfr, 3grs, 5p21). All contain a characteristic α/β Rossmann motif which constitutes a large proportion of the fold. However, there are differences in the additional secondary structures which decrease the SSAP score below 70. At least 30 major fold groups can be identified from an inspection

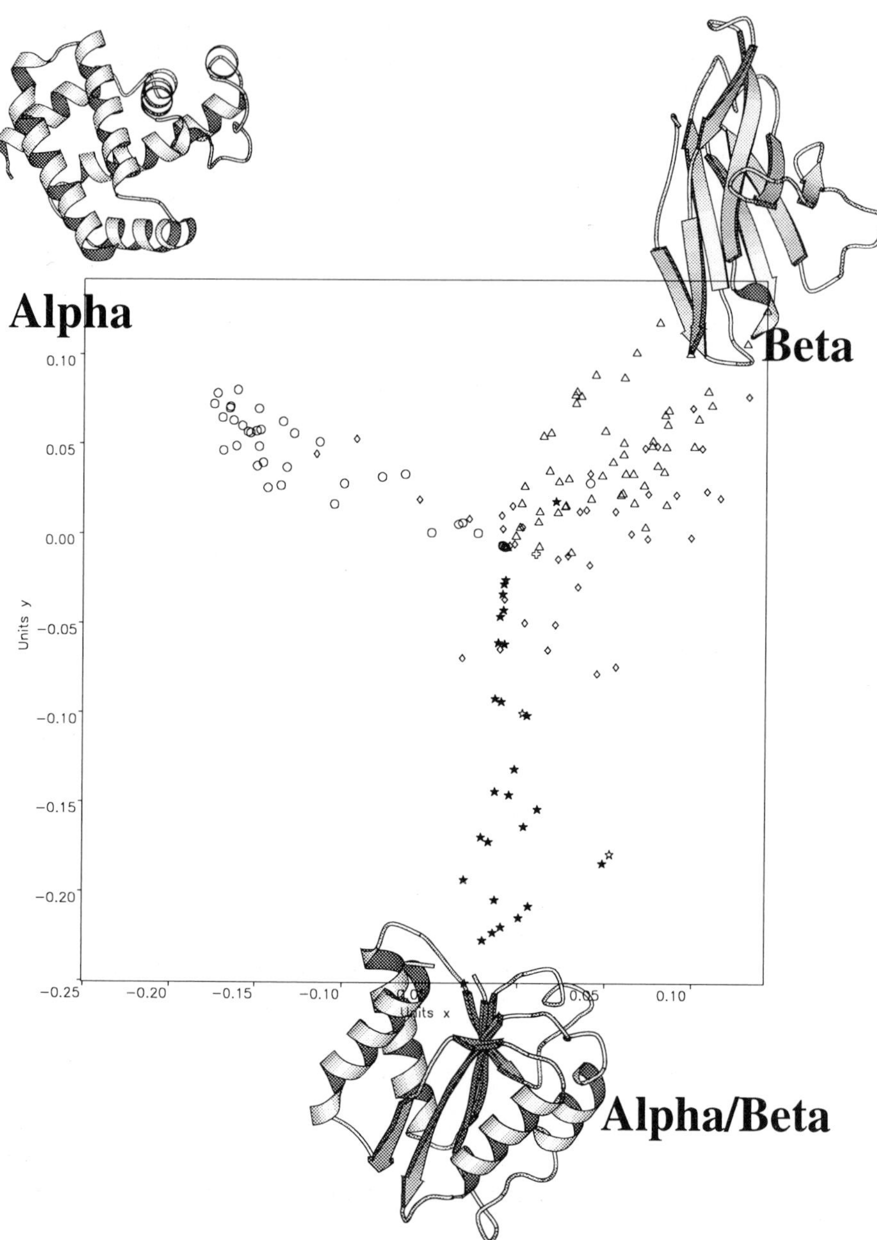

Figure 3. Multidimensional scaling plots generated from the pairwise SSAP homology matrix for the 35Seq family representatives. Symbols correspond to the four classes of protein structure, mainly alpha (○), mainly beta (△, alternating alpha/beta (◇, and non-alternating alpha plus beta (★).

of the TOPS representations (see *Figure 4*). Although this step is not automatic nor necessary for generating fold libraries, the similarities revealed may help to understand the restraints on tertiary conformations.

2.4 Number of folds

Many groups performing data bank searches and classifications have revealed structural similarities between protein pairs having different sequences/functions and previously thought to be unrelated (see *Table 1*, *Figure 5*). This has prompted speculation on the total number of folds possible. Chothia (16) has suggested that there may be fewer than 1000 folds. This estimate was based on the fact that one-third of the sequences determined by the human genome project match known sequences and a quarter of known sequences can be assigned a structure. Assuming 150 unique folds in the current data bank this gave $4 \times 3 \times 150 = 1440$ folds. This falls to ~ 1000 folds if you assume that better sequence alignment methods would have matched more sequences.

We have adopted an approach based solely on the observed occurrence of matching structure which have unrelated sequences. If the number of folds (N) is limited by physical constraints (15) then it should be possible to calculate N from the number of chance matches currently observed. We used a non-redundant data set which contained no pairs of proteins having more than 25% sequence similarity and no pairs displaying both topological (SSAP score > 70) and functional similarity. The number of pairs matching with a SSAP score above 70 but having no functional or sequence similarity can then be determined and used in a statistical calculation (see Equation 1).

$$\frac{\text{Number of matching pairs}}{(\text{Number of non-homologous sequences})^2} = \frac{1}{N}. \qquad [1]$$

Originally we assumed that all folds are equally populated. However, some (e.g. α–β sandwiches and TIM barrel folds) showed far more matches (see *Figure 6*), resulting in a value for N below the number of currently known folds! We therefore suggested these fold 'clusters' represented stable frameworks which can tolerate diverse sequences and functions. If they were adjusted for, we obtained a high probability of there being between 400–700 folds (*Figure 7*).

Whatever the exact number of folds, it is clear that the number of small, compared to the number of protein sequences. Therefore, the approach of fold recognition from sequence becomes increasingly important in structure prediction.

3. Evaluating structural models

Novotny *et al.* (46) demonstrated the inability of standard atomic force fields to detect misfolded proteins using an elegantly simple experiment. Their test

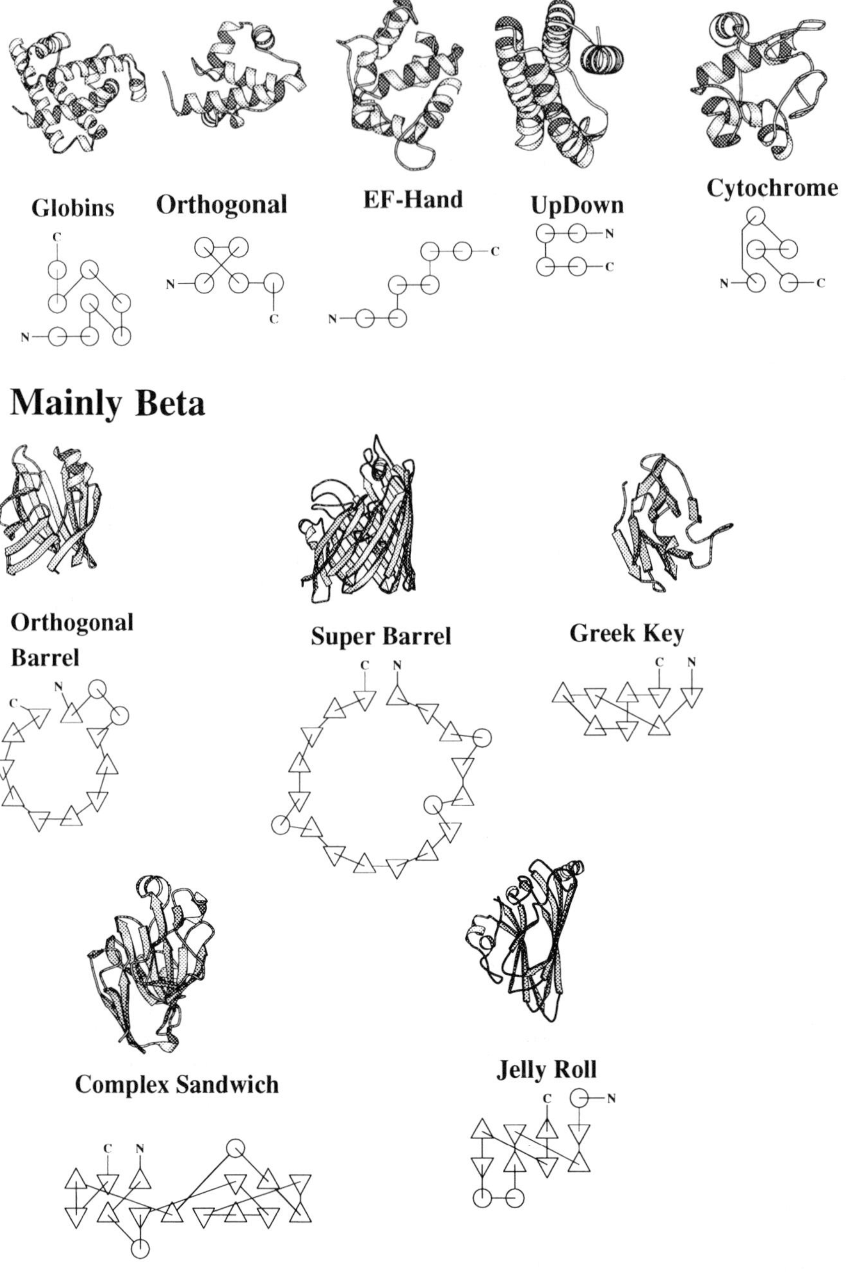

Figure 4. Schematic TOPS representations (45) and MOLSCRIPT diagrams (77) for some of the major fold groups identified by clustering the Brookhaven PDB data bank (3).

Fold Groups - Alternating Alpha/Beta

TIM Barrel

Doubly Wound

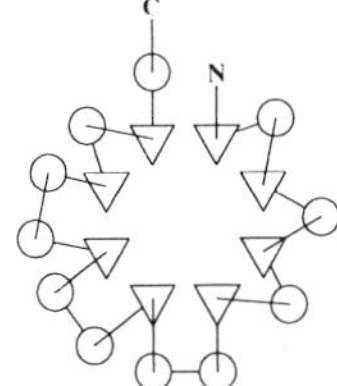

Alpha plus Beta

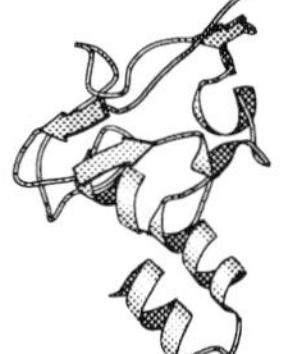

Split beta-alpha-beta Sandwich

Meander Sandwich

Metal Rich

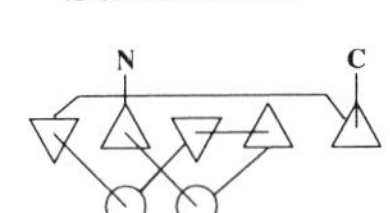

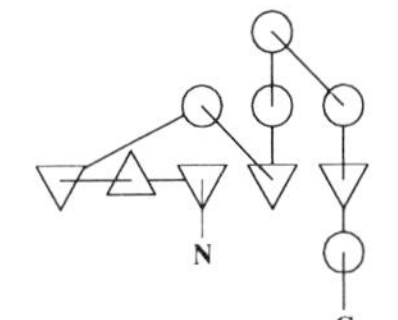

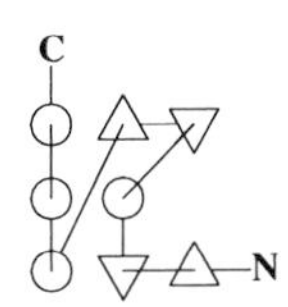

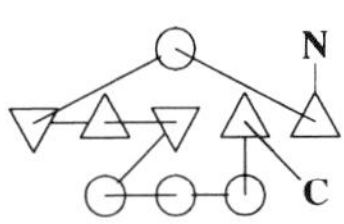

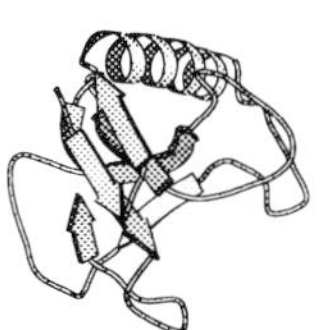

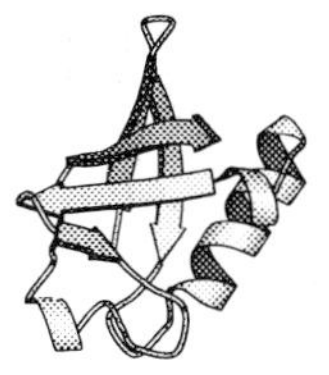

Open Roll

OB Roll

UB Roll

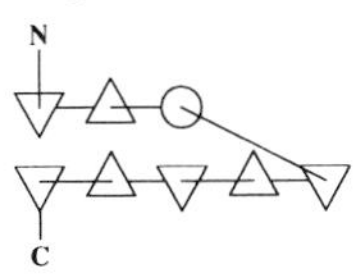

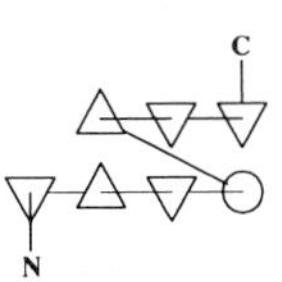

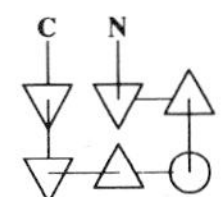

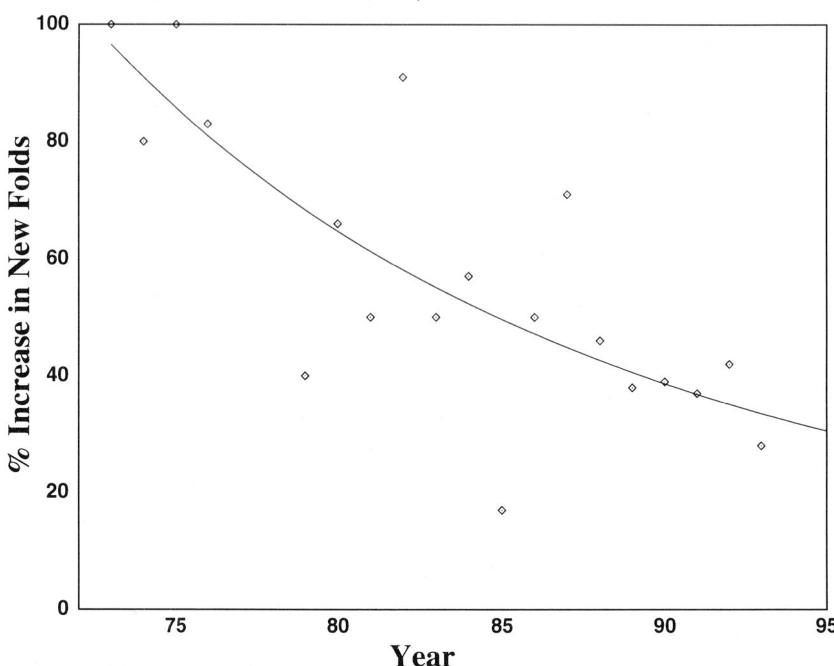

Figure 5. Rate of new fold discovery as a percentage of the number of non-homologous (< 25% sequence identity) folds.

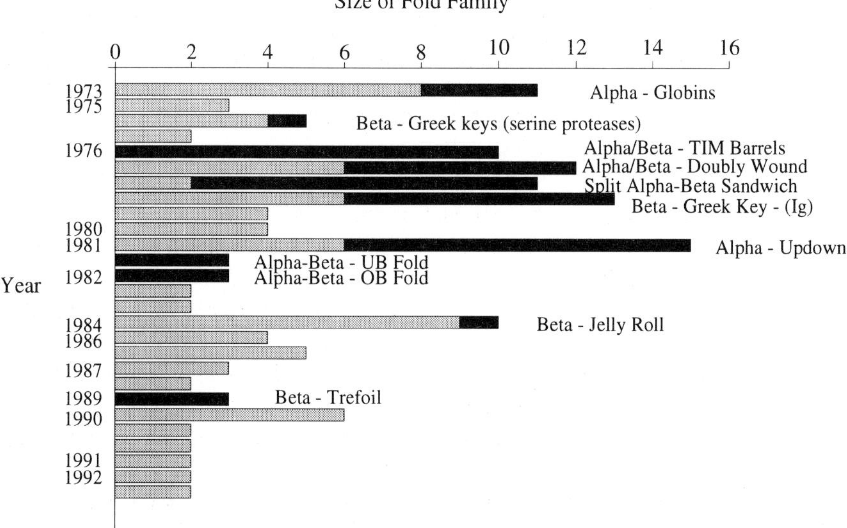

Figure 6. Population of different fold families in the PDB data bank. Grey shaded rectangles show the number of non-homologous representatives (< 25% sequence identity) and black rectangles show the number of non-homologous representatives having no functional similarity. The family is shown against the year when the first representative structure was determined.

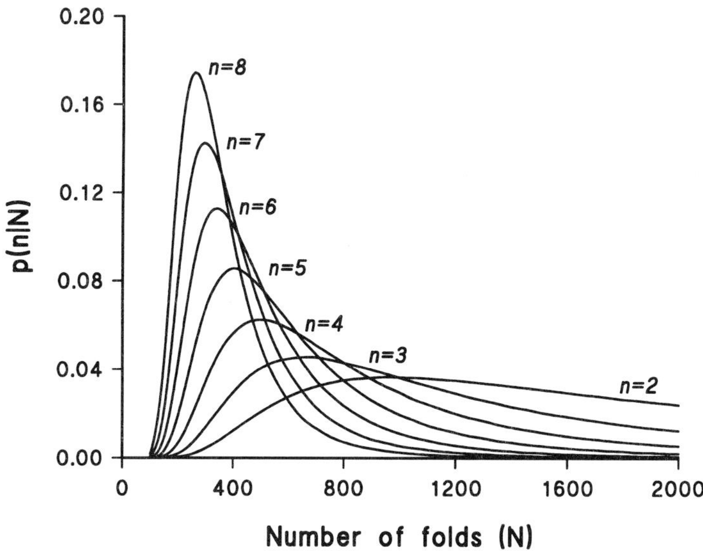

Figure 7. Given a number of observed hits it is useful to consider the probability of this number of hits being produced from fold populations of different sizes. In other words we should like to know the conditional probability of the underlying fold population being of size N given that we have observed n hits. To estimate these probabilities a simple Monte Carlo simulation was run. For each of the 208 protein sequences a fold was randomly chosen from N possibilities. To take into account the observed fold clusters, the first 23 folds were selected with equal probabilities of 0.7/23 whilst the remaining $N-23$ folds were selected with probabilities of $0.3/(N-23)$. The number of identical pairs was then tallied for sequences assigned to the $N-23$ unclustered folds. This experiment was repeated a million times for different value of N over the range 50 to 2000. If $f(n)$ is the total number of all experiments producing n hits, then the conditional probability $p(N|n)$ is estimated as $g(n,N)/f(n)$.

problem was very simple, and yet is a good illustration. In this study, the sequences of myohaemerythrin and an immunoglobulin domain of identical length were exchanged. Both the two native structures, and the two 'misfolded' proteins were then subjected to energy minimization using the CHARMM (47) force field. The results were somewhat surprising in that it was impossible to distinguish between the native and misfolded structures on the basis of the calculated energy sums. Novotny *et al.* (46) correctly surmised that the reason for this failure was the neglect of solvation effects in the force field. In a later study by Novotny *et al.* (48), the force field was modified to approximate the effects of solvent and in this case the misfolded structures could be identified. The work of Novotny *et al.* encouraged several studies into effective methods for evaluating the correctness of protein models, which will now be briefly discussed.

Using a simple solvation energy model alone, Eisenberg and McLachlan (49) were able to distinguish correct models from misfolded models. By calculating a solvation free energy for each amino acid type and calculating the degree of solvent accessibility for each residue in a given model structure, the correctly folded models were clearly distinguished from the misfolded.

Baumann *et al.* (50) also used a solvation term to recognize misfolded protein chains, along with a large number of other general statistical properties of sequences which form stable protein folds. Holm and Sander (51) have proposed another solvation model based on atomic contacts, which appears to be very able at detecting misfolded proteins, even those proteins which have shifts of their sequence on their correct native structure. Interestingly enough a sequence–structure mismatch can quite easily occur not just in theoretically-derived models, but even in crystallographically-derived models. For example one of the xylose isomerase structures in the current Brookhaven database has in part a clearly mistraced chain. Such errors can be detected by use of a suitable solvation-based model evaluation procedure.

Several groups have used statistically-derived pairwise potentials to identify incorrectly folded proteins. Using a simplified side chain definition, Gregoret and Cohen (52) derived a contact preference matrix and attempted to identify correct myoglobin models from a set of automatically generated models with incorrect topology, yet quite reasonable core packing.

Hendlich *et al.* (53) used a set of potentials of mean force, first described by Sippl (54), to not only correctly reject the misfolded protein models of Novotny *et al.* (46), but also to identify the native fold of a protein amongst a large number of decoy conformations generated from a database of structures. In this latter case, the protein sequence of interest was blindly fitted to all contiguous structural fragments taken from a library of highly resolved structures, and the contact energy terms summed in each case. For example, consider a protein sequence of 50 residues being fitted to a structure of length 100 residues. The structure would offer 51 possible conformations for this sequence, starting with the sequence being fitted to the first 50 residues of the structure, and finishing with the sequence being fitted to the last 50. Taking care to eliminate the test protein from the calculation of potentials, Hendlich *et al.* correctly identified 41 out of 65 chain folds. Using factor analysis, Casari and Sippl (55) have found that the principal component of their potentials of mean force is a hydrophobic potential of simple form. This principal component potential alone is found to be almost as successful as the full set of potentials in identifying correct folds. Recently Sippl (56) has shown that a combination of the original pairwise potentials and a recently added solvation potential is useful for detecting errors in crystallographic structures.

In a very similar study to that performed by Hendlich *et al.*, Crippen (52) used a simple contact potential to identify a protein's native fold from all continguous structural fragments of equal length extracted from a library of structures. The success rate (45 out of 56) in this case was higher than that of Hendlich *et al.* due to the fact that the contact parameters in this case were optimized against a 'training set' of correct and incorrect model structures.

An extended study by Maiorov and Crippen (58) has improved upon these results, with the improved contact function correctly identifying virtually all chain folds defined as being 'compact'.

Recently, Bryant and Lawrence (59) showed how a statistically rigorous model could be constructed for residue–residue interactions in protein structures. In this case the energy parameters of the empirical force field were considered as parameters in a statistical model, and were optimized in order to maximize the difference between the energy of a correctly threaded protein structure, and a large number of alternative threadings.

Both the work of Hendlich *et al.* and Crippen demonstrates a very restricted example of fold recognition, whereby sequences are matched against suitable sized contiguous fragments in a template structure. A much harder recognition problem arises when more complex ways of fitting a sequence to a structure are considered, i.e. by allowing for relative insertions and deletions between the object sequence and the template structure. Suitable treatment of insertions and deletions is essential to a generalized method for protein fold recognition.

4. Fold recognition methods

4.1 Matching one fold to the sequence data bank—the inverse protein folding problem

4.1.1 Bowie *et al.* (1991)

Bowie *et al.* (60) match sequences to a fold by describing the fold in terms of the *environment* of each residue location in the structure. The environment is described in terms of local secondary structure (three states: α, β, and coil), solvent accessibility (three states: buried, partially buried, and exposed), and the degree of burial by polar rather than apolar atoms. The environment of a particular residue thus defined is found to be more highly conserved than the identity of the residue itself, and so the method is able to detect more distant sequence–structure relationships than purely sequence-based methods. The method has also been applied to the evaluation of protein models (61). The authors describe this method as a 1D–3D profile method, in that a 3D structure is encoded as a 1D string, which can then be aligned using traditional dynamic programming algorithms. Bowie *et al.* have applied the 1D–3D profile method to the inverse folding problem and have shown that the method can indeed detect remote matches, but in the cases shown the hits have still retained some sequence similarity with the search protein, even though in the case of actin and the 70 kDa heat shock protein the sequence similarity is very weak (62). Environment-based methods appear to be incapable of detecting structural similarities between extremely divergent proteins, and between proteins sharing a common fold through convergent evolution—environment only appears to be conserved up to a point (63). Consider a buried polar residue in one structure that is found to be located in a polar environment. Buried polar residues tend to be functionally important residues, and so it is not surprising then that a protein with a similar structure but with an entirely different function would choose to place a hydrophobic residue at this position in an apolar environment. A further problem with environment-based methods is that they are sensitive to the multimeric state of a protein. Residues buried in a subunit interface of a multimeric protein will not be buried at an equivalent position in a monomeric protein of similar fold.

4.1.2 Godzik *et al.* (1992)

Godzik *et al.* (64) have developed a fold recognition method that combines aspects of the residue environment method of Bowie *et al.* and the optimal threading approach (see later). Their search template, against which sequences are matched, is provided by a 'structural fingerprint' derived from the protein's contact map and the buried/exposed pattern of residues. After identifying possible matches using a simple string matching algorithm, a lattice-based Monte Carlo algorithm is used to test the stability of the proposed model. The reported successes for this method include matches between globins and phycocyanins, the $(\alpha\beta)_8$-barrels, and between members of the copper binding protein family.

4.2 Matching one sequence to a library of folds

The fold recognition methods described so far have been examples of the inverse protein folding problem. Despite the interesting theoretical implications of inverse protein folding, to practising biochemists and molecular biologists interested in identifying possible structures for a newly-characterized gene product these methods are not directly relevant. Roughly speaking a researcher is 20 times more likely to have a novel sequence 'under the microscope' than a novel structure, and so methods for determining compatible structures for given sequences are of vital importance.

4.2.1 Bowie *et al.* (1990)

The method described by Bowie *et al.* (18) can be thought of as a precursor to their inverse protein folding work, and despite the fact that the results were not quite comparable to the more recent methods, the future prospects were clearly in view. The first stage of this method involves the prediction of residue accessibility from multiple sequence alignments, which is itself another interesting recent development (see later). In essence, alignment positions with high average hydrophobicity and high conservation are predicted to be buried and relatively polar variable positions predicted to be exposed to solvent. The degree of predicted exposure at each position of the aligned sequence family is then encoded as a string. This string is then matched against a library of similarly encoded strings, based, however, not on predicted accessibilities but on *real* accessibilities calculated from structural data. Several successful recognition attempts were demonstrated using this method. Of particular note was the matching of an aligned set of Ef Tu sequences with the structure of flavodoxin. The similarity between Ef Tu and flavodoxin can only be detected by sensitive structure comparison methods (44), so this result is quite impressive.

4.2.2 Finkelstein and Reva (1991)

Finkelstein and Reva (19) have used a simplified lattice representation of protein structure for their work on fold recognition. The problem they consider is that of matching a sequence to one of the 60 possible eight stranded β-sandwich topologies. Each strand has three associated variables: length, position in the sequence, and spatial position in the lattice Z direction. The force field used by Finkelstein and Reva includes both short-range and long-range components. The short-range component is simply based on the beta–coil transition constants for single amino acids, similar in many respects to the standard Chou–Fasman (3) propensities. The long-range interaction component has a very simple functional form. For a pair of interacting (contacting) residues, it is defined simply as the sum of their solvent transfer energies as calculated by Fauchere and Pliska (65).

The overall energy of the eight strands in this simple force field is minimized by a simple iterative method. At the heart of the method is a probability matrix (a three-dimensional matrix in this case) for each of the strands, where each matrix cell represents one triplet of the strand variables, i.e. length, sequence position, and spatial position. The values in each cell represent the probability of observing the strand with the values associated with the cell. The novel aspect of this optimization strategy is that the strands themselves do not physically move in the force field, only the probabilities change. At the start of the first iteration the strand coordinate probabilities are assigned some arbitrary values, either all equal, or set close to their expected values (the first strand is unlikely to be positioned near the end of the sequence for example). A new set of probabilities is then calculated using the current mean field and the inverse Boltzmann equation. As more iterations are executed it is to be hoped that most of the probabilities will collapse to zero and that eventually a stable 'self-consistent' state will be reached. Finkelstein and Reva (19) found that the most probable configurations corresponded to the correct alignment of the eight stranded model with the given sequence, and that when the process was repeated for each of the 60 topologies, in some cases the most probable configuration of the native topology had the highest probability of all.

The simplicity of the lattice representation and the uncomplicated force field are critical to the success of this method. A more detailed interresidue potential would prevent the system from reaching a self-consistent state, and would be left either in a single local minimum or more likely oscillating between a number of local minima. In addition, whilst it is quite practical to represent β-sheets on a lattice, it is not clear how α-helices could be reasonably represented. It will be interesting to see whether this method can be extended to classes of protein structure other than the all-β class. The method does however begin to address an important problem in predicting structure in that it attempts to choose from a set of different topologies with identical secondary structure assignments, answering questions such as if I know my protein has seven strands, what is the most likely arrangement of these strands?'.

4.2.3 Jones *et al.* (1992)

Despite the obvious computational advantages of using residue environments, it is clear that the fold of a protein chain is governed by fairly specific protein–protein and protein–solvent atomic interactions. A given protein fold is therefore better modelled in terms of a 'network' of pairwise interatomic energy terms, with the structural role of any given residue described in terms of its interactions. Classifying such a set of interactions into one environmental class such as 'buried alpha helical' will inevitably result in the loss of useful information, reducing the *specificity* of sequence–structure matches evaluated in this way. Put simply, one amphipathic helix is much like any other amphipathic helix. Ideally, we should like to match a sequence to a structure by considering the plethora of detailed pairwise interactions, rather than averaging them into a crude environmental class. However, incorporation of such non-local interactions into standard alignment methods such as the algorithm of Needleman and Wunsch (41), has hitherto proved difficult.

We have applied a novel dynamic programming algorithm (now commonly known as 'double' dynamic programming) (17) to the problem of aligning a given sequence with the 'real' coordinates of a structure, taking into account the detailed pairwise interactions, a process which we call *optimal sequence threading*—or more commonly today, 'threading' for short. The requirement here to match pairwise interactions relates to the requirement of structural comparison methods. We define here the *potential environment* of a residue i as being the sum of all pairwise potential terms involving i and all other residues $j \neq i$. This is a similar definition to that of the *structural environment* of a residue, as described by Taylor and Orengo (30). In the simplest case, structural environment of a residue i is defined as the set of all inter-C^{α} distances between residue i and all other residues $j \neq i$. Taylor and Orengo propose a novel dynamic programming algorithm for the comparison of residue structural environments, and it is a derivation of this method that we have used for the effective comparison of residue potential environments. *Figure 8* outlines the steps involved in the use of double dynamic programming for aligning sequences with structural templates.

Whilst our method is applicable to any form of pairwise potential, including all those which have been described in this review, we chose to use a set

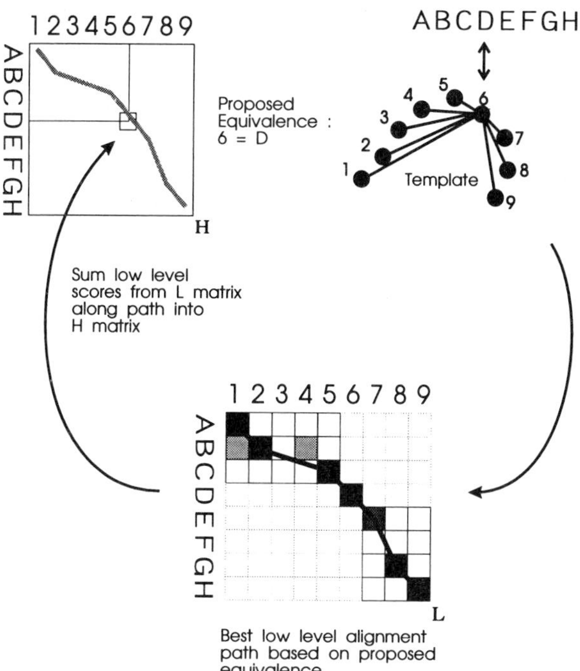

Figure 8. An outline of the double dynamic programming algorithm. For each proposed sequence–structure equivalence an optimal path is calculated based on interactions with the equivalenced residue. Scores are summed along each path into the higher level matrix.

of statistically-derived pairwise potentials similar to those described by Sippl (54). Using the formulation of Sippl, we have constructed short (sequence separation, $k \leqslant 10$), medium ($11 \leqslant k \leqslant 30$), and long ($k > 30$) range potentials between the following atom pairs: $C^{\beta} \rightarrow C^{\beta}$, $C^{\beta} \rightarrow N$, $C^{\beta} \rightarrow O$, $N \rightarrow C^{\beta}$, $N \rightarrow O$, $O \rightarrow C^{\beta}$, and $O \rightarrow N$. For a given pair of atoms, a given residue sequence separation, and a given interaction distance, these potentials provide a measure of energy, which relates to the probability of observing the proposed interaction in native protein structures. Our potentials differ from those proposed by Sippl in the following ways. First, interactions beyond 10 Å are ignored. We have found these interactions to be residue-specific and determined simply by solvation effects. in addition the longer distance interactions are biased towards the larger proteins in the set used to calculate the potentials. In place of the long-distance terms, we substitute a 'solvation potential'. This potential simply measures the frequency with which each amino acid species is found with a certain degree of solvation, approximated by the residue solvent accessible surface area. We define the solvation potential for amino acid residue a as follows:

$$\Delta E^{a}_{\mathrm{solv}}(r) = -kT \ln \left[\frac{f^{a}(r)}{f(r)}\right] \quad [2]$$

where r is the per cent residue accessibility (relative to residue accessibility in GGXGG fully extended pentapeptide), $f^{a}(r)$ is the frequency of occurrence of residue a with accessibility r, and $f(r)$ is the frequency of occurrence of all residues with accessibility r. Residue accessibilities were calculated using the program DSSP (66), applied to Brookhaven coordinate files (40). For multimeric proteins, only the chains explicitly described in the coordinate files were taken into account. A final point to note about our force field is that all pairwise terms involving loop residues are excluded (loop positions are evaluated by the solvation potential alone). Loop residues were excluded due to the fact that loop conformations tend not to be conserved even between closely related proteins, let alone distant relatives.

By dividing the empirical potentials into sequence separation ranges, specific structural significance may be tentatively conferred on each range. For instance, the short-range terms predominate in the matching of secondary structural elements. By threading a sequence segment on to the template of an α-helical conformation and evaluating the short-range potential terms, the possibility of the sequence folding into an α-helix may be evaluated. In a similar way, medium-range terms mediate the matching of supersecondary structural motifs, and the long-range terms, the tertiary packing.

We have found this method capable of detecting quite remote sequence–structure matches. In particular it has proven able to detect the previously mentioned similarity between the globins and the phycocyanins, and is therefore the first sequence analysis method to achieve this. Also of particular note are the results for some $(\alpha\beta)_8$ (TIM)-barrel enzymes and also the β-trefoil folds: trypsin inhibitor DE-3 and interleukin 1β for example. The degree of sequence homology between different $(\alpha\beta)_8$-barrel enzyme families and between trypsin inhibitor DE-3 and interleukin 1β is extremely low (5–10%). As a consequence of this, again, sequence template methods have not proven able to detect these folds. It is therefore clear that new information beyond sequence similarity is being utilized in our method. Although the actin fold was found to be the best match when a hexokinase sequence was matched against the fold library, the separation between actin and the next fold in the list of folds sorted by 'threading energy' was close to zero. We might therefore surmise that the degree of structural similarity between hexokinase and actin represents the current limit to the sensitivity of the method. However, the sensitivity of the method does depend on the secondary structure content of the protein's native fold. For example, the method is more able at detecting similarities between proteins in the all-α class of protein structure than in the αβ class, with the all-β class being the most difficult of all. Also proteins with very large relative insertions and deletions prove difficult to match. Of course, these problems are common to all protein fold recognition methods.

5. Case study

To see how fold recognition might be typically used to identify the structure of a newly-characterized protein sequence, we present here an example of using the optimal threading search program to identify the fold of the N-terminal fragment of the 70 kDa heat shock protein. When the structure for the muscle protein, actin was solved it came as a great surprise to the entire crystallographic community to see that the structure was almost identical to a fragment of the 70 kDa heat shock protein. After superposition, the rmsd between the two structures is found to be only 2.3 Å over 241 C^{α} positions. This structural relationship was not expected despite the fact that many biologists had been searching for evolutionary relationships involving actin.

5.1 Traditional sequence analysis

As a first step in analysing a newly-determined protein sequence it is common practice to search the existing database of protein sequence to identify any similar sequences. In this particular case, for the purposes of predicting the structure of the protein, we are only interested in searching those sequences with a known tertiary structure. In the following section, the following sequence for the heat shock protein fragment was used in all cases:

MSKGPAVGIDLGTTYSCVGVFQHGKVEIIANDQGNRTTPSYVAFTDTERLIGDAAKNQVA
MNPTNTVFFDAKALIGRRFDDAVVQSDMKHWPFMVVNDAGRPKVQVEYKGETKSFYPEE
VSSMVLYKMKEIAEAYLGKTVTNAVVTVPAYFNDSQRQATKDAGTIAGLNVLRIINEPTAA
AIAYGLDKKVGAERNVLIFDLGGGTFDVSILTIEDGIFEVKSTAGDTHLGGEDFDNRMVNH
FIAEFKRKHKKDISENKRAVRRLRTACERAKRTLSSSTQASIEIDSLYEGIDFYTSITRARFEE
LNADLFRGTLDPVEKALRDAKLLDKSQIHDIVLVGGSTRIPKIQKLLQDFFNGKELNKSINP
DEAVAYGAAVNAAILSGDKS

This sequence was aligned against a representative set of sequences taken from entries in the protein structure data bank, using the standard Smith and Waterman (67) dynamic programming alignment algorithm. Dynamic programming methods prove to be the most sensitive methods for pairwise sequence comparisons, but even so, as can be seen in *Figure 9* the expected match is far from the top of the distribution of scores. Consensus template methods (68–72) have been found to be more sensitive than pairwise comparison methods, and so whenever many sequences are available, it is wise to search the sequences with a template or profile based on a multiple alignment. In this case, 41 members of the 70 kDa heat shock protein family were aligned, and a profile constructed according to the method of Gribskov *et al.* (69,72). *Figure 10* shows the distribution of Z-scores (normalized scores) for the heat shock profile search, with the expected match again marked with an arrow.

Despite the fact that in this case traditional sequence analysis techniques have not been successful, it is still highly advisable to start with these methods before resorting to more sophisticated techniques. One of the most significant advantages of sequence-based techniques is that the statistical properties

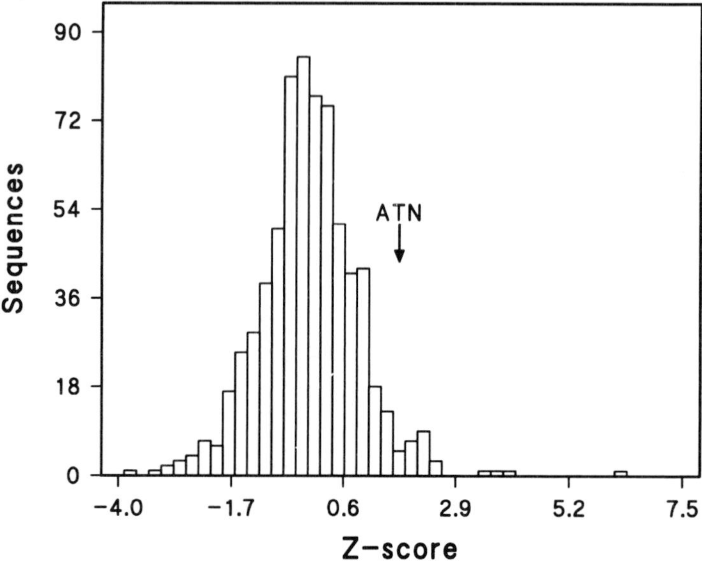

Figure 9. Results from searching a database of 682 sequences with known three-dimensional structure using a single 70 kDa heat shock cognate protein sequence using the alignment method of Smith and Waterman (67).

are well-characterized and given statistically significant results it is easy to make a strong case for an *evolutionary* relationship between the search sequences and the hits extracted from the database.

5.2 Threading

Given an unsuccessful sequence-based search, the fold recognition methods described earlier can be applied. Using the method of Jones *et al.* (17), a search through a library of 187 folds was performed using the heat shock protein sequence. After about 90 minutes the results shown in *Figure 11* were obtained (compare this time to around two to three minutes for the profile and pairwise sequence searches on the same computer system). Note that in this case the match between given sequence and a given structure is quantified in terms of *'energy'*, where low values are best. In this case the expected structural match is ranked at the top of the list (i.e. lowest energy).

5.3 Interpreting the results

The interpretation of the resulting scores is currently a major problem in the use of fold recognition methods. The simplest interpretation is to take the highest scoring (or lowest energy) fold and assume that the search protein adopts an identical topology to the matched protein. In the case of optimal sequence threading it is useful to consider both components of the aggregate scores (the pairwise and solvation terms). On occasions, a protein fold may be identified by a good match in pairwise energy terms but not in solvation

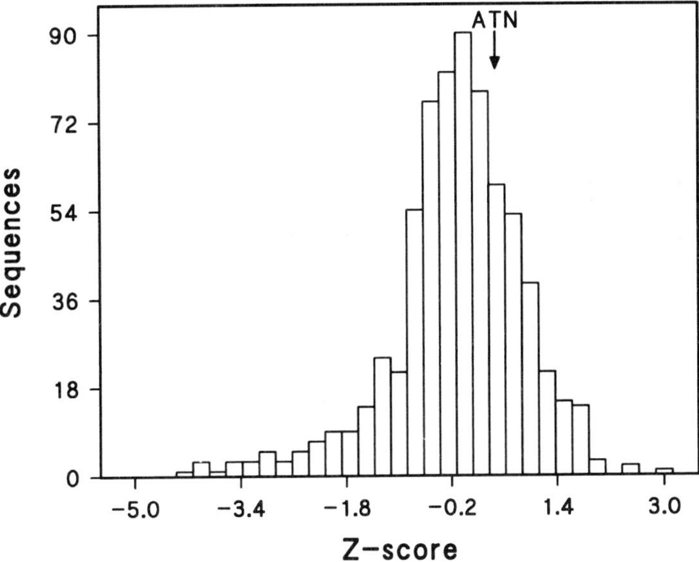

Figure 10. Results from searching a database of 682 sequences with a consensus profile constructed from 20 members of the 70 kDa heat shock cognate protein family using the method of Gribskov *et al.* (69).

terms, and vice versa. The 'take home message' here is that at their present level of sophistication, fold recognition methods can only provide hints as to the correct fold (assuming that the fold is represented in the fold library at all). In favourable cases, the correct fold will make itself very apparent, as in the case of actin and the heat shock protein fragment. In this case a model made from the best scoring fold and the given threading alignment would have been reasonably close (better than 3 Å over the core main chain) to the native structure. Even where a handful of different folds are clustered at the top of the list of scores, and it is not immediately apparent which of them are false positives, a little intelligent application of biological knowledge can sometimes spotlight the more likely candidates. For example, if it is known that the unknown protein binds ATP, do any of the top scoring folds share this function? It is often worth taking, say, the top five hits and returning to purely sequence analytic methods to see if some distant evolutionary relationship can be discerned. A good example of such a combined attack on predicting the structure of a protein was the successful prediction of the structure of amiC (part of the amidase operon of *Pseudomonas aereginosa*), by Wilson *et al.* (73), using a combination of optimal sequence threading and traditional multiple sequence alignment.

5.4 Using the results

The final question to answer is what to do with the resulting information

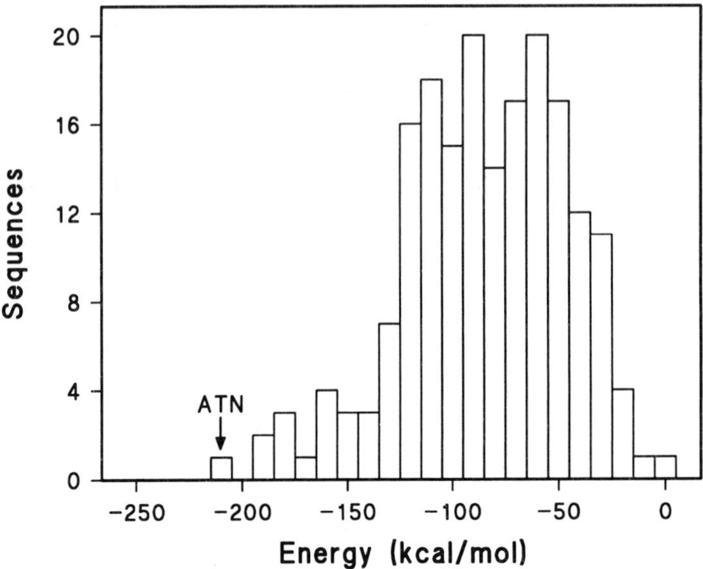

Figure 11. Results from searching a fold library containing 187 folds using a single 70 kDa heat shock cognate protein sequence and the optimal sequence threading method of Jones *et al.* (17).

gleaned from fold recognition. At a minimum the results may suggest further lines of inquiry—perhaps predicting the structural importance of particular residues which might be tested by mutagenesis experiments. These methods will typically provide a number of possible structural models which can be further evaluated by inspection on a graphics workstation, or perhaps automatically by checking for the violation of stereochemical rules. Before long it may prove possible to use these low resolution models as search models in X-ray crystallographic molecular replacement (74) and so allow the rapid solution of a structure where native protein crystals are available but where heavy metal derivatives have proven impossible to obtain.

6. The Asilomar competition

Although the published results for the fold recognition methods look impressive, it is fair to argue that in all cases the correct answers were already known and so it is not clear how well they would perform in real situations where the answers are not known at the time the predictions are made. The results of a very ambitious world-wide experiment have recently been published (78) in a special issue of the journal *Proteins*, where an attempt was made to find out how successful different prediction methods were when rigorously blind-tested. In 1994, John Moult and colleagues approached X-ray crystallographers and NMR spectroscopists around the world and asked them to deposit (in a special data bank) the sequences for any structures they were close to solving. Before these structures were made public, various teams around the world were challenged with the task of predicting them. The results of this experiment were announced at a meeting held at Asilomar in California at the end of 1994, and this ambitious experiment has now become widely known as the Asilomar experiment (or more commonly the Asilomar competition.

The results for the comparative modelling and *ab initio* sections offered few surprises, in that the *ab initio* methods were reasonably successful in predicting secondary structure but not tertiary structure, and homology modelling could predict tertiary structure well, but only when the proteins concerned had very high sequence similarity. The results for the fold recognition section (79), however, showed great promise. Overall, roughly half of the structures in this part of the competition were found to have previously observed folds and yet showed no sequence similarity to other proteins with the same fold. Almost all of these structures were correctly predicted by at least one of the teams. The threading method of Jones *et al.* (17) proved to be the most successful method, with five out of nine folds correctly identified, and with a looser definition of structural similarity, eight out of eleven correct. These results show that, despite their relative early stage of development, fold recognition methods (and threading methods in particular) offer very exciting prospects for prediction of protein tertiary structure in the near future. One point that should be made about the predictions from all of the fold recognition groups was that the sequence–structure alignments were not as accurate as might have been hoped when judged against the alignments obtained from structural superposition of the two structures concerned. This is disappointing, but it is obvious that this will be improved as the methods mature.

7. Software availability

Several of the methods described in this chapter are now generally available either in commercial packages or directly from the authors. A program, called THREADER, which implements the threading approach to protein fold recognition (17,80), has been made widely available to the academic community free of charge, and can be downloaded over the Internet from the World Wide Web address (URL): `ftp://ftp.biochem.ucl.ac.uk/pub/THREADER`.

8. Conclusions and future directions

It is most important to remember that finding a possible fold for a sequence is only the first stage in the process of tertiary structure prediction, and provides the starting point for constructing an all-atom three-dimensional model, using much the same techniques as conventional modelling. However, since the template structure will have very few residues in common with the search sequence, the framework will have changed more than is usual in conventional homology modelling exercises. Again methods to improve modelling of analogous sequences are still being developed, as we learn more about the relationships between such molecules.

The most important reason for predicting structure is to learn more about, and better understand, the function of the protein at the molecular level. The function almost always depends on the specific conformation and location of a relatively small number of residues in the sequences, especially for an enzyme or binding protein. The topology of itself rarely provides insight into biological function except by relationship to other proteins of known function. Thus the topology does not of itself determine the function, which depends rather on the sequence. There are many examples where proteins with the same topology have radically different functions. There are also many examples where proteins with the same topology, but little sequence similarity have functions which are broadly similar, although the details of the 'active site' and the residues involved have changed (e.g. the OB-fold family which bind oligosaccharides and oligonucleotides) (75). Therefore before any conclusions can be drawn from identifying the topology of a sequence, it is necessary to study the sequence for the presence or absence of critical functional residues.

Threading techniques are also valuable for aligning two distantly related sequences which are known to adopt the same topology. In such cases the alignment by sequence is often difficult, if not impossible, and the empirical potentials can be used to find the most probable equivalences.

For tertiary structure prediction, threading is limited by the current library of folds and an obvious future development is, not only to recognize folds, but also to build novel folds, based on topological rules derived from proteins of known structure. Many of the new folds, which have been recently discovered, are made up of the common supersecondary structure motifs, even though their topology has not been observed before. As described above, Finkelstein and Reva (15) built very simplified models of eight stranded β-sandwich structures, with different topologies. In order to change recognition into *ab initio* prediction we still need better potentials, better methods for generating topologies, and better techniques for finding the optimal threading. In addition the kinetic problem of folding a protein is just becoming accessible to experimental determination (76), and this may provide guidance for simulations of folding pathways.

References

1. Thornton, J. M., Flores, T. P., Jones, D. T., and Swindells, M. B. (1991). *Nature*, **354**, 105.
2. Sander, C. and Schneider, R. (1991). *Proteins*, **9**, 56.

3. Chou, P. Y. and Fasman, G. D. (1974). *Biochemistry*, **13**, 212.
4. Garnier, J., Osguthorpe, D. J., and Robson, B. (1978). *J. Mol. Biol.*, **120**, 97.
5. Gibrat, J. F., Garnier, J., and Robson, B. (1987). *J. Mol. Biol.*, **198**, 425.
6. Levin, J. M. and Garnier, J. (1988). *Biochim. Biophys. Acta*, **955**, 283.
7. Biou, V. and Gibrat, J. (1988). *Protein Eng.*, **2**, 185.
8. Sternberg, M. J. E. and Thornton, J. M. (1976). *J. Mol. Biol.*, **105**, 367.
9. Richardson, J. (1981). *Adv. Protein Chem.*, **34**, 167.
10. Chothia, C., Levitt, M., and Richardson, D. (1977). *Proc. Natl. Acad. Sci. USA*, **74**, 4130.
11. Cohen, F. E., Sternberg, M. J. E., and Taylor, W. R. (1981). *J. Mol. Biol.*, **148**, 253.
12. Chothia, C. and Finkelstein, A. V. (1990). *Annu. Rev. Biochem.*, **59**, 1007.
13. Jones, D. T. and Thornton, J. M. (1993). *J. Comput. Aid. Mol. Des.*, **7**, 439.
14. Wodak, S. J. and Rooman, M. J. (1994). *Curr. Opin. Struct. Biol.*, **3**, 247.
15. Finkelstein, A. V. and Ptitsyn, O. B. (1987). *Prog. Biophys. Mol. Biol.*, **50**, 171.
16. Chothia, C. (1993). *Nature*, **357**, 543.
17. Jones, D. T., Taylor, W. R., and Thornton, J. M. (1992). *Nature*, **358**, 86.
18. Bowie, J. U., Clarke, N. D., Pabo, C. O., and Sauer, R. T. (1990). *Proteins*, **7**, 257.
19. Finkelstein, A. V. and Reva, B. A. (1991). *Nature*, **351**, 497.
20. Drexler, K. E. (1981). *Proc. Natl. Acad. Sci. USA*, **78**, 5275.
21. Pabo, C. O. and Sucharek, E. G. (1986). *Biochemistry*, **25**, 5987.
22. Ponder, J. W. and Richards, F. M. (1987). *J. Mol. Biol.*, **193**, 775.
23. Blundell, T. L., Sibanda, B. L., Sternberg, M. J. E., and Thornton, J. M. (1987). *Nature*, **326**, 347.
24. Greer, J. (1985). *Ann. N.Y. Acad. Sci.*, **439**, 44.
25. Chothia, C. and Lesk, A. M. (1986). *EMBO J.*, **5**, 823.
26. Hubbard, T. J. P. and Blundell, T. L. (1987). *Protein Eng.*, **1**, 159.
27. Benner, S. A., Cohen, M. A., and Gonnet, G. H. (1993). *J. Mol. Biol.*, **229**, 1065.
28. Boscott, P. E., Barton, W. G., and Richards, W. G. (1993). *Protein Eng.*, **6**, 261.
29. Rost, B. and Sander, C. (1994). *J. Mol. Biol.*, **235**, 13.
30. Taylor, W. R. and Orengo, C. A. (1989). *J. Mol. Biol.*, **208**, 1.
31. Orengo, C. A., Flores, T. P., Taylor, W. R., and Thornton, J. M. (1993). *Protein Eng.*, **6**, 485.
32. Artymiuk, P. J., Mitchell, E. M., Rice, D. W., and Willett, P. (1989). *J. Inf. Sci.*, **15**, 287.
33. Sali, A. and Blundell, T. L. (1990). *J. Mol. Biol.*, **212**, 403.
34. Overington, J. P., Zhu, Z. Y., Sali, A., Johnson, M. S., Sowdhamini, R., Louie, G., *et al.* (1993). *Biochem. Soc. Trans.*, **21**, 597.
35. Holm, L., Ouzounis, C., Sander, C., Tuparev, G., and Vriend, G. (1992). *Protein Sci.*, **1**, 1691.
36. Holm, L. and Sander, C. (1993). *J. Mol. Biol.*, **233**, 123.
37. Flores, T. P., Orengo, C. A., Moss, D. M., and Thornton, J. M. (1993). *Protein Sci.*, **2**, 1811.
38. Hilbert, M., Bohm, G., and Jaenicke, R. (1993). *Proteins*, **17**, 138.
39. Orengo, C. A., Flores, T. P., Jones, D. T., Taylor, W. R., and Thornton, J. M. (1993). *Curr. Biol.*, **3**, 131.
40. Bernstein, F. C., Koetzle, T. F., Williams, G. J. B., Meyer, E. F., Brice, M. D., Rodgers, J. R., *et al.* (1977). *J. Mol. Biol.*, **112**, 535.
41. Needleman, S. B. and Wunsch, C. D. (1970). *J. Mol. Biol.*, **48**, 443.

42. Hobohm, U., Scharf, M., Schneider, R., and Sander, C. (1992). *Protein Sci.*, **1**, 409.
43. Taylor, W. R. and Orengo, C. A. (1989). *Protein Eng.*, **2**, 505.
44. Orengo, C. A., Brown, N. P., and Taylor, W. R. (1992). *Proteins*, **14**, 139.
45. Flores, T. P., Moss, D. M., and Thornton, J. M., (1993). *Protein Eng.*, **7**, 31.
46. Novotny, J., Bruccoleri, R. E., and Karplus, M. (1984). *J. Mol. Biol.*, **177**, 787.
47. Brooks, B., Bruccoleri, R. E., Olafson, B. D., States, D. J., Swaminathan, S., and Karplus, M. (1983). *J. Comput. Chem.*, **4**, 187.
48. Novotny, J., Rashin, A. A., and Brucolleri, R. E. (1988). *Proteins*, **4**, 19.
49. Eisenberg, D. and McLachlan, A. D. (1986). *Nature*, **319**, 199.
50. Baumann, G., Frommel, C., and Sander, C. (1989). *Protein Eng.*, **2**, 329.
51. Holm, L. and Sander, C. (1993). *J. Mol. Biol.*, **225**, 93.
52. Gregoret, L. M. and Cohen, F. E. (1990). *J. Mol. Biol.*, **211**, 959.
53. Hendlich, M., Lackner, P., Weitckus, S., Floeckner, H., Froschauer, R., Gottsbacher, K., *et al.* (1990). *J. Mol. Biol.*, **216**, 167.
54. Sippl, M. J. (1990). *J. Mol. Biol.*, **213**, 859.
55. Casari, G. and Sippl, M. J. (1992). *J. Mol. Biol.*, **224**, 725.
56. Sippl, M. J. (1993). *Proteins*, **17**, 355.
57. Crippen, G. M. (1991). *Biochemistry*, **30**, 4232.
58. Maiorov, V. N. and Crippen, G. N. (1992). *J. Mol. Biol.*, **227**, 876.
59. Bryant, S. H. and Lawrence, C. E. (1993). *Proteins*, **16**, 92.
60. Bowie, J. U., Lüthy, R., and Eisenberg, D. (1991). *Science*, **253**, 164.
61. Lüthy, R., Bowie, J. U., and Eisenberg, D. (1992). *Nature*, **356**, 83.
62. Bork, P., Sander, C., and Valencia, A. (1992). *Proc. Natl. Acad. Sci. USA*, **89**, 7290.
63. Pickett, S. D., Saqi, M. A. S., and Sternberg, M. J. E. (1992). *J. Mol. Biol.*, **228**, 170.
64. Godzik, A., Kolinski, A., and Skolnick, J. (1992). *J. Mol. Biol.*, **227**, 227.
65. Fauchere, J. L. and Pliska, V. E. (1983). *Eur. J. Med. Chem.*, **18**, 369.
66. Kabsch, W. and Sander, C. (1983). *Biopolymers*, **22**, 2577.
67. Smith, T. F. and Waterman, M. S. (1981). *J. Mol. Biol.*, **147**, 195.
68. Taylor, W. R. (1986). *J. Mol. Biol.*, **188**, 233.
69. Gribskov, M., McLachlan, A. D., and Eisenberg, D. (1987). *Proc. Natl. Acad. Sci. USA*, **84**, 4355.
70. Barton, G. J. and Sternberg, M. J. E. (1990). *J. Mol. Biol.*, **212**, 389.
71. Taylor, W. R. and Jones, D. T. (1991). *Curr. Opin. Struct. Biol.*, **1**, 327.
72. Gribskov, M., Lüthy, R., and Eisenberg, D. (1990). *Methods in enzymology*, Vol. 188, pp. 146–59.
73. Wilson, S. A., Wachira, S. J., Drew, R. E., Jones, D. T., and Pearl, L. H. (1993). *EMBO J.*, **12**, 3637.
74. Brunger, A. T. (1990). *Acta Crystallogr.*, **46**, 46.
75. Murzin, A. G. (1993). *EMBO J.*, **12**, 861.
76. Dobson, (1991). *Curr. Biol.*, **1**, 22.
77. Kraulis, P. J. (1991). *J. Appl. Crystallogr.*, **24**, 946.
78. Lattman, E. E. (1995). *Protein structure prediction: a special issue. Proteins*, **23**, 295.
79. Lemer, C. M. R., Rooman, M. J., and Wodak, S. J. (1995). *Proteins*, **23**, 337.
80. Jones, D. T., Miller, R. T., and Thornton, J. M. (1995). *Proteins*, **23**, 387.

9

The combinatorial approach

FRED E. COHEN and SCOTT R. PRESNELL

1. Introduction

With the advent of protein X-ray crystallography in the 1950s and the more recent use of multidimensional NMR spectroscopy, biologists have become familiar with the detailed organization of protein structures at the atomic level. These structures have fundamentally changed our understanding of mechanistic enzymology and molecular recognition in the immune system. More recently, the ability of structural studies to influence drug discovery and design has become clear (1). In spite of these advances, the rate of protein sequence determination via DNA sequencing vastly exceeds the rate of structure determination. How can this gap between sequence and structure be bridged? With an increase in the number of proteins of known structure, the likelihood that a new sequence will be similar to a known structure increases. Currently, ~ 20% of the sequences identified in genome sequencing efforts are related to well-characterized proteins (2). Whenever possible, homology-based model building provides an accurate and reliable entry into structural information about the sequence of interest (Chapter 6). Unfortunately, most newly-determined sequences share insufficient similarity with the structurally studied group that little if any additional information can be derived unless *de novo* methods are applied. *De novo* methods can take two forms. Energy calculations define a representation of the chain and an energy associated with a particular conformation. Structures of minimum energy, preferably minimum free energy are sought. Detailed representations of the chain that specify the coordinates of all atoms are in principle, the most accurate. However, computational restrictions prevent an adequate search of conformational space with this detailed representation. Extremely simplified representations of the chain, e.g. one sphere per amino acid residue, with conformations confined to those embeddable on a cubic grid or lattice, may overcome the sampling problem but accurate energy functions are difficult to construct and validate. Efforts to endow these simplified representations with sufficiently protein-like behaviour are underway. The second approach is more geometric in nature. The secondary structure of the chain is predicted and rules for the assembly of secondary structure elements are used to construct plausible tertiary structures. Many combinations of α-helices or β-stands are plausible and so many structures result. The likely structures must then be sorted from their unlikely counterparts. The advantages of the *de novo* methods are that they can be applied in all cases, that the result can be a novel, heretofore unknown structure, and that success in building these methods would have important implications for the field of protein design and protein engineering. The major disadvantages are that these methods can be difficult to apply and that they rarely lead to a single predicted structure. The best models are likely to be 4–5 Å rms away from the crystal or NMR structure. This level of accuracy is adequate for many immunology and mutagenesis studies, but is insufficient for enzymology or structure-based drug design.

1.1 Overview of the hierarchical approach

To date, the most useful *de novo* structure predictions have come from the combinatorial geometric approach. In particular, the most successful predictions have been performed on all α-helical proteins. The first step in tertiary structure prediction with this approach is to identify the location of the secondary structure elements (see also Chapter 4). While it is not critical to identify the precise boundaries of the α-helices and β-strands, it is crucial to identify all of the features that contribute to the core of the molecule. Most general purpose secondary structure prediction algorithms are accurate ~ 65% of the time. This is insufficient for subsequent tertiary structure prediction work. In an effort to overcome this limitation, two approaches have been developed. The first asserts that it is possible to identify the folding class of a protein (α/α, α/β, β/β, α + β) and then exploit this knowledge to identify where the secondary structure must lie. For α/α proteins, 80% prediction accuracy has been achieved with a neural network and similar results have been seen with a pattern recognition method (3,4). Another promising method depends upon the alignment of many homologous sequences (5,6). Algorithms have been developed by several groups to exploit the pattern of positional residue variation as well as consensus residue character to improve secondary structure prediction. Some of these methods have been automated and seem to be sufficiently accurate to contribute to meaningful tertiary structure predictions.

Combinatorial packing algorithms have been developed that explore all possible geometrically sensible tertiary arrangements of secondary structure elements. The challenge then shifts to sorting between these alternative structures. At the most simplistic level, structures that would disrupt the connectivity of the chain or create other steric conflicts can be eliminated from further consideration. The molecular surface of the alternative structures can be evaluated and those with unusual ratios of surface area to molecular volume can be excluded (7). Recently, it has become clear that the quality of these methods can be improved by the addition of information derived from the analysis of homologous sequences. For example, residue conservation patterns may point out residues that are likely to be close together in space. The most common example of this feature is the identification of the active site residues as a sequentially distinct but spatially proximal set of highly conserved hydrophilic amino acids.

Experimental studies provide a second logical avenue to sort incorrect alternative structures from their native-like counterparts. Most commonly, these take the form of distance constraints on the spatial separation between sequentially distant residues. For example, interleukin 4 (IL-4) contains three disulfide bridges joining six cysteine residues along the chain (8). From experimental data, the precise bridging of the cysteines is known. Combinatorially generated structures that fail to allow for these three bridges cannot be acceptable models for the correct three-dimensional structure. Although myoglobin lacks disulfide bridges, distance constraints can be placed on the separation of the proximal and distal histidine residues that frame the haem binding pocket. Model structures that could not hold on to the haem group would be poor candidates for the tertiary structure of myoglobin (9).

Recently, Jin *et al.* (10) determined the epitopes for a large group of monoclonal antibodies that bind to conformational determinants on the surface of human growth hormone (hGH) and used these data on the proximity of sequentially distant spatially proximal residues to sort the correct combinatorially generated model structure from a list of ~ 1400 alternatives. Side chain participation in an antibody epitope implies that all of the residues are within 20 Å of one another. Moreover, these epitope side chains are within a pie-shaped wedge of the molecule with a central angle of < 120°. In general, each antibody epitope is consistent with only one-half of the combinatorially generated hGH model structures. Only four structures are compatible with the constraints imposed by a panel of ten monoclonal antibodies. These structures are, on average, 4.2 Å rms away from the known crystal structure of hGH.

2. Secondary structure prediction with tertiary intent

Chapter 4 reviews approaches to the problem of relating a protein amino acid sequence to its secondary structure. Most methods rely on largely local information. This, in part, explains the limited accuracy of these algorithms (~ 65% when applied to individual sequences). In an attempt to overcome these limitations, it is necessary to incorporate tertiary information into secondary structure prediction and perhaps focus on the most regular secondary structures that constitute the hydrophobic core of the molecule. For example, a secondary structure prediction that resulted in an isolated β-strand would present a problem for creating a β-sheet. A more subtle case might yield the prediction of three or five α-helices in a protein that could form a four-helix bundle. Globular proteins have a hydrophobic core and hydrophilic exterior. Predicted secondary structures should be no longer than the diameter of the globule with a distribution of hydrophobic and hydrophilic residues compatible with the preferences of the core and exterior of the molecule. While the best secondary structure prediction may violate some of the conformational tendencies of individual residues or small clusters of residues along the chain, it should facilitate the creation of a tertiary structure that is consistent with the known preferences of globular proteins.

Experimental characterization of folding intermediates suggests that some respectable level of secondary structure is formed early in the kinetic protein folding pathway (11) before chains have reached their final, compact states. A variety of experimental and theoretical studies document the importance of sequentially long-range interactions in protein folding (12). The most reasonable interpretations of both theory and experiment in the area leads to the conjecture that some regular secondary structure is locally specified, and hence may fold quickly, while other regular structure is context-dependent (13).

2.1 Class-dependent secondary structure prediction

Taxonomic studies of protein structure reveals the existence of several well-defined structural classes: α/α (e.g. haemerythrin and myoglobin), β/β (e.g. superoxide dismutase and the immunoglobulins), α/β (e.g. flavodoxin and triosephosphate isomerase), and α + β (e.g. lysozyme and ribonuclease). Within each structural class, the likely independent folding domains have a similar size. In addition, the lengths of the α-helices are comparable. A similar observation obtains for the β-strands. Thus, knowledge of a proteins structural class provides useful constraints for subsequent secondary (14) and tertiary structure prediction.

Prediction of secondary structure content has proven far easier than secondary structure prediction. For example, Sheridan *et al.* (15) developed an algorithm to predict secondary structure content and hence protein structural class from a consideration of protein composition, chain length, and dipeptide patterns that was 85% accurate. Muskal and Kim (16) exploited a tandem neural network to improve the accuracy of protein class prediction even further. In the absence of these computational approaches, circular dichroism spectroscopy offers an experimental route to protein class determination (17).

2.2 Pattern-based secondary structure prediction

Secondary structure prediction techniques are of two types: statistical or numerical-based methods and symbolic or pattern-based methods. From a study of proteins of known structure, Chou and Fasman (18) were the first to develop statistics that approximated the propensity of each of the amino acids to adopt a particular backbone geometry. Garnier *et al.* (14,19) used an information theoretic approach to quantitate the impact of sequential neighbours on the conformation of an individual residue. These two algorithms are two of the most commonly used today in commercial sequence analysis packages. In the seminal works of Lim (20,21) specific sequence patterns of residues in primary sequence were postulated for the formation and continuation of α -helix and β-sheet structure. In addition, specific patterns were also recognized as structure 'breaking'.

2.2.1 Turns

Protein class information can be used to specific advantage in secondary structure prediction. Globular proteins can be represented as a chain of residues constrained within a spherical boundary. Amino acid chain length provides an estimate of the sphere radius required to encapsulate the protein. Finally, the pitch per residue (α-helix = 1.5 Å/residue, β-strand = 3.0 Å/residue) then allows a calculation of a 'link length': the expected number of residues between changes in chain direction (*Table 1*). In the pattern matching work of Cohen *et al.* (22) this link length information was incorporated into the patterns that describe turn location. A different link length was used for each of the three protein structural classes In this work, 'a pattern language for amino and nucleic acid sequences', PLANS, was developed specifically for the task of describing and identifying regular expression patterns of residues in primary sequence. As the strongest turn signals are recognized, those turns are so labelled, and the region around that location is hidden from further consideration as a potential turn site. Such strong turn sites might include regions containing a proline residue or local maxima in average side chain hydrophilicity. Weaker turn signals, such as weakly hydrophobic segments, are recognized in the sequential context of the previously identified strong turns. With this heirarchical algorithm, sensitive turn location is possible (90% of turns are found).

2.2.2 Alpha-helices

Presnell *et al.* (23) also applied these concepts to the pattern-based prediction of α-helices in α/α proteins. Helices can be dissected into three component parts: an amino-terminal capping region (N-cap), the core of the helix, and a carboxy-terminal cap (C-cap) (24). Building on the PLANS system for describing regular expressions, patterns were developed independently to recognize these three helical components. Turn patterns from the previous work were also integrated into the system. While PLANS provides a mechanism for developing and expressing patterns in local sequence it cannot easily be used to describe patterns of patterns, or meta-patterns. For meta-pattern expression in this work, 'a language for the prediction of protein substructures', ALPPS, was developed. ALPPS-based meta-patterns were written

Table 1. Domain size and link length

Class	Domain size (residues)	Radius (Å)	Pitch (Å/residue)	Link length (residues)
α/α	150	20.6	1.5	22[a]
α/β	200	22.7	2.0	19
β/β	100	18.0	3.0	9

[a] In the current set of patterns a link length of 26 residues is used.

to recognize the necessary sequential ordering of the component helix patterns. The pattern-based methods for predicting α-helices performs as well as the method of Garnier *et al.* (14) when all residues are considered equally important in the prediction, but not as well as the most advanced neural net systems. However, secondary structure prediction with the intent of tertiary structure prediction focuses more on the existence and location of the core features of a protein and the core residues of each of those features. When scoring of helical caps is devalued relative to that of helical cores (a scoring technique referred to as trimming), pattern-based methods perform 3% better than the method of Garnier *et al.* (14). Cap trimming increases the scores of neural network-based methods in an analogous fashion.

2.3 Neural networks

Recently, machine learning techniques have been applied to the problem of determining sequence structure correlations. As noted earlier in this chapter, artificial neural networks (ANNs) have proven adept at recognizing and categorizing regular secondary structure.

ANN tools are usually made up of a static network of nodes and arcs, analogous in concept to neurones and synapses in biological neural networks (25). In the most simple case, input nodes are connected to output nodes via weighted links: this form of network is called a perceptron (see *Figure 1*) (26). Network architectures may be made more complex through the inclusion of intermediate or 'hidden' layers that contain mediating nodes. Under other protocols, network arc connections may be dynamically determined.

Before ANNs can be used to recognize a sequence concept, they must first be taught. Training of neural networks involves presenting a series of examples to the input nodes while minimizing the difference between the predicted and actual result measured at the output node(s). The minimization is performed by adjusting the strength of the many arc weights between the input and output nodes, as well as adjusting a threshold on the output nodes. A technique called the 'jack-knife' procedure is often used in training ANNs. In this method, all sequences, save one are used to train the network until a concept converges. The resulting connective weight set is used to evaluate

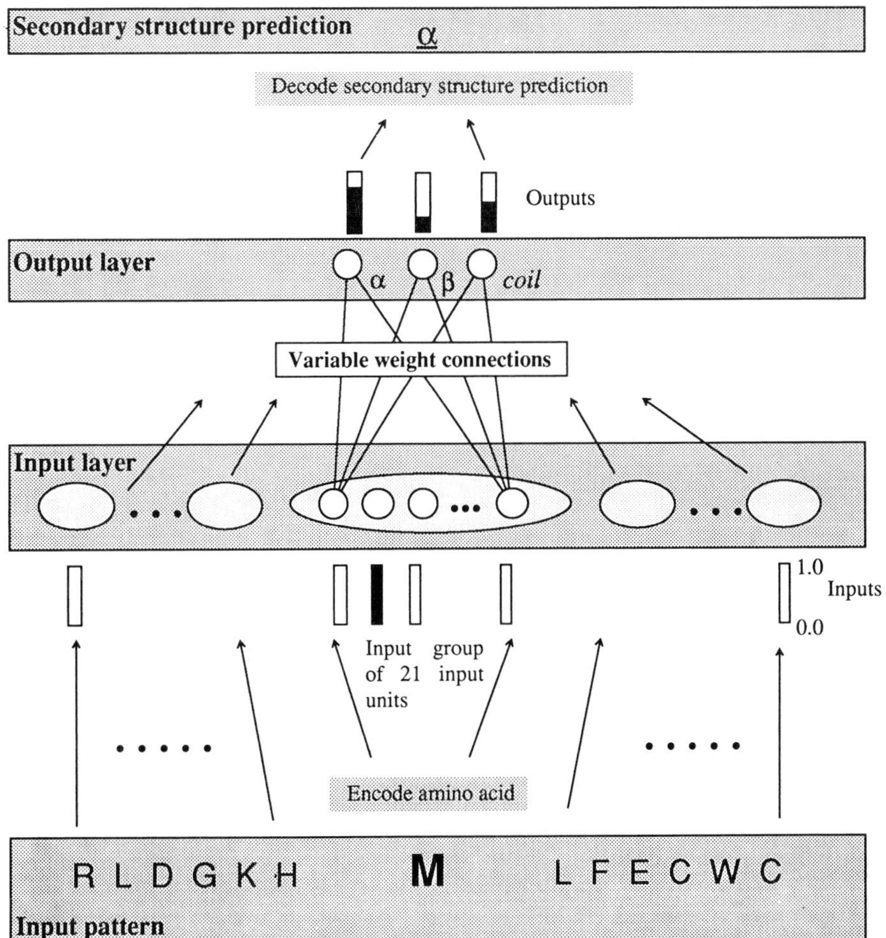

Figure 1. The secondary structure neural network. The input pattern is a sequence of amino acids centred around a central amino acid. Each amino acid is mapped to one input group, which is a collection of 21 units. Each amino acid causes an input of 1.0 to one of the units of its group and an input of 0.0 to the other units. We typically use the six amino acid residues on each side of the central amino acid, for a total of 13 x 21 = 273 input units. There are three output units: helix, strand, and coil. Each input unit is connected to each output unit and the output unit with the greatest output is taken to be the secondary structure prediction for the central amino acid. Additional input units may be accommodated.

the single saved sequence. This procedure is repeated for each sequence in the data set. The resulting weight sets may be averaged, or the best one used in prediction runs of the network. The width of the input layer may be varied in the course of designing a network, but it is fixed during and after training. This width is usually referred to as the 'window' of the ANN: in these systems it is the number of residues that may be considered at any one time. This places a hard limit on the amount of local information considered. Current ANN methods for secondary structure prediction use windows of 13 to 17 residues.

One of the difficulties in the construction and training of ANNs is the issue of memorization. Under ideal conditions, networks should contain data which represents abstract knowledge. However, if the number of independently variable weights is of the same order of magnitude as the number of training examples, the network will have the capability to memorize the specific features of the training examples rather than generalize a concept. Muskal and Kim (16) addressed this problem in their work on ANNs for determining protein secondary structure content, but in general this issue remains an open one in ANN research.

The accuracy obtained in the original work by Qian and Sejnowski (27) on the application of ANNs to secondary structure prediction were significantly better than the classical method of Chou and Fasman (28) but not as successful as the methods available from Garnier *et al.* (14) as compiled in Presnell *et al.* (23). Since the initial work in ANNs applied to sequence structure correlates, new approaches have been proposed and implemented. Kneller *et al.* (3) reconsidered the goal of the networks by training separate networks for the three major protein structural classes in a fashion analogous to Cohen *et al.* (22). In addition, hydrophobic periodicity information was made available to the network. In the case of α/α proteins, supplying the hydrophobic moment as an input resulted in a significant increase (6%) in the predictive power of the system. Periodic information also increased the accuracy of networks trained on α/β proteins, but not for the β/β proteins.

2.4 Utilization of information from homologous sequences

Protein structure evolves more slowly than sequence. Thus, an homologous family of aligned sequences drawn from a phylogenetically large range of organisms should highlight the important features of the sequence while the structurally or funtionally unimportant parts of the sequence should contain amino acid insertions, deletions, or a broad set of chemically diverse mutations. For example, turns should correspond to hydrophilic regions of the aligned sequence with significant variability while secondary structure elements should be subject to more strict sequence conservation, especially in the buried portions of the structure. Active site residues tend to be absolutely conserved hydrophilic residues from a restricted set of catalytically useful side chains. Secondary structure prediction on any individual sequence is likely to err 30–40% of the time at any point along the sequence. However, when these methods are applied to a broad family of sequences, the errors tend to distribute over the entire sequence allowing the useful secondary structure signals to emerge (5,6).

2.5 Interleukin 4: a worked example

The prediction of the structure of interleukin 4 provides a useful example of the secondary structure prediction techniques presented in this chapter. While the use of secondary structure prediction tools can be influenced by personal choice, this example provides a good sense of the common problems encountered in this task.

In this example, sequences from two species, mouse (mIL-4) and man (hIL-4), are examined and exploited through the assumption that their three-dimensional structures will be similar. Initially both sequences are examined for possible turn locations using the pattern-based method of Cohen *et al.* (22) (see *Figure 2*, mIL-4tu and hIL-4tu). Turn prediction allows the separation of the sequence into structural blocks that may be further scrutinized for regular secondary structure elements. In the context of a simple alignment of the two sequences, turns or loop regions are strongly indicated at residues 20, 40, 59, and 123. Turns at 76 and 97 have analogues in the mouse sequence, but appear more variable in exact location. Insertions and deletions in the alignments also indicate possible structural flexibility. From the six turn locations, we can identify six blocks of sequence to consider (the last turn is located very close to the end of the sequence).

Using pattern-based helix prediction methods of Presnell *et al.* (23), five of the blocks contain the signatures of a possible helix (see *Figure 2*, mIL-4ha and hIL-4ha). Five putative helices can be assigned (A–E). The fourth block is devoid of helical signals. The helix capping patterns (mIL-4ca, hIL-4ca; n = N-cap, c = C-cap) are not as strong for the human sequence as has been observed in other cases (23). Helix capping patterns were more pronounced in the mouse sequence. Capping patterns in the centre of blocks are often ignored when helix indications are strong both before and after these signals (blocks D and E). By examining the pattern-based predictions of both sequences and the neural network predictions (mIL-4nn, hIL-4nn) (3), the locations of the helix end-points can be determined. Putative helix extents were identified as: A = 4–17; B = 23–35; C = 42–58; D = 77–86; E = 108–122.

Circular dichroism studies of IL-4 indicated that the protein was likely to contain exclusively α-helical structure. For a variety of reasons, we suspected that IL-4 would be a four-helix bundle. With these constraints in mind, there was an 'extra' core helix predicted for IL-4. Originally, the putative helix B was suspect because of a proline located in the middle of the putative helix in the mouse sequence. When structure building calculations were performed, different groupings of four helices were used to construct likely structures: those groupings that did not include putative helix B, gave the most reasonable structures. Genetically re-engineered chimeric proteins were constructed in which the putative B helix region from the mouse sequence replaced the same region in the human sequence. The result of these experiments also suggested the putative B helix was not critical to the core

```
                Block A          Block B          Block C          4th          Block D                           Block E
         1             15 16             30 31             45 46             60 61             75 76             90 91            105 106            120 121            135
mIL-4ca  ..............c ........cn..... n...........nn. ............... ..c.-n...c..... ............... .-------......- --......n...... ....c.....
mIL-4tu  ...t........... ........t...... ...........t... ............... ..t.-.....t.... ............... .-------......- --..t.......... ..........
mIL-4ha  .........hhhhhh hhhh..........h .hhhhhhh....... ..hhhhhhhhhhhhh hhhh-.....hhhhh hh.hhhhhhhhhhhh h-------......- --...hhhhhhhhhh hhhhhh....
mIL-4nn  ....h..hhhhhhhh hhhhh.........h ..h.hhhh.....hh hhhhhhhhhhhhhhh ....-........hh hhhhhhhhhhhhhhh h-------hh.hhh- --hhh..h.hhhhhh hhhhhhhh..

mIL-4    HIHGCDKNHLREIIG ILNEVTGEGTPCTEM DVPNVLTATKNTTES ELVCRASKVLRIFYL KHGK-TPCLKKNSSV LMELQRLFRAFRCLD S-------SISCTM- --NESKSTSLKDFLE SLKSIMQMDY
hIL-4    --HKCD-ITLQEIIK TLNSLTEQKTLCTEL TVTDIFAASKNTTEK ETFCRAATVLRQFYS HHEKDTRCLGATAQQ FHRHKQLIRFLKRLD RNLWGLAGLNSCPVK EANQS---TLENFLE RLKTIMREKYSKCS
hIL-4nn  --...h-hhhhhhhh hhh.hhhhhhhhhhh hhhhhhhh.....hh hhhhhhhhhhhhhh. h.hh.hhhhhhhhhh hhhhhhhhhhhhhhh hhhhhhh......hh hh.hh---hhhhhhh hhhhhhhhh......
hIL-4ha  --....-..hhhhhh h.............. .hhhhhhh......h hhhhhhhhhhhh... ............... ......hhhhhhhhh hhhhh.......... .....---.hhhhhh hhhhhhhhhhhhh..

hIL-4tu  --....-........ .......t....... ............t.. ............... t.............. ...t........... ...........t... .....---....... .....t.........
hIL-4ca  --....-........ ............... ............... ............... .n............. ............... ..n............ .....---....... ..c............
```

Figure 2. Secondary structure prediction of the mouse (mIL-4) and human (hIL-4) sequence. The ca line contains helical N-cap (n) and C-cap (c) predictions. The tu line locates turns and the ha line locates helical core regions. The nn line contains the neural network prediction for an all-helical protein. The sequences are shown with the usual one letter amino acid code. The blocks indicate an approximate parsing of the sequence into structural units.

Table 2. Interleukin 4 secondary structure: model versus experiment

Helix	Model	NMR	Difference
A	4–17	5–17	+1, 0
C	42–58	41–57	−1, −1
D	77–86	72–90	−5, +4
E	108–122	110–125	+2, +3

structure of the protein. (Much of the original work from this example is discussed in detail in ref. 8.)

Results from recent NMR work (29) show that the predictions for the A, C, D, and E helix were fairly accurate (see *Table 2*). The primary location of the four helices was approximately correct. The most significant difference is in helix D, which was predicted to be one turn too short on either end.

3. Tertiary structure from secondary structure

On its own, information about the location of secondary structure elements provides little useful information on the overall folded structure of the protein. In order to exploit successful secondary structure predictions, a strategy for generating tertiary structures is required. α-Helices and β-sheets have a predictable distribution of ridges and valleys that make some packing arrangements much more likely than others. If one can identify preferential packing sites along the surfaces of these secondary structures (e.g. a hydrophobic patch on an α-helix), then the number of plausible tertiary arrangements of secondary structure elements is more manageable. Even with these constraints, a large number of possible structures remain. Consider the case of four α-helices forming a bundle. Six types of four-helix bundles are known (see *Figure 3*). For each type, there are 4! or 24 possible ways to label the helices and each helix could point up or down (2^4 directions). For this simple case 2304 helical bundle topologies are possible. Plausible variations in the details of helix–helix packing allow this number to grow geometrically.

3.1 Identification of tertiary packing sites on secondary structures

3.1.1 All-helical proteins

The ridges and valleys on the surface of an α-helix can be arranged to create two plausible helix–helix pairing geometries, one at +20° (Type III) and a second at −70°. A review of the second interaction class suggests that it involves two subclasses, one with smaller residues at the interaction centre and an angle of −80° (Type I) and the other with larger residues at the

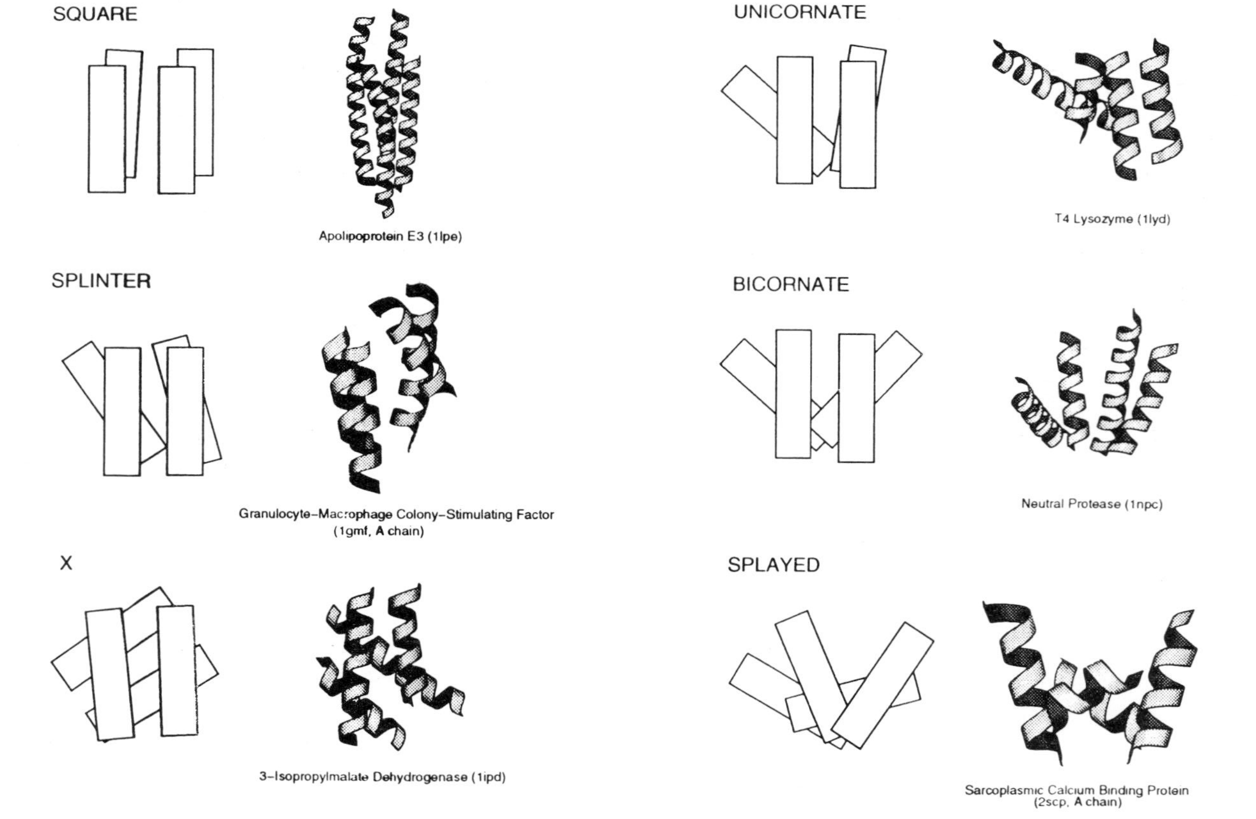

Figure 3. There are six classes of bundles as defined by the pattern of interhelical angles. Prototypes and representative bundles of each type are shown.

interaction centre and an interhelical angle of −60° (Type II). Richmond and Richards (30) developed an algorithm to detect plausible helix–helix packing sites by looking for residue clusters on the face of an α-helix composed largely of hydrophobic residues (see *Figure 4*). The importance of the interaction site could be related to the potential solvent accessible surface area loss on helix–helix packing. Type I, II, and III packing sites are identified with the program AAPATCH. Occasionally, residues spaced by four along the chain (i and $i + 4$) are identified as possible packing sites. In general this is redundant information and the site list can be reduced by eliminating either site without loss of generality (31).

3.1.2 All-beta proteins

Current methods are limited to the study of β-sandwiches (e.g. immunoglobulin-like molecules or superoxide dismutase) and will not handle β-barrels (e.g. serine proteases) or trefoil structures (e.g. interleukin 1, soybean trypsin inhibitor) (32). For each strand, the packing face is identified by determining whether the even residues or the odd residues contain a longer or stronger cluster of hydrophobic residues. Lysine residues are considered hydrophobic

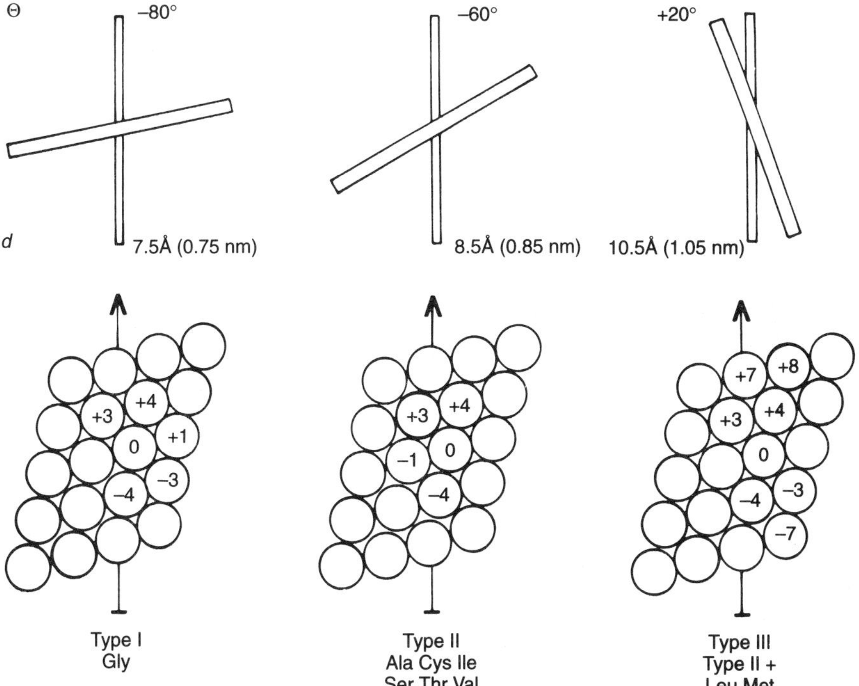

Figure 4. The typical dehedral angle (Θ) and separation (*d*) between a pair of α-helix axes is given for each class. The residues on the surface of an α-helix are represented by circles and a typical pattern of residues for each interaction are described.

if the strand is on the edge of a β-sheet and hydrophilic if the strand is in the middle of the sheet. Strands lacking a pair of hydrophobic amino acids spaced by two residues (i and $i + 2$) are not allowed in the core β-sandwich structure.

3.1.3 Alpha/beta proteins

The α-helix packing sites can be identified in a fashion similar to that used for α/α proteins. The most hydrophobic cluster of four residues forming a diamond on the helical surface ($i, i + 3, i + 4, i + 7$) defines the α/β packing site (33). The β-strands in α/β proteins tend to be protected from the solvent by α-helices on both sides. Thus, the sequence does not suggest a preferential hydrophobic packing face and so all possible strand orderings and sheet hydrogen bonding patterns must be considered.

3.2 Combinatorial searches through conformational space

3.2.1 All-helical proteins

From the list of all possible helix–helix packing sites, the program AAFOLD generates a list of helical pairings that could produce a folded structure subject to four criteria (31):

- separate sites on the same helix cannot be paired
- no two helices can form more than one pair with each other
- all helices must appear in each list
- site pairs are made only between sites of the same class

AABUILD takes the list of helix interaction site pairings and constructs three-dimensional models of the possible structures. While the interhelix packing angle is specified by the packing type, the relative orientation of the helices remains to be determined. For example, the helices could pack in a parallel or antiparallel fashion. Also, the central packing residues do not point directly toward one another. Instead, packing is optimized by rotating each helix by a ± 25° skew angle about the helix axis (34). Thus, each pair of helices can be placed in eight distinct pairings, two directions times two skew angles for each of two helices. Each three-dimensional model is examined and those that disrupt the connectivity of the chain or violate other steric constraints are eliminated from further consideration.

3.2.2 All-beta proteins

Imagine a β-sandwich core composed of two three stranded β-sheets. Given the correct β-strand assignments from a secondary structure prediction, each strand could be tried in all of the six possible locations in either of two directions with at least three distinct strand registrations. Given the importance of hydrogen bonds in β-sheet organization, only structures with substantial strand overlap in each β-sheet are allowed. Moreover, hydrophobic residues on the interior surfaces of the β-sheets tend to follow certain patterns that enhance sheet–sheet packing (see *Figure 5*). Structures lacking these patterns or failing to comply with a number of topological 'rules' about β-sheet structure (e.g. the handedness of the connection between parallel strands in a β-sheet is right-handed) are eliminated from further consideration (32).

3.2.3 Alpha/beta proteins

In general, α/β proteins are composed of a single β-sheet coated by α-helices on both sides. Under certain conditions, the β-sheet can wrap up on itself to form a barrel coated by α-helices on the external face. α/β Barrels are an extremely common folding motif constrained by a large number of topological, geometric, and energetic constraints and better modelled by homology-based methods. For the planar α/β proteins, a combinatorial strategy similar to the one used to sample plausible β-sheet structures is followed. All possible orderings of strands to form a sheet are considered. While there is a

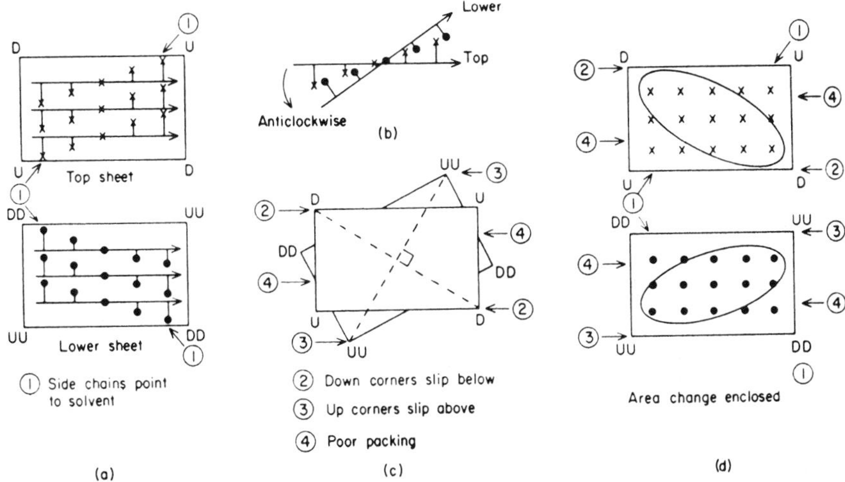

Figure 5. Stereochemistry of β-sheet stacking. (a) Two β-sheets with U and UU indicating the up-pointing regions of the top and lower sheet and with D and DD showing the down-pointing corners of the sheets. The use of U and UU does not suggest a difference in twist between the two sheets. The β-strands run horizontally and the rough direction of the side chains is shown by the vertical line to the cross in the top sheet and to the dot in the lower sheet. (b) Illustrates that when the β-strands in the top sheet lie with an anticlockwise rotation with respect to the strand direction in the lower sheet, the side chains in the two sheets line up parallel. (c) Illustrates the stacking of two sheets with an anticlockwise rotation so the diagonal between the down-pointing corners of the top sheet is perpendicular to the diagonal between the up-pointing corners of the lower sheet. (d) Relates the observed anticomplementary pattern of hydrophobic accessible surface area changes on sheet stacking to the stereochemistry of the interaction. Note that the observed hydrophobic accessible surface area changes cannot be explained by simply superimposing two planar representations for the β-sheets and rotating anticlockwise the β-strand direction of one of the β-sheets.

preference for all of the strands to point in the same direction, this is not necessary. α-Helices are placed following the rule that the connection between parallel strands is right-handed using the diamond-shaped docking site previously described. β-Sheet hydrogen bonding is maximized and a variety of topological constraints are imposed to eliminate unlikely structures from further consideration (33).

4. Sorting between alternative models

The advantage of energy minimization approaches to structure prediction lies in their simplicity, one structure is continuously deformed in search of the best structure. Multiple minima on the energy surface prevents this approach from reaching the 'correct' structure before getting trapped in a local minimum. For the combinatorial structure generation strategy, the multiple minima problem is replaced by the multiple alternative problem. Conformational space is sampled broadly, but the ability to rank order accurately the alternative structures is lacking. While the 'correct' structure is generally a member of the list of combinatorial alternatives, it has proven difficult to avoid 'throwing out the baby with the bath water'. Listed below are a series of methods used to reduce the set of alternative structures while retaining the 'correct' structure.

4.1 Experimental constraints on model structures

Implicit in the success of any tertiary structure prediction is accurate secondary structure prediction. Circular dichroism (CD) and Fourier transform infrared (FTIR) spectroscopy can be used to determine secondary structure content, although they are not helpful in defining the intrachain location of α-helices and β-strands. Thus, if a protein is predicted to contain 50% α-helical structure and no β-strands, CD or FTIR spectroscopy can be used to confirm that this is a sensible prediction or reveal that this is an incorrect starting point (7).

Disulfide bridges are the most common chemical cross-links in protein structures. When present, they define the spatial aposition of residues separated in sequence along the chain. While model structures that allow the correct disulfide bridges to form are not necessarily correct, models that are inconsistent with correct disulfide pairing can be eliminated from further consideration (8).

In a similar way, the existence or creation of metal binding sites can be used to eliminate incorrect structures from consideration. In proteins of known structure, molecular biologists have inserted a pair of loops containing the sequence Asp–His–Asp. These engineered molecules can chelate a divalent cation (e.g. Zn^{2+}) if the loops come together in space. From a list of alternative models, loop insertion experiments can be designed that could separate plausible structures from their implausible counterparts. For example, if only one-half of the models place a loop *i* near loop *j* and the appropriately mutated protein gains the ability to bind Zn^{2+}, then the other one-half of the models can be eliminated. Myoglobin provides another example of this strategy. The globins contain two histidines that interact with the iron atom at the centre of the haem. Thus, models of apomyoglobin that fail to create a haem sized cavity between the proximal and distal histidines are poor models for myoglobin. While combinatorial helix packing calculations suggested 20 alternative structures for apomyoglobin, only two were compatible with the haem constraint and these differed from each other by < 0.3 Å rms and from the crystal structure by 4.4 Å rms (9).

Experimental constraints can cause problems if the structural implication of the experiment is misinterpreted. The structure of tryptophan synthase was modelled in advance of the solution of the crystal structure. Limited proteolysis of tryptophan synthase resulted in an α-fragment of 188 residues that was stable and contained approximately equal amounts of α-helix and β-sheet structure judged by CD. Thus, tryptophan synthase was felt to be a two domain structure. Secondary structure prediction revealed six β-strands and α-helices, and tertiary structure prediction yielded a planar α/β structure with an all-parallel β-sheet for the first 188 residues. Crystallography demonstrated a structure containing 268 residues that formed an α/β barrel with ten helices and eight β-strands forming an all-parallel β-sheet. The limited proteolysis experiment cleaved not at the domain boundary, but in the middle of the α/β barrel domain. The fact that the resulting α-fragment was stable remains most peculiar. However, the interpretation of the proteolysis data as a valid domain boundary resulted in an error that doomed the subsequent secondary and tertiary structure prediction (35).

4.2 Theoretical constraints on model structures

Model structures incorporate sequence information in the identification of secondary structure packing sites. However, current combinatorial construction algorithms treat all side chain groups as if they are isosteric. While this is computationally efficient, it fails to consider side chain packing, side chain/side chain interaction preferences, and side chain/solvent interaction preferences. In an attempt to overcome these difficulties and to further reduce the list of plausible alternative structures, a series of constraints based on theoretical or empirical grounds can be implemented.

The geometric properties of real side chains can distort the quality of the packing interface produced by idealized models of secondary structure interactions. QPACK was developed to assess the quality of packing in model built structures (36). A simplified side chain representation is used that places one sphere at the centroid of aliphatic side chains. For aromatic side chains, spheres are centred at the β-carbon and ring centre(s). The spheres begin as points and are allowed to grow at a rate proportional to their ideal radii.

Sphere growth is terminated when spheres collide with their counterparts on neighbouring side chains. The final side chain volumes obtained in the model structures can be compared to those obtained on average for each amino acid in a series of real structures. In general, model quality correlates with the appropriateness of the side chain volumes.

It is also possible to look at detailed side chain/side chain contacts in the model structures and compare these with the distributions observed in proteins of known structure. Residue–residue interaction energy tables derived from these distributions have been determined by a number of investigators using a variety of related techniques. In general, high quality model structures have appropriately low energy values while their misfolded counterparts are higher energy structures.

Finally, solvent accessibility calculations provide an important check on the quality of model structures. First, some model structures have an inappropriately large solvent accessible surface given their molecular weight. From the work of Teller (37), it is known that the total area for an extended chain is 1.45 M where M is the molecular weight, while the folded chain accessible area is 11.12 $M^{2/3}$. Some combinatorially generated model structures have more surface area than is expected from a folded chain and can be eliminated from further consideration. Secondly, it is possible to evaluate the per residue solvent accessibility and ask if the distribution of fractional accessibility is sensible for each amino acid type. For example, if arginine residues are consistently buried yet phenylalanine residues are frequently exposed, then the model structure is unlikely to be correct. For comparison purposes, distributions of average side chain fractional accessibility have been tabulated by Rose and colleagues (38).

Recently, a variety of empirical and semi-empirical tools have been developed to assess the quality of model-built structures. In general, these methods can identify incorrect structures but cannot be used select only the correct structure from a list of plausible alternatives. In principle, energy calculations should be able to sort between alternative structures and optimize the best model structure. To date, limitations on the convergence properties of these algorithms and perhaps limits with the accuracy of the underlying potential functions have frustrated these more direct approaches to the multiple alternative structure problem.

4.3 Interleukin 4: a worked example

Beginning with the secondary structure prediction of *Table 2*, AAPATCH identified 19 packing sites (Type I: 49; Type II: 10, 11, 13, 24, 28, 29, 30, 32, 48, 49, 51, 80, 119; Type III: 14, 48, 52, 109, 116). AAFOLD and AABUILD generated 90403 compact four-helical bundles. Only 311 satisfied the three disulfide bridge distance constraints (Cys3–Cys127, Cys24–Cys65, and Cys46–Cys99). These structures were then rank ordered by solvent accessible surface contact area. The most compact structures were all antiparallel four-helix bundles with two overhand connections. The best structure is shown schematically in *Figure 6*. A more detailed stereo diagram with the loop regions added is presented in *Figure 7*. The top seven structures were right-handed antiparallel four-helix bundles with two overhand connections. The eighth structure was a left-handed antiparallel four-helix bundle with two overhand connections. While the models of the right-handed structures seemed to create a better hydrophobic core, the left-handed structure bears a closer resemblance to the NMR structure (29). The rms deviation between this model and the NMR structure is 4.6 Å.

5. Conclusion

When confronted with a new sequence, the existence of an homologous protein sequence of known structure provides the most rapid route to a useful

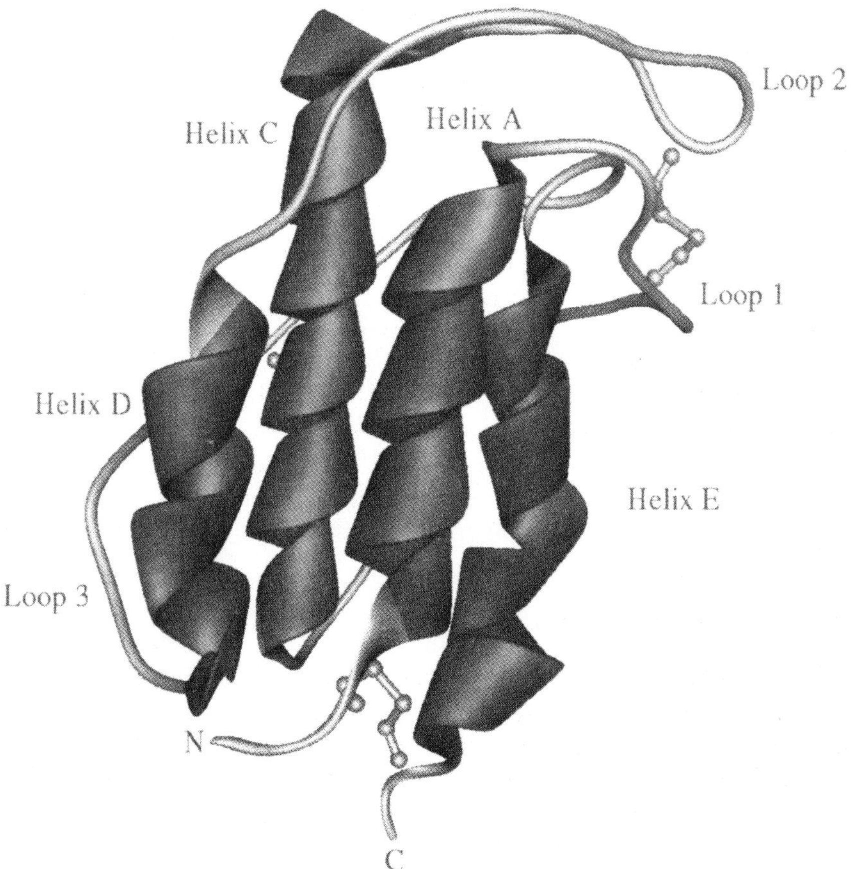

Figure 6. Ribbon diagram of the complete predicted structure of human IL-4.

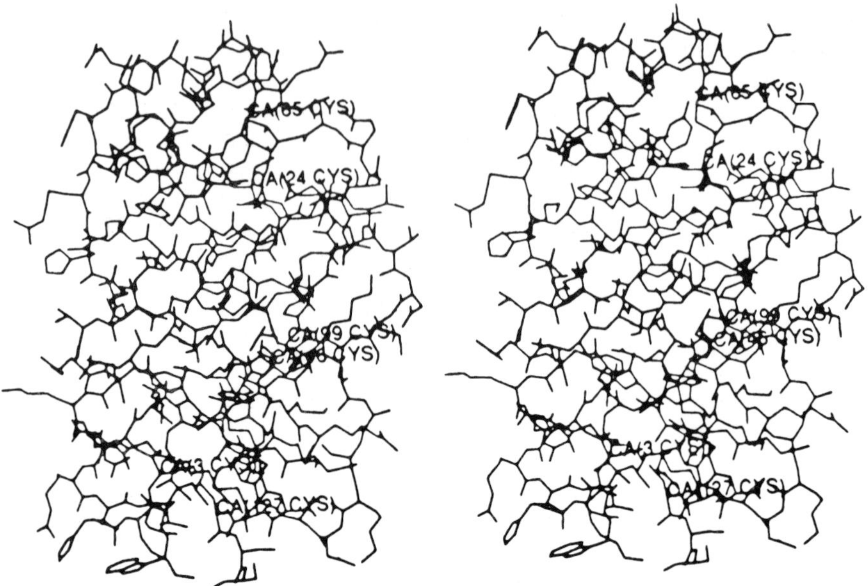

Figure 7. Stereo diagram of the complete predicted structure of human IL-4.

model of the structure of the new sequence. In the absence of this information, *de novo* methods may be used to create an approximate model of a protein tertiary structure. Secondary structure prediction remains a difficult problem, but the existence of sequence information for a family of homologous proteins can be helpful. Combinatorial algorithms exist that effectively explore the geometrically sensible juxtapositions of secondary structure elements to produce an approximate tertiary structure. Additional biochemical and biophysical information can be used to reduce the list of plausible structures. Theoretical models combined with sufficient experimental information can be used to generate useful intermediate resolution models of proteins of unknown structure. Computer algorithms to carry out this work are available on request from the authors.

References

1. Kuntz, I. D. (1992) *Science* **257,** 1078.
2. Pascarella, S. and Argos, P. (1992) *Protein Eng.* **5,** 121.
3. Kneller, D. G., Cohen, F. E., and Langridge, R. (1990) *J. Mol. Biol.* **214,** 171.
4. Presnell, S. R., Cohen, B. I., and Cohen, F. E. (1993) *CABIOS* **9,** 373.
5. Benner, S. A. and Gerloff, D. (1991) *Adv. Enzyme Regul.* **31,** 121.
6. Russell, R., Breed, J., and Barton, G. (1992) *FEBS Lett.* **304,** 15.
7. Cohen, F. E., Kosen, P. A., Kuntz, I. D., Epstein, L. B., Ciardelli, T. L., and Smith, K. A. (1986) *Science* **234,** 349.

8. Curtis, B. M., Presnell, S. R., Srinivasan, S., Sassenfeld, H., Klinke, R., Jeffrey, E., *et al.* (1991) *Proteins: Struct. Funct. Genet.* **11,** 111.
9. Cohen, F. E. and Sternberg, M. J. E. (1980) *J. Mol. Biol.* **137,** 9.
10. Jin, L., Cohen, F., and Wells, J. (1994) *Proc. Natl. Acad. Sci. USA* **91,** 113.
11. Hughson, F. M., Wright, P. E., and Baldwin, R. L. (1990) *Science* **249,** 1544.
12. Dill, K. A. (1990) *Biochemistry* **29,** 7133.
13. Cohen, B. I., Presnell, S. R., and Cohen, F. E. (1993) *Protein Sci.* **2,** 2134.
14. Garnier, J. R., Osguthorpe, D. J., and Robson, B. (1978) *J. Mol. Biol.* **120,** 97.
15. Sheridan, R. P., Dixon, J. S., and Venkataraghavan, R. (1985) *Biopolymers* **24,** 1995.
16. Muskal, S. M. and Kim, S.-H. (1992) *J. Mol. Biol.* **225,** 713.
17. Johnson, W. C. (1992) In *Methods in enzymology* (ed. J. N. Abelson and M. I. Simon), Vol, 210, pp. 426–47.
18. Chou, P. Y. and Fasman, G. D. (1974) *Biochemistry* **13,** 222.
19. Gibrat, J. F., Garnier, J. R., and Robson, B. (1987) *J. Mol. Biol.* **198,** 425.
20. Lim, V. I. (1974) *J. Mol. Biol.* **88,** 873.
21. Lim, V. I. (1974) *J. Mol. Biol.* **88,** 857.
22. Cohen, F. E., Abarbanel, R. M., Kuntz, I. D., and Fletterick, R. J. (1986) *Biochemistry* **25,** 266.
23. Presnell, S. R., Cohen, B. I., and Cohen, F. E. (1992) *Biochemistry* **31,** 983.
24. Richardson, J. S. and Richardson, D. C. (1988) *Proteins* **4,** 229.
25. McClelland, J. L. and Rumelhart, D. E. (1988) *Explorations in parallel distributed processing*. MIT Press, Cambridge, MA.
26. Rosenblatt, F. (1958) *Psychol. Rev.* **62,** 559.
27. Qian, N. and Sejnowski, T. J. (1988) *J. Mol. Biol.* **202,** 865.
28. Chou, P. Y. and Fasman, G. D. (1978) *Annu. Rev. Biochem.* **47,** 251.
29. Smith, L. J., Redfield, A. C., Boyd, A. J., Lawrence, G. M., Edwards, R., Smith, R. A., *et al.* (1992) *J. Mol. Biol.* **224,** 899.
30. Richmond, T. J. and Richards, F. M. (1978) *J. Mol. Biol.* **119,** 537.
31. Cohen, F. E., Richmond, T. J., and Richards, F. M. (1979) *J. Mol. Biol.* **132,** 275.
32. Cohen, F. E., Sternberg, M. J. E., and Taylor, W. R. (1980) *Nature (London)* **285,** 378.
33. Cohen, F. E., Sternberg, M. J. E., and Taylor, W. R. (1982) *J. Mol. Biol.* **156,** 821.
34. Harris, N. L., Presnell, S. R., and Cohen, F. E. (1994) *J. Mol. Biol.* **236,** 1356.
35. Hurle, M. R., Matthews, C. R., Cohen, F. C., Kuntz, I. D., Toumadje, A., and Johnson, W. C. (1987) *Proteins: Struct. Funct. Genet.* **2,** 162.
36. Gregoret, L. M. and Cohen, F. E. (1990) *J. Mol. Biol.* **211,** 959.
37. Teller, D. (1976) *Nature* **260,** 729.
38. Rose, G., Geselowitz, A., Lesser, G., Lee, R., and Zehfus, M. (1985) *Science* **229,** 834.

10

Modelling protein conformation by molecular mechanics and dynamics

MARY E. KARPEN and CHARLES L. BROOKS III

1. Introduction

Molecular mechanics and dynamics are used extensively in protein structure prediction, complementing experimental approaches and database ('knowledge-based') analyses. These methods are able to predict conformation at the atomic level. This high degree of resolution, however, carries a large computational cost. Hence limits are often placed on system size, the extent of conformational space searched, and the length of simulations. None the less, molecular mechanics and dynamics approaches are able to address many important protein structure prediction problems.

Protein structure prediction is a multifaceted problem which encompasses a wide range of scientific questions. The classic prediction problem is that of folding an amino acid sequence into its native tertiary conformation. Other prediction problems of a more limited scope are also of interest. For example, predicting protein dynamics given a static protein structure is often crucial to understanding protein structure and function (1). Another example is the prediction of local regions of protein conformation, important in site-directed mutagenesis and homology model building.

Molecular mechanics and dynamics simulations have been applied to all of these prediction problems, using approaches which reduce the computational costs. Current computational resources do not allow *ab initio* protein folding given only the knowledge of the amino acid sequence, but this folding can be accomplished via molecular dynamics with additional information such as distance restraints (for example, see refs 2–5). The additional information restrains the search to relevant conformational space, reducing the computational complexity of the task. A restrained search space is also used when predicting local protein conformation, where only the region of interest is allowed to move. Certain aspects of protein structure can be modelled using simpler atomic systems, such as peptides (6,7). In addition to restraints on search space and system size, the length of a simulation is also chosen to contain computational cost. Current computer resources allow protein motion to modelled into the 10–100 nanosecond time-scales using molecular dynamics.

Two methods form the basis of molecular mechanics: energy minimization and molecular dynamics. Both methods use a potential energy function to describe atomic interactions, which are modelled using relatively simple mathematical forms of bonded and non-bonded interactions. Dynamics are calculated using Newton's equations of motions.

Due to the size of biological macromolecules, quantum mechanical treatments of protein conformation and dynamics are not feasible in all but the most localized instances (8–10). For many applications, valuable insights into protein structure can be gained from classical mechanics models. Furthermore, these models often provide a more accurate representation of conformational energetics than presently available from quantum chemical approaches (8). Solvated systems, which require simulations of large numbers of atoms, can currently only be treated using classical mechanics methods.

In energy minimization, a minimum in the potential energy function is determined in an iterative fashion given an initial starting conformation. The potential energy of a large system of atoms can be minimized in this way, using relatively small amounts of computational power. The method is limited in the information it can provide in that no protein dynamics are considered. Also, the resulting minimum may be a local minimum rather than a true global minimum, a classical problem with all minimizations. Hence the minimum found depends on the initial conditions of the system.

In molecular dynamics (MD), the derivative of the potential energy function with respect to atomic position is used to determine the forces on each atom in the system. The acceleration of each atom is then determined from these forces using Newton's equations of motion. Given an initial atomic position and velocity, integration of the acceleration is carried out to obtain a new atomic position. In this way molecular dynamics simulations compute the time evolution of an atomic system.

Molecular dynamics can be used in conjunction with energy minimization to avoid getting trapped in local minima on the potential energy surface; simulated annealing, used extensively in crystallographic refinement, is one such example (11,12). Molecular dynamics plays an even more fundamental role in protein structure prediction. Only limited knowledge is gained from a static treatment of protein conformation or states via energy minimization. Protein motion is often intrinsically involved in the function of a protein (1) and hence molecular dynamics simulations are required. Molecular dynamics is also used as an efficient 'ensemble sampler' in free energy simulations, i.e. as a means to generate a large number of relevant conformations.

Free energy simulations are becoming an increasingly important tool in protein structure prediction. A minimum in potential energy, even if it is the global minimum, may not correspond to a minimum in the free energy of a system since entropic terms are ignored. In order to include entropy, free energy calculations require a large number of protein states from which statistics can be accumulated. To this end, molecular dynamics are performed to generate a large number of conformations, needed for the convergence of the thermodynamic properties. These simulations are very computationally intensive and are currently applied mainly to peptide systems or local regions of protein structure (including drug binding sites).

Since the late 1970s, molecular mechanics have been used to study biomolecular systems (13). As computational power has increased, so has the breadth of molecular mechanics applications; many reviews exist that examine various aspects of the field (1,6–9,14–18). In the spirit of 'a practical approach', here we have chosen a set of informative applications which represent a range of protein structure prediction problems. These are described briefly below, and in more depth in the following sections.

1.1 Protein dynamics

Proteins are inherently dynamical molecules, a point which is sometimes overlooked given the most common protein structure is a single average structure obtained from X-ray crystal diffraction studies (13). Conformational dynamics have been found in many cases to affect directly the function of a protein (1,8,14,19) and we examine molecular dynamics simulations which support this finding.

We start with a brief review of general aspects of protein dynamics, and present studies using various methods to identify correlated motion within a protein. Correlated motion is found to occur within structural units, such as secondary structures, and can lead to conformational change. We then present several studies which demonstrate the importance of protein motion to function. First, we discuss several simulations of the HIV-1 protease, which give information on functionally important collective motion (20–22). We then examine a recent simulation of the enzyme acetylcholinesterase, important in neurone synaptic junctions (23,24), where a long, narrow channel to the active site in the crystal structure did not appear to allow the high catalytic rate observed for this protein. Molecular dynamics simulations revealed a 'back door' path to the active site, which dynamically opens to potentially allow passage of products, reactants, and solvent, thereby effectively increasing the reaction rate.

1.2 Peptides

An important and useful approach to problems in computational chemistry is to use a smaller, simpler molecular system for more intensive computational effort and either extrapolate the results to a larger system or use the results to 'bootstrap' to a more complex system, where the problem is treated using a simpler computational model (9). This approach has been used to examine the relationship between protein secondary structure and sequence (6,7), where peptides are used as models for initial protein folding of secondary structure elements. We will review several studies performed in our group, which aim to calculate the free energy surface of various peptides as a function of conformation (25–28). Comparisons of results from different amino acid sequences show how sequence affects one of the most common protein conformations, the α-helix. Results from these simple systems have implications for helix–coil transitions in proteins.

1.3 Protein stability

Site-directed mutagenesis is a powerful tool in molecular biology, used not only to understand natural protein structures but also in protein design and engineering. Prediction of mutation effects a priori can aid the experimentalist in identifying key residues for mutation as well as assisting in understanding mutation results at the atomic level. We examine free energy simulations aimed at predicting protein stability, using examples from T1 ribonuclease (29), subtilisin (30), lysozyme (31,32), and barnase (33). Results from these simulations show how simple analyses of interaction energies can be misleading as to the effect of mutations. Free energy simulations as applied to protein stability are relatively new and include approximations which have not been fully tested (30). We feel this is a valuable and informative approach in spite of current limitations.

1.4 Tertiary protein folding—refinement

Many tools are evolving which have great promise in predicting the topology of a protein structure (see Chapters 8 and 9). In order to obtain atomic level details from these models, a 'refinement' process is needed, similar to refinement used in X-ray crystallography and NMR structure determination. Much work is needed in this field. As a first step, we briefly discuss an example of structure refinement via restrained molecular dynamics (2).

Molecular mechanics and dynamics are also playing an increasingly large role in protein structure prediction based on a homologous protein structure. Loop building in particular has benefited from simulated annealing methods (34–36). We refer the reader to these references and to Chapters 6 and 7 for details of loop building techniques. Another important application of molecular mechanics to protein structure prediction is in protein–ligand binding studies, discussed in Chapter 11.

In the following sections, we first detail the methods used in molecular mechanics studies. We then present the applications to give the reader a feel for the power of the molecular mechanics techniques as well as their current limitations. We end the chapter with our view of where molecular mechanics techniques can lead us in the near future in the quest for protein structure prediction.

2. Methods

We review here the basic methods of energy minimization and dynamics calculations as used in the modelling of protein (and peptide) structure. As noted in the introduction, these methods permit atomic level 'resolution' of protein structure to be investigated. However, the computational cost and the inherent limitations of these techniques are important considerations in assessing the use, and evaluating the outcome, of molecular mechanics and dynamics calculations.

In general, it is useful to view molecular mechanics and dynamics as a means of exploring the structure–energy relationships in biological molecules. As such, the limitations of the methodologies arise from two primary sources. One is the accuracy of the representation, i.e. the potential energy function. The other is the limitation in sampling of the 'important' regions of conformational space of the system of interest. We will first discuss the general form of the potential energy functions used in molecular mechanics calculations on biomolecules and then present the salient features of the methods to manipulate molecular structures on these energy surfaces, e.g. energy minimization and molecular dynamics techniques.

2.1 Potential energy functions

At the core of a molecular mechanics calculation is the representation of the energy given the configuration of a particular biomolecule (the positions of all the constituent atoms). The potential energy function encapsulates this information in a form which is a compromise between ease of evaluation and accuracy. It is empirical in nature, deriving the constants representing bond stretching, angle bending, torsional rotations, and non-bonded interactions predominantly through comparisons with experimental data. In general, the simplest functional form which captures the basic physics of these interactions is currently employed in many potential functions in common usage, e.g. AMBER, CHARMM, GROMOS, and Discover. The functional form of this potential energy function (U) for N atoms, each atom, i, at position r_i is:

$$
\begin{aligned}
U(\boldsymbol{r}_1, \boldsymbol{r}_2, \ldots, \boldsymbol{r}_N) = {} & \frac{1}{2} \sum_{\textbf{bonds}} K^i_{bond}\,(b^i - b^i_0)^2 \\
& + \frac{1}{2} \sum_{\textbf{angles}} K^i_{angle}\,(\theta^i - \theta^i_0)^2 \\
& + \sum_{\textbf{torsion}} K^i_{torsion}\,[1 + \cos(n\phi^i - \delta^i)] \\
& + \sum_{\textbf{non-bonded pairs}} \frac{A_{ij}}{r^{12}_{ij}} - \frac{B_{ij}}{r^{6}_{ij}} + \frac{q_i q_j}{\varepsilon r_{ij}}
\end{aligned}
\qquad [1]
$$

In this expression the empirically derived terms K^i_{bond}, b^i_0, K^i_{angle}, θ^i_0, $K^i_{torsion}$, n, and δ^i correspond, respectively, to the bonded force constant, the reference bond distance, the valence angle force constant, the reference valence angle, the dihedral angle (torsional) potential force constant, the periodicity of the torsional potential (e.g. rotations about a single C–C bond normal have three minima corresponding to $g+$, $g-$, and t, and hence $n = 3$), and finally a phase for the torsional potential. The remaining constants determine the non-bonded forces: A_{ij} and B_{ij} determine the van der Waals forces, and the electrostatic forces are determined by the electrostatic charges, q_i and q_j, and the dielectric constant, ε. See Cornell *et al.* (9) for a review of potential energy functions.

In many instances, modellers have used different representations of the 'dielectric' by modifying the value and form of ε used. In particular, the grossest features of solvation by water can be mimicked by use of a dielectric constant of 78 for the electrostatic interactions. Special forms, such as $\varepsilon = 4r_{ij}\varepsilon_0$ (ε_0 is the vacuum dielectric constant) and other functions of the inter-atomic separation, have been used to represent 'environmental' influences in different circumstances (37,38).

We note that the non-bonded parameters are the most important terms in the potential in determining the overall energetic aspects of folded proteins and peptides. These terms are also the most difficult to parameterize. However, the non-bonded parameters and charges present in the force fields mentioned above have all been extensively tested, and these force fields yield roughly equivalent representations of the molecular interactions. Despite a similar overall agreement of the force fields in total, it is not advisable to mix parameters from differing force fields since the parameters for electrostatics, van der Waals interactions, and torsions are quite strongly coupled. Thus, barriers between adjacent minima, as well as the relative energy of the minima (e.g. α-helix versus extended conformations), can be substantially altered in 'mixed' potentials.

2.2 Energy minimization

We now turn to a description of the methods of molecular mechanics. As just noted, we can classify these methods as to how they explore the structure–energy relationship in biologically interesting molecules.

Energy minimization methods are perhaps the simplest molecular mechanics technique. These methods search very localized regions of conformational space, around the initial conditions given to the search, and identify minima in the energy function. Thus they provide a limited search of conformations and are directed, due to their algorithmic design, to the nearest local minimum. An important feature to recognize about energy minimization methods is that they provide no guarantee of identifying the global energy minimum. Thus, the outcome of an energy minimization is strongly dependent on the initial state conformation of the system. For these reasons, minimization is most useful when used in conjunction with broader search-based methods, such as molecular dynamics or Monte Carlo-based approaches.

The algorithmic basis for energy minimization is to, in a 'smart' way, vary the conformation of a molecule such that the energy of the system decreases. To accomplish this, energy minimization algorithms use information about the derivative of the potential energy with respect to the atomic coordinates. The simplest approach is that of steepest descents. This algorithm evaluates the gradient of the potential energy with respect to the molecular coordinates and adjusts the coordinates along the negative gradient direction. Specifically, if $\boldsymbol{r}$ represents the full configuration of a protein molecule and $-\nabla_r U$ is the gradient of the potential energy with respect to that configuration (note that $-\nabla_r U$ is the force on the atomic coordinates of the protein molecule), then the steepest descents algorithm moves the current configuration to a new one as:

$$\boldsymbol{r}_{n+1} = \boldsymbol{r}_n - \alpha \boldsymbol{s}_n \qquad [2]$$

where $\boldsymbol{s}_n$ is the unit vector in the direction of the potential energy gradient (i.e. $\boldsymbol{s}_n = \nabla_r U / |\nabla_r U|$). The parameter α determines the size of the step along the negative gradient vector, and hence influences how many steps are required to reach the minimum. In general, convergence to the minimum is best when α is relatively small (on the order of 0.01 Å–0.1 Å). Since all steps are along the steepest descents direction, it is clear that this algorithm will lead directly to the nearest minimum. Furthermore, because of the simplicity of the 'search' algorithm used here, steepest descents minimization is most useful in early stages of structure 'regularization' (in early stages of model building where there may be many non-bonded overlaps) where large forces are present and small steps can rapidly relieve highly energetic interactions.

A number of related first derivative methods are also used for energy minimization, for example the conjugate gradient methods and the Powell methods. These algorithms differ from steepest descents primarily in their representation of the gradient information used in determining the direction of moving the coordinates. These methods have features which often make them more desirable for use over the simpler steepest descents approach. Reviews, and references to the original literature, can be found in refs 8,15,39.

Second derivative information can also be used to provide more detail on the local nature of the potential energy surface, and hence the 'best' direction to move. These methods come generally under the heading of Newton–Raphson based methods. Since they use second derivative information, they are generally more expensive, computationally, to use. Also, because many of these algorithms require storage and manipulation of the full second derivative matrix (of order $3N \times 3N$, where N is the number of atoms in the system) the memory requirements inhibit their use on large biological macromolecules (8).

As noted in the beginning of this section, energy minimization methods permit only local searching of the energy–structure relationship for biological molecules. Therefore, the greatest use of these methods is in the conditioning of structures from experimental data. For example, in initiating a modelling study on a protein or DNA molecule for which a crystallographic or NMR structure exists, one uses minimization methods to 'relax' the experimental structure on to the empirical energy surface. Also, minimization can be used in modelling studies associated with homology building.

2.3 Molecular dynamics

Molecular dynamics (MD) simulation methods have become the 'workhorse' for exploration of structure–energy relationships in biological molecules. These methods are based upon the numerical solution of the classical Newtonian equations of motion. They are used extensively for structure refinement in model building (homology models, or refinement against experimental data), conformational searching, and sampling as required in free energy calculations. The basis of MD algorithms is the expression of the potential energy derivative as a finite difference step in time. Thus, a commonly employed MD algorithm, the Verlet algorithm (40), is schematically represented as:

$$\boldsymbol{r}_{n+1} = 2\boldsymbol{r}_n - \boldsymbol{r}_{n-1} + \Delta t^2 \boldsymbol{F}_n/m \qquad [3]$$

where $\boldsymbol{r}_n$ is the coordinate vector for a particle and $\boldsymbol{F}_n$ is its force vector at the discretized point in time i, Δt is the time step used in propagating the positions, and m is the mass of the particle ($\boldsymbol{F}/m$ is the acceleration vector). Note that these equations give an MD simulation in the '*NVE*' ensemble (i.e. constant number of atoms N, volume V, and energy E).

The stability of this MD algorithm is controlled by the choice of time step. The time step should be about ten times faster than the highest frequency motion. Typically, stable protein dynamics trajectories (the path of all of the molecular coordinates through 'configuration space') are achievable for time steps less than two or three femtoseconds (41). This puts a significant limitation on the length of time (and consequently the amount of conformational space) which can be sampled in a molecular dynamics trajectory. Simulations of protein molecules extending over periods of several nanoseconds (i.e. 10^6 iterations) are becoming routine. Some recent simulations have sampled a single molecular trajectory for approximately 100 nsec (42).

The 'radius of convergence' (for use of larger time steps) of the dynamics algorithm can be increased by projecting out high frequency motions using methods such as the SHAKE algorithm (43) or more recently developed schemes (44,45). The SHAKE algorithm constrains internal degrees of freedom, such as high frequency bond vibration involving hydrogens, permitting time steps of 2–3 fsec. The other approaches allow, in principle, much larger time steps. It still remains to be demonstrated whether or not these new approaches are competitive with conventional algorithms in MD. Nevertheless, they will likely prove important in the general repertoire of conformational sampling methods.

It is also possible to introduce the control of temperature into molecular dynamics. This feature is responsible for many of the useful connections between the molecular trajectories and statistical thermodynamic quantities of interest in free energy calculations, as well as in the 'staging' of heating and cooling cycles used in structure refinement. A detailed discussion of temperature 'thermostatting' is beyond the scope of the present review. However, references exist which discuss various aspects of these approaches (8,46). For our purposes here, let it suffice to note that it is possible to carry out molecular dynamics simulations at a fixed temperature T (thereby performing calculations in the canonical—constant NVT—ensemble) as well as with the use of predetermined schedules for heating/cooling of the system. The latter is the basis of simulated annealing methods, which themselves are a mainstay of structure search, refinement, and model building.

2.4 MD protocols

Below we outline basic protocols used in molecular dynamics.
Typical variables specified for an MD run:

- length of time step
- boundary conditions: none, periodic (47), stochastic (8,48)
- treatment of non-bonded interactions, e.g. cut-offs versus Ewald (6,49)
- frequency of updating the non-bonded list (for non-bonded cut-offs)
- ensemble simulated (NVE, NVT, etc.)
- temperature and method of thermostatting (for NVT simulations)
- pressure and method of maintaining constant pressure (for NPT simulations)
- starting atomic positions
- starting atomic velocities (usually assigned from a Boltzmann distribution at the appropriate temperature)
- form of restraints (if required)

Protocol 1. A typical molecular dynamics protocol

1. Apply appropriate restraints, if necessary.
2. Relieve initial strain via energy minimization. Do a sufficient number of iterations so that the potential energy of the system plateaus (1000–5000 steps is typical).

Protocol 1. *Continued*

3. Equilibrate the system, i.e. do several steps of MD until measurable thermodynamic variables equilibrate (i.e. reach a plateau). The thermodynamic variables should behave as expected given the ensemble simulated: for *NVE* simulations, energy should be conserved; for *NVT*, the temperature should be approximately constant (depending on the thermostatting), etc. At equilibrium, the fluctuations in potential and kinetic energies should be relatively constant.
4. Collect coordinates (and velocities, if needed) during 'production' dynamics. Do the number of MD steps required for the problem, storing the coordinates at regular intervals. For free energy simulations, do not process coordinates from every step, as the conformations should be 'independent' observations.
5. Analyse the MD data from step 4.

Protocol 2. A protocol for building a 'box' of solvent

1. Build a lattice of solvent molecules that is somewhat less dense than required.
2. Perform MD on the order of tens of picoseconds to move the solvent off-lattice. Check the appropriate radial distribution functions (47) to assure the lattice effects have been removed and the system has equilibrated.
3. Apply external pressure to slowly shrink the box to the appropriate density. Make sure the system equilibrates.

Protocol 3. A protocol for solvating a solute

1. Minimize the solute *in vacuo* or using an *r*-dependent dielectric constant.
2. Obtain an equilibrated box of solvent (note that the 'box' may be any shape, though it must be space-filling if using periodic boundary conditions). The box should be large enough to hold the solute and several layers of solvent (37). If using periodic boundary conditions and non-bonded cut-offs, make sure the box is large enough so the solute does not 'see' solutes in neighbouring image volumes.
3. Centre the solute in the box and remove solvent molecules that are within 2.6 Å of solute atoms.
4. Equilibrate the system by first minimizing and then doing several steps of MD.

2.5 Free energy calculations

At the core of free energy calculations are two observations:

(a) The sum of thermodynamic state variables (energy, entropy, free energy, etc.) around a closed thermodynamic cycle is zero.

(b) The difference in thermodynamic state variables between two well-defined states is independent of the pathway taken to move from one state to the next.

Point (a) can be demonstrated by considering the change in stability of a protein due to point mutations at a specific site. If N_f denotes the folded state of a protein with its 'native' sequence, N_u the unfolded state, and M_f and M_u the folded and unfolded states of a specific mutant protein derived by changing (or deleting) one (or more) of the amino acids in the native protein, then we may write two thermodynamic processes that can be used to construct the relative stability based on the free energy of unfolding ΔG_u^N and ΔG_u^M.

$$N_f \overset{\Delta G_u^N}{\Rightarrow} N_u$$

$$\text{and } M_f \overset{\Delta G_u^M}{\Rightarrow} M_u \qquad [4]$$

The relative stability of the native protein sequence compared to the mutant protein is $\Delta\Delta G = \Delta G_u^N - \Delta G_u^M$. Connecting the left- and right-hand sides of these relationships into a closed cycle yields:

$$\begin{array}{ccc} N_f & \overset{\Delta G_u^N}{\Rightarrow} & N_u \\ \Delta G_m^F \Uparrow & & \Downarrow -\Delta G_m^U \\ M_f & \underset{-\Delta G_u^M}{\Leftarrow} & M_u \end{array} \qquad [5]$$

The overall free energy change (or, for that matter, the change in any thermodynamic state variable) is zero, i.e. $\Delta G_u^N - \Delta G_m^U - \Delta G_u^M + \Delta G_m^F = 0$. This also yields the relationship:

$$\Delta\Delta G = \Delta G_u^N - \Delta G_u^M = \Delta G_m^U - \Delta G_m^F, \qquad [6]$$

which is the basis of free energy perturbation calculations (see below).

The free energy differences for unfolding, ΔG_u^N and ΔG_u^M, can often be measured by experiment. An extensive search of conformational space is required for protein folding or unfolding, however, so these differences are much more computationally expensive to calculate than the mutational free energy differences, ΔG_m^U and ΔG_m^F, which involve the non-physical (alchemical) change in the underlying chemical structure. Because of point (b) above, the free energy change is path-independent, and any computationally convenient path is as good as any other. Thus, the experimental and chemical perturbation results can be directly compared as in Equation 6. This is the general framework for calculations of relative free energy changes between differing chemical types (native and mutant proteins in their folded conformation) as well as the basis of conformational free energy calculations.

Point (b) above implies that we can define any 'mapping' between the two states of interest. First consider the alchemical (or *chemical perturbation*) pathway to computing $\Delta\Delta G$s. The potential energy function representing a protein in its folded configuration can be expressed as U_N and U_M for the native and mutant protein states, respectively. Also, we can define a pathway which maps between the two states via a hybrid potential energy function $U_\lambda = (1 - \lambda)U_N + \lambda U_M$. This potential represents the native folded sequence when $\lambda = 0$ and the mutant when $\lambda = 1$. From statistical mechanics (17,50,51), the free energy difference between two states of differing λ values is expressed as:

$$\Delta G_{\lambda_1\lambda_2} = -k_B T \ln \langle \exp \left[\frac{(U_{\lambda_1} - U_{\lambda_2})}{-k_B T} \right] \rangle_{\lambda_1} \qquad [7]$$

or equivalently as:

$$\Delta G_{\lambda_1\lambda_2} = \int_{\lambda_1}^{\lambda_2} \langle \frac{dU_\lambda}{d\lambda} \rangle d\lambda \qquad [8]$$

and the free energy change associated with mutating the native protein in the folded state is arrived at from the above expressions for the case that $\lambda_1 = 0$ and $\lambda_2 = 1$. In practice, this calculation is carried out in several 'windows', or incremental changes in λ_1 and λ_2, and the total free energy change is computed as the sum of the incremental pieces. The general rule of thumb is that the λ step size is chosen such that the free energy difference between windows is less than $2k_B T$.

Equation 8 is the 'thermodynamic integration' form of the equation, and is the basis of the class of methods known as slow growth, where the λ variable is slowly forced from the initial value to the final, changing by a small increment during each step of the dynamics simulation. The overall free energy change is then simply related to the sum of the integrand throughout the MD simulation.

The calculation of conformational free energies too can be cast into this framework. The relationship between the free energy of a biopolymer in a particular conformation and the frequency with which that conformation occurs in a statistically representative sample of structures is given by:

$$A(\xi) = -k_B T \ln [\rho(\xi)] \cong -k_B T \ln [h(\xi)] + C \qquad [9]$$

where $A(\xi)$ is the Helmholtz free energy (*NVT* ensemble) as a function of the conformational coordinate ξ (also known as the potential of mean force),

$\rho(\xi)$ is the canonical distribution of ξ, and $h(\xi)$ is a histogram of ξ which forms a finite simulation approximant to $\rho(\xi)$. C is an arbitrary constant which is related to the overall (and unknown) normalization of the true distribution function. From this relationship one can develop two approaches to computing conformational free energy changes in biopolymers.

The first is exactly equivalent to what is done for chemical 'perturbations'. The coordinate, however, is no longer λ, which describes the degree of potential energy function hybridization between two chemical species, but is instead one or more conformational degrees of freedom. A molecular dynamics simulation is carried out with the value of these degrees of freedom fixed at the reference values. The energy difference between this conformation and another nearby conformation is computed and averaged as the argument of the exponent given above, i.e.

$$\Delta A_{\xi_1\xi_2} = -k_{\mathrm{B}}T \ln \langle \exp \left[\frac{(U_{\xi_2} - U_{\xi_1})}{-k_{\mathrm{B}}T} \right] \rangle_{\xi_1}. \tag{10}$$

Using this formula one can 'bootstrap' one's way through a range of ξ and hence construct the underlying potential of mean force, $\Delta A(\xi)$, to within an arbitrary constant which shifts the entire surface.

Alternatively, the potential of mean force can be computed by calculating the histogram in the coordinate of interest (i.e. the reaction coordinate, ξ) during the course of a molecular dynamics trajectory. However, since barrier crossing is often infrequent in such simulations, one must add a biasing potential that allows the system to move into (and through) regions of high free energy. In biased, or umbrella, sampling molecular dynamics, one uses a biasing potential, $U^*(\xi_i)$, to keep the conformation in a particular region of conformational space centred around ξ_i. From a series of molecular dynamics simulations centred around differing values of ξ_i, histograms in ξ are computed, $h_i^*(\xi)$. These histograms yield biased distributions of the conformational coordinate ξ since they were computed from a biased molecular dynamics simulation. Therefore, the bias must be removed using the exact relationship:

$$h_i^{\mathrm{unbiased}}(\xi) = C_i \exp \left[\frac{U^*(\xi_i)}{k_{\mathrm{B}}T} \right] h_i^*(\xi). \tag{11}$$

In this expression the arbitrary constant C_i is used to 'shift' the (unbiased) distributions, which result from different biased samplings (windows), to yield the optimal overlap in regions of the histograms with common values of ξ. The resultant 'spliced' histogram is constructed to cover the entire region of interest in the conformational coordinate and the potential of mean force follows from this (52,53).

Conformational free energy functions (or surfaces if the conformation is dependent on more than one variable) provide insights into the conformational control of biopolymer function. Examples where the use of conformational free energy simulations are helpful include understanding the origin of helix capping effects in peptides, the formation of secondary structure in peptides, and the folding of proteins.

3. Applications

3.1 Protein dynamics

Several excellent reviews on protein dynamics give extensive background in the types of motion exhibited by proteins and how these motions are observed experimentally and theoretically (1,8,14–16). Listed below some important aspects of protein dynamics are summarized.

(a) Bond length stretching is the highest frequency protein motion, taking place at femtosecond time-scales at room temperature (1). Other high frequency motion includes bond angle bending and constrained dihedral angle motion such as motion in the backbone dihedral angle associated with the peptide bond, which is highly constrained due to the double bond character of the peptide bond (1). These conformational motions represent some of the earliest processes associated with the dissociation of ligands from myoglobin.

(b) Unlike the 'hard' degrees of freedom described above, the ϕ and ψ backbone dihedral angles are relatively unconstrained and are called 'soft' degrees of freedom. Many side chain dihedral angles are also soft degrees of freedom. Unhindered motion in these angles (as at the protein surface) is of high amplitude and relatively low frequency (picoseconds to nanoseconds). Hindered motion, as in the tightly packed protein core, is similar to motion in the hard degrees of freedom. Dihedral angle transitions can occur in the protein core, but this motion typically requires the collective motion of several angles (1).

(c) Although motion along the soft degrees of freedom dominate protein dynamics, the hard degrees of freedom significantly reduce barriers to motion in the soft degrees of freedom (1). Hence theoretical models allowing movement in only the soft degrees of freedom are limited in their ability to reproduce protein motion (8).

(d) Molecular dynamics simulations indicate that conformational change occurs one dihedral angle at a time or as a peptide flip, where ψ of residue i and ϕ of residue $i + 1$ change in an anti-correlated fashion, which rotates the peptide bond but causes little change in the C^{α} positions (54–57). Hence, conformational changes are brought about by a series of sequential dihedral angle changes. A large change in conformation involves the movement of anywhere from only a few dihedral angles to small shifts in several internal degrees of freedom (1).

(e) In general, atomic motion within a protein is anisotropic and anharmonic

(58). Hence, methods which assume all motion is harmonic (normal mode analyses) give limited information about predicted protein motion (59), since atomic motion is confined to a single, hyperparabolic minimum. Molecular dynamics simulations allow anisotropic and anharmonic motion, but analysis of long time-scale motion is limited due to practical limits on simulation time.

(f) Protein dynamics simulations give detailed information on atomic motions into the nanosecond range, which are often difficult to obtain experimentally. Many important protein motion events take place at longer than nanosecond time-scales, including secondary structure formation (< millisecond), conformational changes, and tertiary structure formation (milliseconds to seconds). Promising experimental (e.g. pulse labelled NMR) and theoretical (e.g. free energy simulations) methods are being developed to study long time-scale protein motion, though much work remains (1).

(g) Protein dynamics in simulations are significantly affected by solvent effects (8). Recent increases in computational power have allowed many full protein simulations to include explicit solvent, which most accurately models solvent effects (22,56,60,61).

3.1.1 Analysing MD trajectories

Several methods have been used to analyse dynamical motions in proteins. The display of thermodynamic quantities (e.g. total energy, temperature) as a function of time gives information on whether the simulation has reached equilibrium (at least on the time-scale considered). The actual behaviour of these functions will depend on the ensemble sampled (e.g. *NVE*, *NVT*, or *NPT*). The root mean square (rms) deviation in the coordinates, as a function of time, from the starting conformation can indicate whether the system is undergoing structural changes, and time series of internal coordinates give detailed information on the structural changes. Another method for analysing MD trajectories is the time correlation of atomic motion (8,62,63).

One may also be interested in how fluctuations in one region of a molecule correlate with fluctuations in other regions. Concerted movements within a protein, such as occur in conformational changes and protein folding, require correlated motion. Protein regions which have strong local interactions may show correlated motion during equilibrium motion. Studying correlated equilibrium motion can identify these structural units and point to potential sites for concerted motion during non-equilibrium processes such as conformational changes and protein folding. Below we describe several studies which examine correlated behaviour of protein dynamics.

Rojewska and Elber (64) found helix dynamics in a four-helix bundle are dominated by rigid helix motions, where atomic motion within an α-helix was highly correlated and relative motion between helices was due to relatively discrete free energy minima, producing a set of 'packing modes'. Ichiye and Karplus (65) used the equal-time cross-correlation matrix for atomic fluctuations (8) to find correlated motion between different regions in a protein. The elements of this matrix are:

$$C(i,j) = \frac{\langle \Delta \boldsymbol{r}_i \cdot \Delta \boldsymbol{r}_j \rangle}{\langle \Delta \boldsymbol{r}_i^2 \rangle^{1/2} \langle \Delta \boldsymbol{r}_j^2 \rangle^{1/2}}, \qquad [12]$$

where $\Delta \mathbf{r}_i$ is the displacement of atom i from its average position (after global rotation and translation from the reference structure is removed), and $\langle \ldots \rangle$ denotes the average overall conformations. They found that even in the small bovine pancreatic trypsin inhibitor molecule two regions of correlated motion emerge, the 'top' region containing the two stranded β-sheet and two neighbouring peptide strands, and a 'middle' region containing a 3_{10} helix, one side of the α-helix, and a core of hydrophobic residues.

Amadei *et al.* (59) used an interesting approach to identify important modes of protein motion. They analysed a molecular dynamics simulation of lysozyme by constructing the covariance matrix of the positional fluctuations:

$$C = \left\langle (\boldsymbol{x} - \langle \boldsymbol{x} \rangle)\ (\boldsymbol{x} - \langle \boldsymbol{x} \rangle)^T \right\rangle, \qquad [13]$$

using a $3N$-dimensional vector $\boldsymbol{x}$ of all atomic coordinates. They then diagonalized this matrix, resulting in $3N$ eigenvalues and eigenvectors. In the new, generalized coordinate set defined by the eigenvectors, a small number of eigenvalues sum to contribute to most of the positional fluctuations. This smaller 'essential' subspace of the original $3N$ Cartesian space thus describes most all of the high amplitude, anharmonic motion. Motion in the remaining coordinates is highly constrained and contributes little to positional fluctuations (see also ref. 8).

Amadei *et al.* found the 'essential' motion of large amplitude and low frequency was mainly due to backbone rather than side chain atom motion. This motion is correlated over many atoms and they hypothesize that only this motion is relevant to protein function. In contrast, the physically constrained motion produces local, uncorrelated atomic fluctuations that have little bearing on protein function. A promising extension of this analysis is to limit molecular dynamics to only the essential subspace, which would greatly reduce the required search space and potentially allow long time-scale motion to be studied (59).

3.1.2 HIV-1 protease dynamics studies

Several groups have investigated the dynamics of the HIV-1 protease, which is essential for viral replication and therefore a target of drug design. Several crystal structures of this dimeric protein and its relatives have been solved, and the results suggest dynamics play an important role in protein function. In particular, a flexible 'flap' region is evident which is involved in substrate

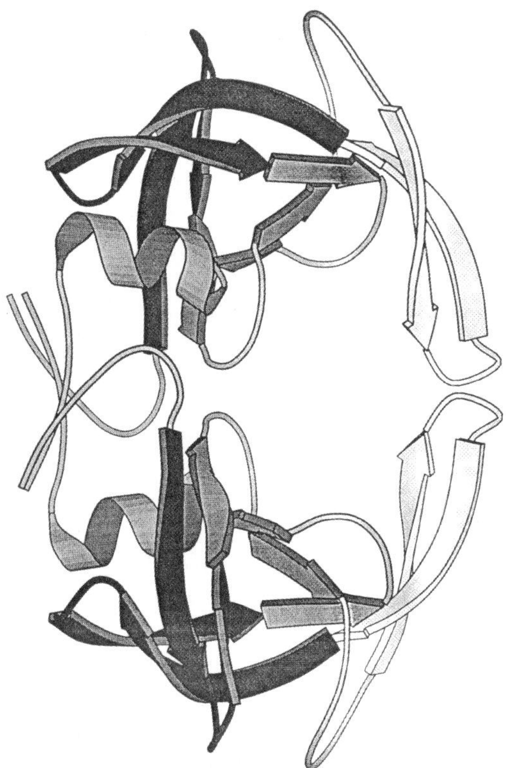

Figure 1. Ribbon diagram of the HIV-1 protease dimer (84). The white region denotes the flexible flap, and the dark grey region denotes the cantilever and fulcrum during dimer motion. The figure was created using MolScript (86).

binding (*Figure 1*). Below we discuss salient features from a number of MD simulations carried out to study various aspects of motion in HIV-1 protease.

Harte *et al.* (21) produced dynamical cross-correlation maps for the HIV-1 dimer from the cross-correlation matrix given in Equation 12. Several regions of through space correlations were identified. From these analyses, Harte *et al.* concluded that the motion of the flap region is anti-correlated with motion in a 'cantilever' and a 'fulcrum' region (*Figure 1*). Correlations in motion between the two monomers were also observed, so that motion in one monomer is communicated to the other monomer. These simulations help explain experimental observations of inactive mutants with mutations in the fulcrum/cantilever region.

Venable *et al.* (20) undertook a similar study, except they simulated dynamics of an HIV-1 monomer rather than a dimer. This was motivated by experimental efforts to stabilize the monomer over the dimer, which would be aided by a more accurate representation of the monomer. Inhibition of dimerization is expected to specifically disrupt HIV-1 protease function.

Venable *et al.* simulated the dynamics of the HIV-1 monomer in a sphere of water molecules, using a single monomer from the dimeric crystal structure as the starting conformation. Marked deviations from the starting structure occurred in the simulation, so that after 160 psec there was ~ 3.0 Å rms difference from the dimeric crystal structure versus ~ 1.5 Å expected if the dynamical structure corresponded to the crystal structure (20,21). The deviations were largest at the active site, flap region, and the termini, which all form the dimer interface. Venable *et al.* also looked at dihedral angle transitions and found the local conformation at the tip of the flap varied very little, but extensive changes in conformation occurred at the base of the flap at residues 45 and 55, indicating these residues act as hinges for rigid flap motion. From dynamical cross-correlation maps, it is seen that the correlated flap/cantilever motion no longer occurs, so that the absence of the second monomer effectively decouples the flap/cantilever motion.

An HIV-1 protease mutant, Asn 88 → Gln, inactivates the enzyme even though it is distant from the active site and dimerization interface. To investigate the cause of inactivation, Harte and Beveridge (22) simulated the dynamics of this mutant, using the native HIV-1 structure with Asn 88 replaced by Gln as the starting structure. In the 100 psec simulation time, the structure diverged from the native structure, and several hydrogen bonds at the dimer interface were disrupted. While this study using free dynamics can not thermodynamically determine the reason for inactivation, it gives supporting evidence that inactivation is due to dimer destabilization. The authors note, however, that the simulation did not achieve structural equilibrium (as determined from rms differences in atomic positions from the starting structure, which did not plateau with time), so that the structure may have found, with time, a stable dimeric conformation. Also, the effects of the mutation on the unfolded protein were not considered, which can affect protein stability.

3.1.3 Effect of dynamics on reaction rates

Acetylcholinesterase (AChE) breaks down the neurotransmitter acetylcholine, clearing the synaptic junction for the next synaptic firing. This enzyme is inhibited by the drug tacrine, found to relieve symptoms in some Alzheimer's patients, who have been shown to have low levels of acetylcholine (66,67). Both acetylcholine and tacrine bind at the active site (68), which in the X-ray crystal structure is in a deep gorge with an average diameter of 4.4 Å, smaller than the diameter of the substrate or the inhibitor (23). The catalytic rate is fast, however, and the reaction is nearly diffusion controlled. Hence the question arises of how the substrates and products are able quickly to enter and leave the active site.

Axelsen *et al.* (23) studied eight different crystal structures of AChE to find evidence for the solution to this paradox, and found the active site gorge geometry changed relatively little in the different structures, even in the presence of different ligands. Hence the average structure was quite stable with few crystal to crystal differences, and no evidence of alternative pathways to the active site were found. Indeed, it was known that ligands freely exchanged within the protein crystal even though a symmetry-related AChE molecule blocked the gorge entrance in the crystal.

Axelsen *et al.* performed molecular dynamics simulations on the AChE molecule in an attempt to find potential conformational changes that would facilitate ligand access to the AChE active site. They included the symmetry-related AChE molecule at the gorge entrance in the simulations.

Axelsen *et al.* discovered that certain atomic positions at the active site diverged significantly from their crystal structure positions early on in the dynamics run. Various means were tried to correct these deviations (changing force field parameters, including solutes from the crystallizing liquor) to no avail. The authors then re-examined the electron density maps and discovered that several peaks originally assigned to water could actually be a DECA molecule (used in AChE purification) and two NH_4^+ molecules bound in the active site gorge. Repeating the molecular dynamics simulation with these additional molecules in the active site gorge decreased the rms atomic deviation for the active site triad residues from 1.6 Å to 0.7 Å. In addition, they observed an improvement in the R-factor, which measures the how well the atomic model fits the diffraction data. Hence, they showed molecular dynamics results can be used by experimentalists to evaluate the validity of the experimental data.

During the above simulation, no major conformational changes were observed, and the role of protein dynamics in facilitating the movement of ligands to and from the active site could not be determined. A second MD study by Gilson *et al.* (24) points to a possible mechanism for this facilitation. They found a conformational change occurred within the first 119 psec of the simulation, which opened a back door to the AChE active site at Trp 84 in the region of a 'thin' gorge wall.

The back door opening was brief, but the fact that it opened at all in this short simulation suggests that within the 0.1 msec time it takes to hydrolyse the substrate it is possible (though not conclusive) that many opening events could occur. The opening required relatively small movements of only three residues, and hence would be expected to occur with some ease. These results suggest site-directed mutagenesis experiments which would alter the motion of the back door residues and would hence test the back door hypothesis (24).

To summarize, dynamics studies of the AChE molecule greatly assisted in predicting both its structure and dynamics. Additional molecules in the crystal structure were located, a potential route that would enhance the reaction rate was identified, and suggestions to experimentally test the predicted structure and dynamics were made. This is an excellent example of how computational studies can complement experimental studies.

3.2 Peptides

Many simulations have used peptides as simple models for protein structure, where specific questions of atomic level influences on protein conformation can be addressed. Factors stabilizing α-helices have been particularly well studied, as helix hydrogen bonding interactions are local, taking place between residues i and $i + 4$. Thus a short peptide can form a complete helical unit. In addition, much experimental data on helix–coil transitions are available to compare with theoretical results. Below we review some results from our laboratory on calculating conformational free energy surfaces for helix initiation and propagation. Our aim is to give some useful rules of thumb for generating converged free energy surfaces.

Tobias and Brooks (25) calculated the free energy surface for helix initiation and propagation using an all-alanine helix. Alanine was used as the 'generic' residue, as it is the simplest amino acid with a conformational degree of freedom similar in extent to all other amino acids, with the exception of glycine (which has greater conformational freedom due to the absence of a side chain) and proline (which has less conformational freedom due to the constraints imposed by its pyrrolidine ring). Studies of peptides with other sequences used the appropriate alanine peptide's free energy surface as the reference for calculating changes in free energy differences.

Tobias and Brooks used umbrella sampling to calculate the free energy surface for helix unfolding for the blocked peptide H_3CCO-(ala)$_3$-$NHCH_3$ in TIP3 water (69). They used the distance between the two atoms forming the helix hydrogen bond as the reaction coordinate; at distances near 2 Å, one turn of an α-helix forms, and at longer distances (~ 9 Å) the peptide is in an extended conformation (*Figure 2a*). As suggested from the free energy surface (*Figure 3*, solid line), there is very little free energy difference between the helix and extended states. Components of the free energy can also be calculated using energy decomposition. For example, in this calculation the change in peptide–peptide energy for helix unfolding is -7.6 kcal mol^{-1} and the change in peptide–solvent energy is 10.9 kcal mol^{-1}. See Tobias and Brooks (25) for details on carrying out these decompositions.

From these calculations, the authors conclude that peptide–peptide interactions, particularly the electrostatic component, favour the extended conformation over the single helical turn. Entropy also appears to favour the extended over the helix state. Solvent–peptide interactions, in contrast, favour the helix over the extended conformation. These effects cancel each other out, so that the free energy difference between helix and extended states is about zero. The thermodynamic results can be rationalized on the atomic level based on peptide dipole interactions (see ref. 25 for details).

These results underscore the importance of including solvent in the simulation. *Figure 3* shows a similar study for the peptide H_3CCO-pro-(ala)$_2$-$NHCH_3$, where proline replaces alanine in the first position (27). Here, the

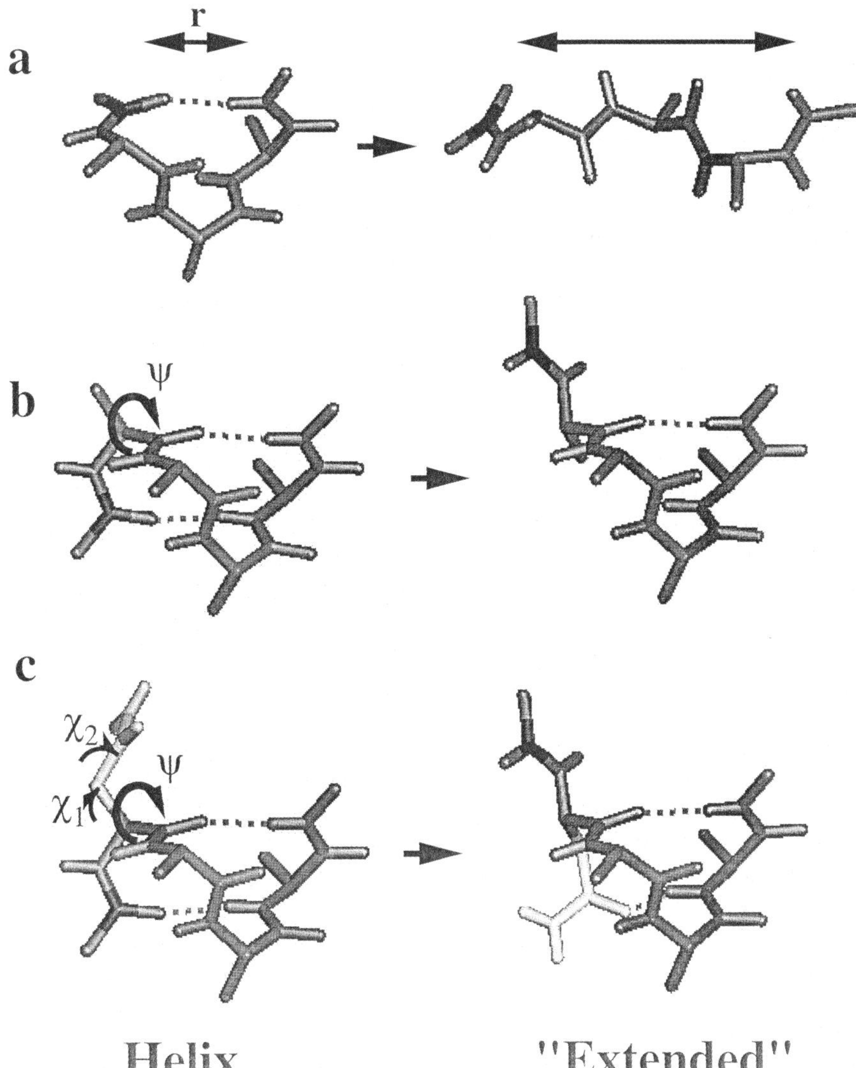

Figure 2. Reaction coordinates for peptide free energy simulations. The dark grey denotes the N-terminal carbonyl carbon, and the dotted lines denote hydrogen bonds. Only the amide hydrogens are shown. All peptides are blocked with H_3COC- at the N-terminus and -$NHCH_3$ at the C-terminus. (a) The reaction coordinate, r, for unfolding one turn of an α-helix in an alanine peptide. (b) The reaction coordinate, ψ, for unwinding the N-terminal turn of a two-turn alanine α-helix into an extended conformation. (c) As in (b), but with the N-terminal alanine replaced by asparagine. Note that the Asn side chain (white) replaces backbone carbonyl in the N-terminal helix hydrogen bond when Asn is in the extended state. The two side chain dihedral angles χ_1 and χ_2 are indicated.

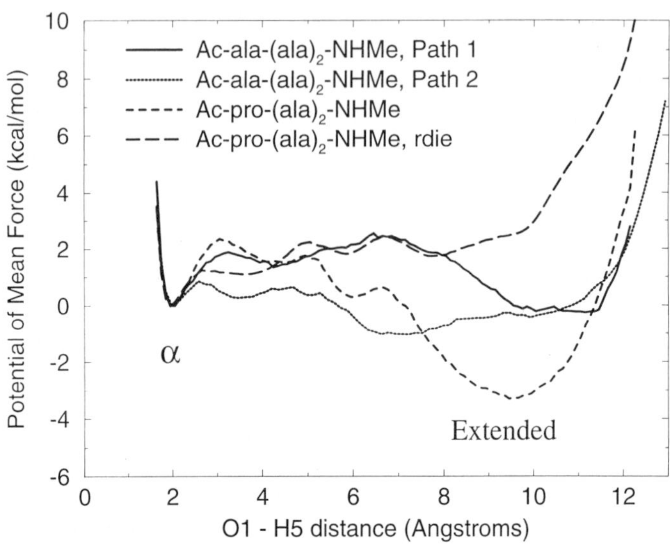

Figure 3. Free energy surfaces for the helix unfolding reaction for various peptides and solvents. The reaction coordinate is the distance between the N-terminal carbonyl oxygen (O1) and the C-terminal amide proton (H5). All simulations were in explicit TIP3 water solvent with periodic boundary conditions except for the curve labelled 'rdie', where an *r*-dependent dielectric $\varepsilon = r_{ij}\varepsilon_0$ was used. The curves are arbitrarily aligned with the α-helix well at 0 kcal mol^{-1}. Restrained MD was performed at 1–2 Å increments of O1–H5 distance. The resulting fragments of free energy surfaces (calculated by Equations 8 and 10) were spliced together to form a smooth surface (52). See refs 25, 27, 71 for details.

free energy surface calculated with an *r*-dependent dielectric ($\varepsilon = r_{ij}\varepsilon_0$) is compared with the surface calculated in a fully solvated system. The α-helix minimum is increased relative to the extended peptide minimum in the *r*-dependent dielectric surface, since there are no water molecules competing for hydrogen bonds with the peptide backbone (70).

The reaction coordinate dictates to some extent the path a molecule will take from one conformation to another. It is important to remember, however, that two simulations may take different paths even though they follow the same reaction coordinate; the conformations can vary in a dimension that is independent of the reaction coordinate. For example, in helix unwinding experiments (25,71), two paths were observed in going from one turn of a helix ('ααα') to an extended conformation ('βββ'). One simulation progressed through 'ααα–ααβ–αββ–βββ' ('Path 1' in *Figure 3*) and another 'ααα–βαα–ββα–βββ' ('Path 2' in *Figure 3*). The free energy difference of the extended versus the helix turn was virtually identical, as expected since free energy differences are independent of the paths taken. Hence it is difficult to draw path-dependent conclusions (e.g. reaction rates) from restrained dynamics, both because one is choosing a path a priori via a choice of reaction coordinate, but also because the path may vary in coordinates orthogonal to the reaction coordinate.

In the above studies a single reaction coordinate was used. Often times a single coordinate is not adequate to describe the free energy surface of a system. This is illustrated below in a study we undertook to quantify the effect of asparagine in the N-cap position of an α-helix. Asparagine is often found in the position preceding the helix N-terminus, called the N-cap position (72). We studied the effect of this residue on N-terminal helix propagation, and compared the results to work by Tobias and Brooks (26), which examined helix propagation in an all-alanine peptide.

Tobias and Brooks (26) used the ψ dihedral angle of the first N-terminal alanine as the reaction coordinate, rotating this angle through 360° to map out the free energy surface as a function of ψ (*Figure 2b*). At ψ values near −45°, a second helical turn is added N-terminal to an existing helical turn (which is restrained during all of the simulations to maintain the turn). At values near 120°, ψ causes the N-terminal residue to have an extended conformation, so only a single turn of the helix exists. The ϕ angle changes little between extended and helical conformations, so no additional constraints on ϕ are needed. The resulting alanine free energy surface is shown in *Figure 4*.

We attempted a similar approach, using umbrella sampling, with the first alanine replaced by an asparagine (*Figure 2c*). When ψ of Asn is near −45°, two turns of an α-helix are formed, as in the all-alanine case. When ψ is near 120°, the conformation is extended, but a hydrogen bond is formed between

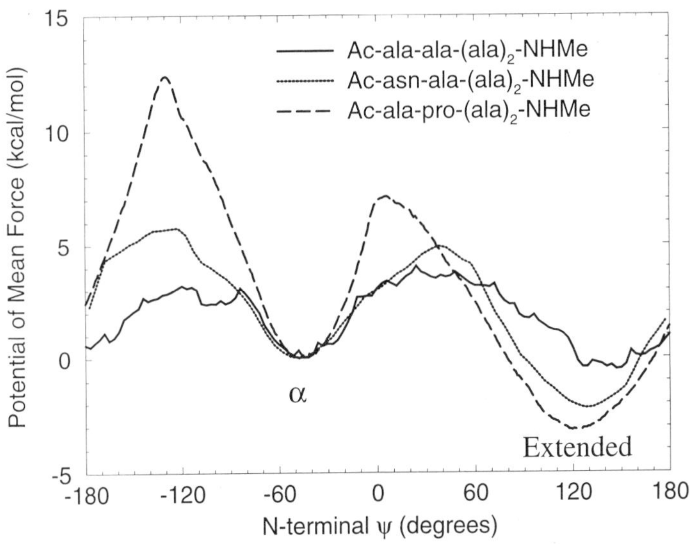

Figure 4. Free energy surfaces for N-terminal helix propagation in water. The curves are arbitrarily aligned with the two-turn helix well (α) at 0 kcal mol^{-1}. General methodology: restrained MD was performed at 6°–15° increments of ψ. See refs 26–28 for details.

the Asn side chain and the N-terminus of the helix, hence 'capping' the helix, stabilizing the helix N-terminus (see below).

The addition of the Asn side chain adds two more soft degrees of freedom when compared with Ala (*Figure 2c*). If during the simulation these angles freely rotate *or* do not change, overlapping conformations then occur in each neighbouring window, a requirement for constructing a valid free energy surface. In our case, χ_2 freely rotated, but χ_1 would rotate to a new conformation during the equilibration period, and two neighbouring windows would not necessarily have overlapping conformations. We found it necessary to include χ_1 as an additional reaction coordinate, and calculated a two-dimensional free energy surface using the weighted histogram analysis method (WHAM) (52) to optimally splice the windows together (*Figure 5*).

The source of this 'χ_1 flip' problem is evident in *Figure 5*. There are two free energy minima for N-capping conformations, one at ψ, χ_1 of Asn = 120°, 180° and a second at ψ, χ_1 = 160°, 60°. As ψ progresses from 120° to 160°, the χ_1 angle switches abruptly to maintain the N-cap hydrogen bond. This is also

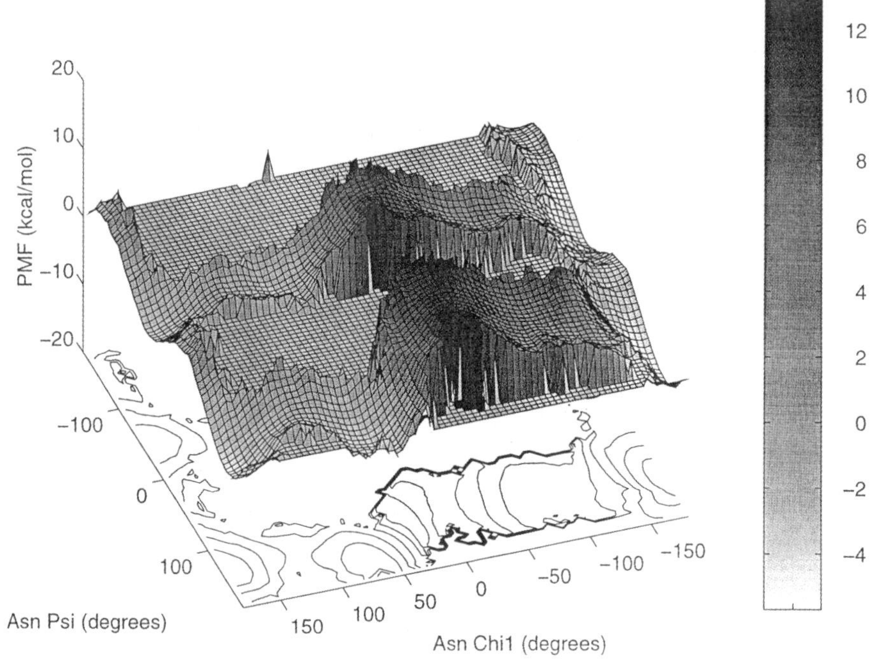

Figure 5. Free energy surface as a function of Asn ψ and χ_1 angles. The minima at ψ, χ_1 values of (120°, 180°) and (160°, 60°) represent capping structures. The two-turn α-helix has a minima at ψ = −45°. The surface was created via umbrella sampling with ψ and χ_1 angles restrained to various values. The C-terminal helix turn was restrained to a hydrogen bond distance of 2 Å, and the N-terminal ϕ was restrained to 80° in all windows. The flat regions at 0 kcal mol^{-1} were not sampled. See ref. 28 for details.

seen in X-ray crystal structures of proteins, where N-caps with ψ angles near 120° more often have χ_1 angles near 180°, and those with ψ angles near 160° more often have χ_1 angles near 60° (data not shown). Hence these two angles are correlated and the reaction coordinate must take both into account. It is not always necessary to have a multidimensional reaction coordinate in such instances; it may be possible to define a smoothly changing composite variable to serve as a single reaction coordinate connecting two or more conformations of interest.

The choice of the direction taken by the reaction coordinate (i.e. increasing or decreasing) may also influence convergence of the free energy surface. In the N-terminal propagation simulations described above, one can rotate the ψ angle either clockwise or counter-clockwise. When the ψ angle was rotated so the angle decreased from 180°, the ϕ angle would 'flip' to positive angles near the barrier at −120°. This sampling problem did not occur when the ψ angle was increased from −180° (the counter rotation). A Ramachandran plot based on the alanine dipeptide free energy surface shows why this is so (*Figure 6*). When ψ is decreased, ϕ approaches a branch point near $\psi = -120°$, such that once the peptide progresses to a conformation with a positive ϕ value it is difficult to return to negative values, as ψ decreases, due to barriers. Hence, the conformation reached at −180° is trapped in a high energy con-

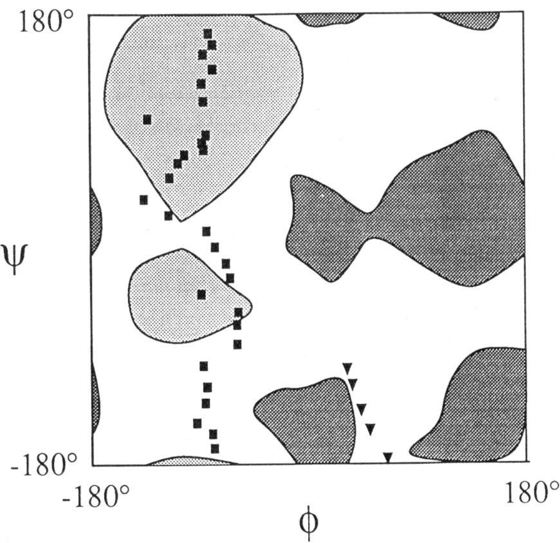

Figure 6. Approximate free energy surface for alanine dipeptide as a function of the ϕ and ψ backbone dihedral angles, simplified from the alanine dipeptide in water free energy surface (85). The light grey areas represent free energy wells, and the dark grey areas represent free energy peaks. The squares denote a pathway increasing in ψ, and the triangles denote the path taken when ψ is decreased towards −180°, leading to positive ϕ angles.

formation and is not the same as that at 180°, a requirement for valid calculation of the free energy surface. In contrast, when ψ is increased, the ϕ angle may make a brief excursion toward positive values, but barriers push the conformation back to negative ϕ values, so the conformation does not become trapped in the positive ϕ region.

Hence, choosing a reaction coordinate requires some prior knowledge of the free energy surface, obviously not available at the start of a simulation. Careful consideration of possible reaction coordinates can significantly reduce sampling problems and result in valid, reproducible free energy differences.

Now that we have discussed some of the issues involved in constructing conformational free energy surfaces, we review the results. The Asn free energy surface for helix capping (*Figure 5*) can be integrated over χ_1, giving the free energy as a function of ψ only. This surface is compared to that for N-terminal propagation of an all-alanine peptide in *Figure 4.* N-terminal propagation with proline in the first position of the turn is also given. Both proline in the N1 position and Asn in the N-cap position inhibit N-terminal helix propagation, so that $\Delta\Delta G^{A\rightarrow P}_{\beta\rightarrow\alpha}$ is 3 kcal mol^{-1} and $\Delta\Delta G^{A\rightarrow N}_{\beta\rightarrow\alpha}$ 2 kcal mol^{-1}, respectively ($\beta \rightarrow \alpha$ denotes propagating the helix a second turn). We found proline did not enhance helix initiation (*Figure 3*) or C-terminal propagation (not shown) when compared with alanine, and the best explanation for its high frequency of occurrence in the N-terminal position is due to its helix terminating effect. This effect originates from constraints placed on the residue preceding the proline residue, due to the presence of the pyrrolidine ring (data not shown). Asparagine in the N-cap position, also found with high frequency in protein structures (72), stabilizes the helix termini by hydrogen bonding to amide protons at the N-terminus. Hence both proline in the N1 position and asparagine in the N-cap position act as helix stop signals.

Hermans' group has performed a similar series of studies of the sequence effects on helix stability (73,74). Rather than simulating the process involving conformational change, they make use of chemical perturbation, where one residue is mutated into another during the simulation, often called alchemy. The relative free energy difference $\Delta\Delta G$ can be obtained via either method allowing the results to be directly compared. Our results are similar to those found by Hermans' group, even though the peptides used are somewhat different. A similar process of chemical perturbation in proteins is discussed in the next section.

3.3 Protein stability

Several groups (29–33) have used chemical perturbation to characterize the effect of site-directed mutagenesis on protein stability. Because of the potential importance of this application, we discuss the underlying assumptions, limitations, and strengths of current methods to compute mutational free energy differences.

The free energy cycle for protein stability as a function of mutation is given in Equation 5. The free energy difference $\Delta\Delta G$ is calculated via chemical perturbation, $\Delta\Delta G = \Delta G_m^U - \Delta G_m^F$, which requires several major assumptions. In order for the free energy differences to converge, the mutant and wild-type states must have few differences in their respective potential energy functions (e.g. only small differences in atomic charge or the number of side chain atoms). In the folded state, only relatively small, local conformational changes between the wild-type and mutant are allowed. Another variable affecting convergence is the modelling of the unfolded state.

Whereas the folded state has a well-defined conformation, i.e. a narrow well on the free energy surface, the unfolded state corresponds to a broad basin encompassing a large number of conformations. Adequate sampling of these conformations carries a prohibitively large computational cost, so simple models for the unfolded protein must be used. If one assumes the unfolded protein has little persistent structure, the effect of the rest of the peptide chain on a particular residue averages out with time and does little to affect the thermodynamics. Only close neighbours, whose interactions persist with time, would influence differences in thermodynamics due to 'mutation'. This is supported by statistical analyses of structures in the Protein Data Bank (75) that find the influence of neighbouring residues on local conformation (the only type of conformation expected in the unfolded protein) decreases rapidly with through-sequence distance (76). Because it is assumed only short-range interactions have influence in the thermodynamics of the unfolded state, short peptides have been used to model the unfolded protein. It is assumed a short peptide containing the residue to be 'mutated' will exhibit mutational free energy differences similar to that for a full unfolded protein simulation.

Although it is difficult to access the validity of this assumption, comparison of results from simulations and experiments have shown very good agreement on free energy differences, even though a variety unfolded state models have been used (*Figure* 7). This implies either that the unfolded state is well represented by a peptide model, or that the mutation mainly affects the folded state. The greatest difference in $\Delta\Delta G$ between theory and simulation is for an Arg → His mutation (31), in which a substantially large number of atoms are mutated and a positive charge redistributed.

Yun-yu *et al.* (30) review several limitations of current methods for determining protein stability via simulations. Here we wish to address the question they raise of whether it is better to model the environment of a system (e.g. boundary conditions, non-bonded cut-off distances) as well as possible, or to match the environments of the folded and unfolded simulations so systematic errors will cancel when calculating the free energy difference. This issue arises because typically the environment of the peptide can be modelled with greater accuracy (e.g. using periodic boundary conditions) than that of the more computationally demanding folded protein (e.g. using stochastic boundary conditions).

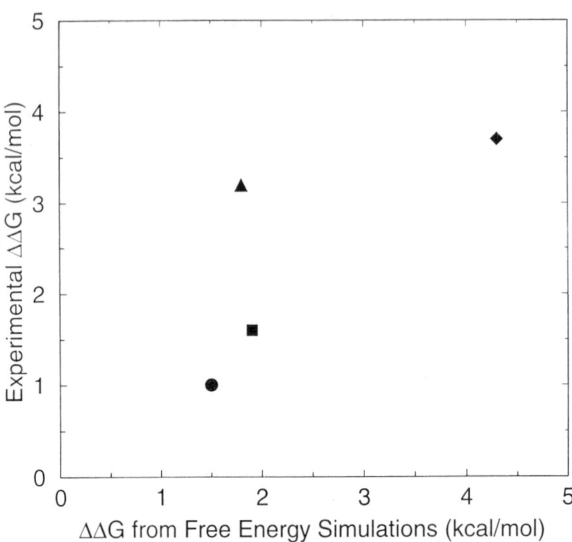

Figure 7. Comparison of calculated and experimentally-derived relative protein stabilities as a function of mutation. Data are from ◆ Prevost *et al.* (33), ▲ Tidor and Karplus (31), ■ Dang *et al.* (32), ● Yun-yu *et al.* (30).

In determining $\Delta\Delta G$ for protein stability via chemical perturbation, the systematic error term is $(\varepsilon_{M_u} - \varepsilon_{N_u}) - (\varepsilon_{M_f} - \varepsilon_{N_f})$, where ε is the systematic error for the system denoted by the subscripts. Any environmental artefact that is mutation-independent will cancel within the unfolded or the folded state calculation. If both these states have the same environment, then any additional conformation-independent artefacts will also cancel. This is not possible if unfolded and folded states have different environments. Increased accuracy in modelling the environment of the unfolded state reduces the magnitude of ε_{M_u} and ε_{N_u}, however, and it is not clear which is the better trade-off to make.

While calculating protein stability in barnase as a function of an Asn 215 → Ser mutation, Yun-yu *et al.* identified a convergence problem which lead them to question the general validity of current protein stability calculations. They performed two separate chemical perturbation simulations of barnase in the folded state, which differed by relatively small changes in simulation protocol. Residue 215 changed conformation early in the first simulation, giving a different hydrogen bonding pattern than expected for Ser in the end state. Reducing the initial rate of change in λ resulted in the expected conformation in a second simulation. Yun-yu *et al.* found that the electrostatic components of the free energy differences for the two simulations varied more widely than they expected, particularly when the contributions of protein–protein and protein–solvent free energy differences were determined separately. They argue that since similar paths were taken, these quantities should be similar.

We argue that the non-convergence problem was identifiable, and could thus be avoided. The path taken by a peptide through conformational space can affect the convergence of the free energy surface in chemical perturbation studies, just as it does in conformational free energy studies. Since a sudden change in conformation occurred in one of the Yun-yu *et al.* simulations, undersampling becomes an issue. There may not be enough 'connecting' conformations to accurately determine the difference in free energy as a function of λ. Careful analysis of the trajectory taken through conformational space is required to determine whether the free energy surface is correctly converged.

Undersampling issues aside, it is not unexpected that two simulations which have different ending structures show relatively large differences in free energy components. Unlike the total free energy difference, these components are path-dependent, and the path depends not only on λ, but on all of conformational space. Also, we have found in our studies that large differences in components can cancel to produce small differences in total relative free energy; for example entropy and energy are often correlated, a side chain participating in a strong non-bonded interaction does not exhibit the freedom of movement that a minimally interacting side chain does. When calculating the difference in free energy between the two states, differences in energy and entropy terms tend to cancel producing only small differences in free energy, although component analysis would show relatively large differences. Hence, we do not feel Yun-yu *et al.* results suggest that the determination of protein stabilities via chemical perturbation are currently unreliable.

Several general results emerge from protein stability simulations. The common practice of merely calculating interaction energy differences between the crystal structures of the wild-type and mutant proteins can be misleading as to the source of experimentally measured stability differences, since entropy effects or effects in the unfolded state can be important (31). In addition, free energy studies show stability differences can result from a large number of small differences in interactions, which would be difficult to discern in an interaction energy calculation.

It seems unlikely that the agreement between experiment and simulation shown in *Figure 7* is merely due to serendipity. Although protein stability calculations are still in their infancy, the results have been very encouraging given the many unknowns underlying the required assumptions. With the rapid advances in computer power, and the development of parallel computing (77–79), great improvements in the accuracy of these calculations should be possible.

3.4 Tertiary protein folding—refinement

Techniques are fast being developed to fold proteins using simplified models (3,80–83). Molecular mechanics will be extremely useful in moving from simplified representations to full atom models. Several efforts using molecular dynamics in the full atom refinement are underway (4,5), and here we focus on work by Vieth *et al.* (2).

Skolnick's group has developed methods to fold proteins on a lattice to predict backbone topology (3,80). The resulting structures are defined by C^{α} coordinates and coordinates for side chain centres of mass. From lattice simulations of the GCN4 leucine zipper dimer, Vieth *et al.* (2) chose five of the lowest energy structures as starting structures for further refinement. Their rms difference in C^{α} coordinates from the crystal structure ranged from 2.3 Å to 3.7 Å. A full backbone and side chain representation was generated from each structure using commercial software.

These starting structures were subjected to energy minimization, followed by solvation with a 6 Å shell of water. This system was again minimized, and several rounds of simulated annealing were performed to find minimum structures. Simulated annealing is composed of a series of molecular dynamics cycles, first at high temperatures followed by a slow cooling period and data collection at low temperatures. The high temperature dynamics effectively 'flattens' the free energy surface, allowing barriers to be traversed more freely in the search for a low energy structure. Slow cooling 'anneals' the structure, where it adopts the minimum energy conformation in that particular region. While this method has no guarantee of finding the global minimum, it has proven effective and efficient in finding low energy minima.

For each starting structure, a set of independent simulated annealing refinement cycles between the temperatures of 100 K and 800 K were performed, and the resulting low energy structures were 'averaged'. This mean structure was minimized, solvated, and further refined, and the whole cycle repeated until the mean structures 'converged' by having only small variations in cycle to cycle backbone rms values. All final mean structures from all starting conformations were again averaged to arrive at a final 'best guess' at the tertiary structure.

Averaging the resulting structures, to arrive at a 'best guess' structure for comparison to experimentally-derived structures, makes sense when one realizes that the native structure in solution is in reality a collection of structures, all within the native structure well on the free energy surface. Most experimentally-derived structures are the average of this multitude of conformations. In addition, if the free energy well can be approximated by a hyperparabola, the average structure will be near the minimum.

The resulting GCN4 dimer agreed remarkably well with the crystal structure, with a backbone rms difference of 0.81 Å, and a full atom rms difference of 2.29 Å. When the crystal structure is itself similarly refined, the resulting structure shows similar rms differences. These results are extremely encouraging in our goal of attaining full atom models from limited three-dimensional data.

4. Summary

Molecular mechanics and dynamics are in theory capable of solving the protein folding problem. In practice, however, additional information is needed to restrain the search space since *de novo* protein folding via MD is not computationally feasible. Many sources for this additional information are being developed and include experimental data, peptide simulations, topology prediction, and homology studies. With these additional sources of information, we may soon be able to routinely predict three-dimensional protein structure at atomic resolution.

Acknowledgements

M. E. K. gratefully acknowledges the generous support from the W.M. Keck Center for Computational Biology and the NIH postdoctoral fellowship grant GM14525. C. L. B. acknowledges partial support from the NIH through GM37554, GM48807, RR06009.

References

1. S. F. Sneddon and C. L. Brooks III. (1991). In *Molecular structures in biology* (ed. R. Diamond, T. F. Koetzle, K. Prout, and J. S. Richardson), pp. 114–63. Oxford University Press, Oxford.
2. M. Vieth, A. Kolinski, C. L. Brooks III, and J. Skolnick. (1994). *J. Mol. Biol.* **237**, 361.
3. J. Skolnick, A. Kolinski, C. L. Brooks III, A. Godzik, and A. Rey. (1993). *Curr. Biol.* **3**, 414.
4. C. W. G. van Gelder, F. J. J. Leusen, J. A. M. Leunissen, and J. H. Noordik. (1994). *Proteins* **18**, 147.
5. P. E. Correa. (1990). *Proteins* **7**, 366.
6. C. L. Brooks III and D. A. Case. (1993). *Chem. Rev.* **93**, 2487.
7. J. Hermans. (1993). *Curr. Opin. Struct. Biol.* **3**, 270.
8. C. L. Brooks III, M. Karplus, and B. M. Pettitt. (1988). *Proteins: a theoretical perspective of dynamics, structure, and thermodynamics* (ed. I. Prigogine and S. A. Rice). Advances in Chemical Physics, Vol. LXXI. Wiley, New York.
9. W. D. Cornell, A. E. Howard, and P. Kollman. (1991). *Curr. Opin. Struct. Biol.* **1**, 201.
10. M. J. Field, P. A. Bash, and M. Karplus. (1990). *J. Comput. Chem.* **11**, 700.
11. A. T. Brunger, J. Kuriyan, and M. Karplus. (1987). *Science* **235**, 458.
12. S. R. Wilson and W. L. Cui. (1990). *Biopolymers* **29**, 225.

13. M. Karplus. (1986). *Ann. N. Y. Acad. Sci.* **482**, 255.
14. H. Frauenfelder, S. G. Sligar, and P. G. Wolynes. (1991). *Science* **254**, 1598.
15. J. A. McCammon and S. C. Harvey. (1987). *Dynamics of proteins and nucleic acids*. Cambridge University Press, Cambridge.
16. M. Karplus, A. T. Brunger, R. Elber, and J. Kuriyan. (1987). *Cold Spring Harbor Symp. Quant. Biol.* **LII**, 381.
17. W. F. van Gunsteren and H. J. C. Berendsen. (1990). *Angew. Chem. Int. Ed. Engl.* **29**, 992.
18. H. J. C. Berendsen. (1991). *Curr. Opin. Struct. Biol.* **1**, 191.
19. D. Joseph, G. A. Petsko, and M. Karplus. (1990). *Science* **249**, 1425.
20. R. M. Venable, B. R. Brooks, and F. W. Carson. (1993). *Proteins* **15**, 374.
21. W. E. Harte Jr., S. Swaminathan, and D. L. Beveridge. (1992). *Proteins* **13**, 175.
22. W. E. Harte Jr. and D. L. Beveridge. (1993). *J. Am. Chem. Soc.* **115**, 1231.
23. P. H. Axelsen, M. Harel, I. Silman, and J. L. Sussman. (1994). *Protein Sci.* **3**, 188.
24. M. K. Gilson, *et al.* (1994). *Science* **263**, 1276.
25. D. J. Tobias and C. L. Brooks III. (1991). *Biochemistry* **30**, 6059.
26. D. J. Tobias, S. F. Sneddon, and C. L. Brooks III. (1991). In *Advances in biomolecular simulations* (ed. R. Lavery, J. L. Rivail, and J. Smith), Vol. 239, pp. 174–99. American Institute of Physics, New York.
27. M. E. Karpen and C. L. Brooks III. Unpublished results (1995).
28. M. E. Karpen and C. L. Brooks III. Unpublished results (1995).
29. S. F. Sneddon and D. J. Tobias. (1992). *Biochemistry* **31**, 2842.
30. S. Yun-yu, *et al.* (1993). *Protein Eng.* **6**, 289.
31. B. Tidor and M. Karplus. (1991). *Biochemistry* **30**, 3217.
32. L. X. Dang, K. M. Merz Jr., and P. A. Kollman. (1989). *J. Am. Chem. Soc.* **111**, 8505.
33. M. Prevost, S. J. Wodak, B. Tidor, and M. Karplus. (1991). *Proc. Natl. Acad. Sci. USA* **88**, 10880.
34. C. A. Laughton. (1994). *Protein Eng.* **7**, 235.
35. U. Rao and M. M. Teeter. (1993). *Protein Eng.* **6**, 837.
36. V. P. Collura, P. J. Greaney, and B. Robson. (1994). *Protein Eng.* **7**, 221.
37. P. J. Steinbach, R. J. Loncharich, and B. R. Brooks. (1991). *Chem. Phys.* **158**, 383.
38. R. L. Ornstein. (1990). *J. Biomol. Struct. Dyn.* **7**, 1019.
39. B. R. Brooks, *et al.* (1983). *J. Comput. Chem.* **4**, 187.
40. L. Verlet. (1967). *Phys. Rev.* **159**, 98.
41. J. L. Scully and J. Hermans. (1993). *Mol. Simul.* **11**, 67.
42. P. E. Smith, B. M. Pettitt, and M. Karplus. (1993). *J. Phys. Chem.* **97**, 6907.
43. J.-P. Ryckaert, G. Ciccotti, and H. J. C. Berendsen. (1977). *J. Comput. Phys.* **23**, 327.
44. N. Gronbech-Jensen and S. Doniach. (1994). *J. Comput. Chem.* **15**, 997.
45. G. Zhang and T. Schlick. (1994). *J. Chem. Phys.* **101**, 4995.
46. C. L. Brooks III. (1993). *Curr. Opin. Struct. Biol.* **3**, 92.
47. A. P. Allen and D. J. Tildesley. (1989). *Computer simulations of liquids.* Oxford University Press, Oxford.
48. C. L. Brooks III, A. Brunger, and M. Karplus. (1985). *Biopolymers* **24**, 843.
49. R. J. Loncharich and B. R. Brooks. (1989). *Proteins* **6**, 32.
50. R. W. Zwanzig. (1954). *J. Chem. Phys.* **22**, 1420.
51. D. L. Beveridge and F. M. Dicapua. (1989). *Annu. Rev. Biophys. Biophys. Chem.* **18**, 431.

52. E. M. Boczko and C. L. Brooks III. (1993). *J. Phys. Chem.* **97**, 4509.
53. J. P. Valleau and G. M. Torrie. (1977). In *Modern theoretical chemistry* (ed. B. J. Berne), Vol. 5, Ch. 4 and 5. Plenum Press, New York.
54. M. Levitt. (1983). *J. Mol. Biol.* **168**, 621.
55. R. Czerminski and R. Elber. (1989). *Proc. Natl. Acad. Sci. USA* **86**, 6963.
56. I. Chandrasekhar, G. M. Clore, A. Szabo, A. M. Gronenborn, and B. R. Brooks. (1992). *J. Mol. Biol.* **226**, 239.
57. M. E. Karpen, D. T. Tobias, and C. L. Brooks III. (1993). *Biochemistry* **32**, 412.
58. T. Ichiye and M. Karplus. (1987). *Proteins* **2**, 236.
59. A. Amadei, A. B. M. Linssen, and H. J. C. Berendsen. (1993). *Proteins* **17**, 412.
60. C. L. Brooks III. (1992). *J. Mol. Biol.* **227**, 375.
61. C. L. Brooks III. (1995). *Curr. Opin. Struct. Biol.* **5**, 211.
62. S. Swaminathan, T. Ichiye, W. F. van Gunsteren, and M. Karplus. (1982). *Biochemistry* **21**, 5230.
63. M. Levitt. (1983). *J. Mol. Biol.* **168**, 621.
64. D. Rojewska and R. Elber. (1990). *Proteins* **7**, 265.
65. T. Ichiye and M. Karpus. (1991). *Proteins* **11**, 205.
66. J. T. Coyle, D. L. Price, and M. R. DeLong. (1983). *Science* **219**, 1184.
67. S. A. Eagger, R. Levy, and B. J. Sahakian. (1991). *Lancet* **337**, 989.
68. M. Harel, *et al.* (1993). *Proc. Natl. Acad. Sci. USA* **90**, 9031.
69. W. L. Jorgensen, J. Chandrasekhar, J. Madhur, R. W. Impey, and M. L. Klein. (1983). *J. Chem. Phys.* **79**, 926.
70. S. F. Sneddon, D. J. Tobias, and C. L. Brooks III. (1989). *J. Mol. Biol.* **209**, 817.
71. C. L. Brooks III and L. Nilsson. (1993). *J. Am. Chem. Soc.* **115**, 11034.
72. J. S. Richardson and D. C. Richardson. (1988). *Science* **240**, 1648.
73. J. Hermans, A. G. Anderson, and R. H. Yun. (1992). *Biochemistry* **31**, 5646.
74. R. H. Yun and J. Hermans. (1991). *Proteins* **10**, 219.
75. F. C. Bernstein, *et al.* (1977). *J. Mol. Biol.* **112**, 535.
76. J.-F. Gibrat, B. Robson, and J. Garnier. (1991). *Biochemistry* **30**, 1578.
77. C. L. Brooks III, W. S. Young, and D. J. Tobias. (1991). *Intl. J. Supercomput. Appl.* **5**, 98.
78. S. E. DeBolt and P. A. Kollman. (1993). *J. Comput. Chem.* **14**, 312.
79. S. L. Lin, J. Mellor-Crummey, B. M. Pettitt, and G. N. Phillips Jr. (1992). *J. Comput. Chem.* **13**, 1022.
80. A. Kolinski and J. Skolnick. (1993). *J. Chem. Phys.* **98**, 7420.
81. T. Dandekar and P. Argos. (1994). *J. Mol. Biol.* **236**, 844.
82. D. G. Covell. (1994). *J. Mol. Biol.* **235**, 1032.
83. M. J. Smith-Brown, D. Kominos, and R. M. Levy. (1993). *Protein Eng.* **6**, 605.
84. A. Wlodawer, *et al.* (1989). *Science* **245**, 616.
85. A. G. Anderson and J. Hermans. (1988). *Proteins* **3**, 262.
86. P. J. Kraulis. (1991). *J. Appl. Crystallogr.* **24**, 946.

11

Docking ligands to proteins

BRIAN K. SHOICHET

1. Introduction

A docking method seeks to find ways of fitting two molecules together in favourable configurations. This simple goal is actively pursued because of its central role in two very practical areas of research: structure-based drug discovery and the molecular modelling of biological function. Sadly, our understanding of how molecules recognize one another is still rather poor. Both the importance of the field, and our limited grasp of its intricacies, have lead to a great number of proposals for solving what is often called 'the docking problem.' None of them is complete. For clarity's sake, I will discuss ways to approach docking problems as if there was a right way, and we knew what we were doing. Sometimes there isn't and we don't. Buyer beware.

Having begun with a caveat, I hasten to add that we have learned a great deal in the last ten years. We can automatically locate regions on proteins that are candidates for ligand binding ('hot-spots'). The components of a ligand–protein complex can be reliably and accurately reassembled, if they are in the proper (bound) conformations. Mounting evidence suggests that docking can discover novel lead inhibitors that bind in the micromolar concentration range, though many difficulties remain (1,2). Structure-based elaboration of novel leads can produce inhibitors that bind in the nanonmolar and picomolar concentration ranges (3,4).

Ambiguities attend like surly waiters on most docking calculations. To overcome this difficulty, I will emphasize problems for which docking can suggest experimentally testable hypotheses. Wherever possible, I shall outline controls for docking calculations. These are often possible and always worthwhile.

I will familiarize you with approaches to three common goals of docking programs:

- fitting a small molecule into a protein
- docking two proteins together
- novel inhibitor discovery using molecule databases

For any of these you will need:

- an atomic resolution structure of your protein (*Table 1*)
- an atomic resolution structure of your ligand or ligands (*Table 1*)
- a scientific-style computer workstation with high resolution graphics
- a docking computer program (*Table 2*)

2. Docking programs

There are a number of different docking computer programs available to the reader. I will illustrate most examples using just one of these, the Dock program, introduced by Kuntz and colleagues (5–7). It is appropriate to begin, however, with a survey of the field.

All docking programs must solve three fundamental problems:

- where, in a relatively large protein site, to fit the ligand
- what conformations of ligand and protein best complement each other
- how to evaluate the energies of the various complexes

Table 1. Molecular databases

Type of molecule	**Source**	**Comments**
Protein or DNA	Protein Data Bank (PDB) (47) Protein Data Bank, Chemistry Department, Brookhaven National Laboratory, Upton, NY 11973, USA e-mail: pdb@bnl.gov	Atomic resolution protein and DNA structures, mostlyfrom X-ray and NMR experiments
Small molecule	Cambridge Structural Database (CSD) (38) Dr Olga Kennard FRS, Cambridge Crystallographic Data Centre, University Chemical Laboratory, Lensfield Road, Cambridge CB2 1EW, UK	X-ray determined structures 10^5 compounds
Small molecule	Available Chemical Directory (ACD), MDL Information Systems, Inc, San Leandro, CA 94577, USA	Computer-generated structures commercially available, 1.5 x 10^5 compounds
Small molecule	Marketing Department, Chemical Abstracts Service, 2540 Olentaey River Road, PO Box 3012, Columbus, OH 43210–0012, USA	Computer-generated structures of literature compounds, approx. 6 x 10^6 compounds
Small molecule	Triad and Iliad Dr Paul A. Bartlett, Department of Chemistry, University of California, Berkeley, CA 94720, USA; e-mail: paul@fire.cchem.berkeley.edu	Computer-generated structures hydrocarbon scaffolds, tricyclic and acyclic, respectively, 4×10^5 and 1×10^5 compounds

Table 2a. Some docking programs

Program	Algorithm	Small ligand docking?	Protein–protein docking?	Database search?	Ligand design	Conformation search	Scoring
Rendezvous	Grid search	Yes	Yes	No	No	Yes	Force field
Soft docking	Grid search	Yes	Yes	No	No	No	Polar/apolar contact
Grow	Fragment	Yes	No	No	Yes	Yes	Force field
Concepts	Fragment	No	No	No	Yes	Yes	Force field
Ludi	Fragment	Yes	No	Limited	Yes	Possible	Force field
Hook	Fragment	Possilbe	No	No	Yes	Possible	Force field
AutoDock	Kinetic	Yes	Yes	No	No	Yes	Force field
Dock	Descriptor	Yes	Yes	Yes	No	No	Force field
Caveat	Descriptor	Yes	No	Yes	No	No	Steric fit
Clix	Descriptor	Yes	ND	Yes	No	No	Force field

Table 2b. Corresponding authors for docking programs

Program name	Corresponding author
Rendezvous Batiment	Joel Janin, Universite Paris-Sud, Laboratoire de Biologie Physicochimique, 433, 91405 Orsay Cedex, France Shoshana Wodak, Universite Libre de Bruxelles, 67 Rue des Chevaux, Rhode-St-Genese, 1640, Belgium
Soft docking	Sung-Hou Kim, Department of Chemistry, University of California, Berkeley, CA 94720, USA
Grow	Joeseph B. Moon or Jeffrey Howe, The Upjohn Company, Computational Chemistry, 301 Henrietta Street, Kalamazoo, MI 49001, USA
Concepts	David Pearlman, Vertex Pharmaceuticals Inc., Cambridge, MA 02139–4211, USA
Ludi	Hans-Joachim Bohm, BASF AG, Central Research, D-6700 Ludwigshafen, Germany Distributed by Biosym Technologies Inc., 1515 Rt 10, Suite 1000, Parsippany, NJ 07054, USA
Hook	Martin Karplus, Department of Chemistry, Harvard University, Cambridge, MA 02138, USA
AutoDock	Arthur J. Olson, The Scripps Research Institute, Department of Molecular Biology, MB5, 10666 North Torrey Pines Road, La Jolla, CA 92037, USA
Dock	Irwin D. Kuntz, Department of Pharmaceutical Chemistry, School of Pharmacy, University of California, San Francisco, CA 94143, USA
Caveat	Paul Bartlett, Department of Chemistry, University of California, Berkeley, CA 94720, USA; e-mail: `paul@fire.cchem.berkeley.edu`
Clix	M. C. Lawrence, CSIRO, Division of Biomolecular Engineering, 343 Royal Parade, Parvill, Victoria, 3052, Australia

Solving these problems rigorously requires elaborate energy calculations (8) and consideration of many more conformational and configurational degrees of freedom than is now computationally feasible for a docking method. To get reasonable answers in reasonable amounts of time, docking algorithms simplify the problem. Common approximations include modelling explicit waters by a dielectric continuum, using enthalpy as a proxy for free energy, and treating intrisincally flexible ligands and proteins as rigid objects.

Docking programs may be grouped into four families, depending on how they address these problems, and what simplifications they use (2). These families are: descriptor methods, grid methods, fragment methods, and kinetic methods (*Table 2*). All begin with the structure of a protein and that of a ligand, or ligands, and ask: where does the ligand bind, and, for inhibitor discovery, which ligand binds best?

2.1 Docking with descriptors

The protein is first analysed for regions of likely complementarity. These 'hot-spots' are places on the protein surface where a ligand atom might fit well; they describe the binding region. The idea is to match ligand atoms to receptor hot-spots, and by so doing generate orientations of the ligand in the protein. For any given ligand many orientations are sampled. Dock, for instance, might try several thousand orientations for a small ligand while a protein ligand might sample tens or even hundreds of thousands of orientations. Orientations are evaluated for goodness of fit, typically through use of a molecular mechanics-type energy function. Though not exhaustive, descriptor methods are fast and can often sample densely in a particular region of the protein. They rely on being able to identify the hot-spots well, a point to which I shall return. Most descriptor methods treat ligands and proteins as rigid objects, though this is not inherent to the algorithms (see for instance ref. 9). Descriptor programs have been used extensively for single ligand docking and for inhibitor discovery (Dock, Caveat, and Clix programs).

2.2 Grid search

Grid searches fit the ligand into the protein by rotating and translating the ligand in discrete steps while holding the protein rigid. Grid searches have the advantage of always getting to the neighbourhood of the correct solution, which cannot be said of the descriptor matching methods. They often take a long time. The accuracy of grid methods is limited to the resolution of the step size used in the search, but the higher the accuracy the longer the search. Most grid methods treat the ligand and protein as rigid objects, though some allow for conformational relaxation (10). These programs have been used extensively for single ligand docking problems (10,11,49).

2.3 Fragment

Fragment methods identify regions of high complementarity on a protein surface by docking functional groups independently into proteins. By breaking ligands into fragments, many of the configurational and conformational issues in docking disappear. This is done at the expense of connectivity information, which fragment methods can in principle gain back by reconnection algorithms at the end of the calculation. Fragment methods can be used for molecular elaboration of existing inhibitors. They are useful for novel inhibitor design when working with molecules whose synthetic chemistry is as modular as the computer-generated fragments, such as peptides and oligonucleotides. The Concepts (12), Ludi (13), and Grow programs (14) are good examples of this idea, with the latter having been experimentally tested.

2.4 Kinetic

Kinetic docking techniques sample potential surface, using molecular dynamics or simulated annealing, to fit ligands to proteins. Like the grid methods, kinetic approaches are conceptually appealing because they mimic how we imagine ligands encounter proteins in solution. An advantage of kinetic methods is that they merge the configurational and conformational aspects of the docking problem smoothly. A disadvantage is that the complex topography and multiple minima of molecular potential surfaces often lead to long run times and minima traps. They have been used extensively in single ligand docking, see especially the method of Goodsell and Olson (15) and its application by Stoddard (16), and the recent method of Abagyan (50).

3. Preparing your input

All docking projects begin with the structure of the ligand and the protein. One has important decisions to make regarding the reliability of these structures and which regions on them to target most heavily.

3.1 Using protein structures

Atomic resolution protein structures have three sources: X-ray crystallography, nuclear magnetic resonance (NMR), and homology model building. The last of these is described in Chapters 6 and 7; my only caution here is that model-built structures are the riskiest to employ, but see Ring *et al.* (17) for their use in inhibitor design, and Bax *et al.* on protein–protein docking (18). I will focus on what regions of experimental structures require consideration before docking.

While experimental structures give you coordinates for most or all atoms in a protein, they are not equally trustworthy or useful. For example, most X-ray structures include bound waters and ions along with protein atoms. Many of these waters and ions, unlike the protein atoms, are displaceable by a ligand. Some, however, are tightly bound and are best considered constitutive parts of the protein. Most docking programs have no way of knowing which is which; you must decide which of these displaceable groups to include and which to exclude before the docking program sees your protein. Another common problem is that some protein atoms are ill-determined. Some appear to be poorly constrained in the structure (high thermal factors), others are ill-resolved (poor electron density, few NOE peaks). You must decide which atoms are reliable before beginning a docking calculation. In happy circumstances the structures of several different protein conformers will be available to you; which should you use?

Eliminate as many waters and ions as possible to have the largest protein or protein ligand surface possible to dock to. In most docking calculations performed in the Kuntz group in the last several years, *all* waters and ions were eliminated. This is the simplest approach, but it has several times lead to problems (7,19). Nevertheless, I recommend deleting all waters and ions in a potential binding site except those that fulfil at least one of the following criteria:

- they occur in three or more different states of the protein structure (e.g. different conformers, different crystal forms, different ligands bound)
- they have unusually low thermal factors
- they have three or more polar interactions with protein atoms
- they have been implicated in the mechanism of the protein

Poorly determined protein atoms can lead to the same sorts of difficulties as waters and ions; they preclude ligand binding to the volume that they occupy. Nevertheless, we usually include all the atoms present in the published structure. It is occasionally useful to remove the following atoms from the binding site of a protein (20):

- atoms with occupancies of zero
- atoms with thermal factors greater than 60 $Å^2$
- poorly constrained regions of NMR structures showing large deviations from model to model

Zero occupancy atoms can be removed without much angst, but the thermal factor and NMR constraint criteria are riskier. Deleting protein atoms should be resorted to only when one suspects that they are interfering with a docking calculation (20).

There are occasionally several structures of the same protein. These may have been solved in complexes with different ligands, or in different crystal forms. Use the experimental form that most closely resembles the state that you expect the protein to be in when it binds the ligand. For novel inhibitor discovery consider using *all the* protein conformations, sequentially. This will not seriously increase the computation time and will improve one's chance of getting a highly complementary ligand–protein fit. Any well-determined protein structure is a good target for inhibitor discovery.

3.2 Targeting binding sites

Any docking project will benefit from knowing, in advance, what parts of the protein surface are most likely to fit the ligand, and what parts can be ignored. For a descriptor method, such as Dock, such information is critical.

3.2.1 Sphere descriptors

There are two computational ways to identify these hot-spots: geometrically and energetically. An experimentally-based method uses the location of a bound ligand. Dock has traditionally relied on a geometric method that maps clefts or ridges on the surface of a protein (5) (*Protocol 1*), though ligand coordinates have also been used to map the site (7). Spheres are calculated to fit, but not intersect, local pockets in the molecular surface of a protein (*Figure 1*). Such pockets are good places to try to fit a ligand atom in docking;

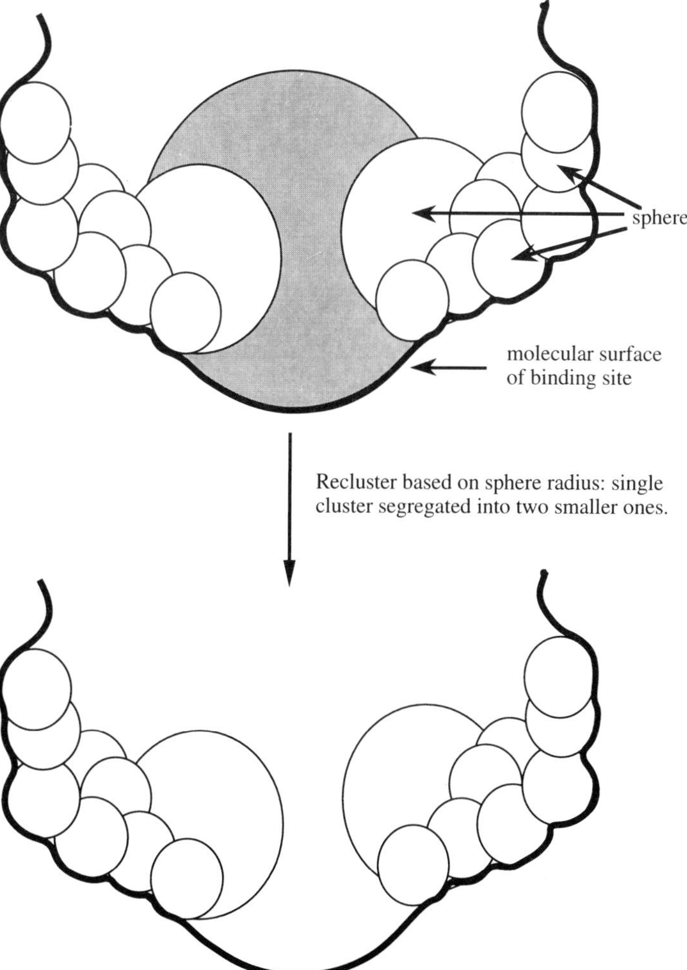

Figure 1. Dock spheres and the subclustering algorithm. Subclustering splits up large binding sites by segregating groups of spheres based on their radii. Here, a large radius sphere (shaded) is removed from the sphere set, creating two subsites.

they are a positive image of the negative mould presented by a protein surface to a ligand. The spheres are produced as intersecting clusters—there will usually be several clusters on the surface of a protein, each defines a putative binding site in that surface. A typical size for a binding site cluster used in docking is 30–60 spheres. The more spheres in a cluster the larger the site that Dock will explore and the longer it will take to do so. Use the smallest number of spheres, or any type of descriptor, as possible. If there are more than 100 spheres in your putative binding site, divide the clusters into smaller groups of spheres and dock to each one separately. Reclustering into smaller sphere sets can be done with an editor, or by using the program Cluster, which is distributed with Dock for this purpose.

Protocol 1. Describing the binding site with docking spheres

1. Make a copy of the protein coordinates on the disk.
2. Remove most water molecules, ions, and bound ligands. Occasionally ligands help define the target sites (e.g. co-factors); leave these in.
3. You can generate spheres to a predefined site or to the entire protein, using the clusters so produced to define the site. The second method can be used to identify non-active site binding regions (21), but use the first if you know what region of the protein you wish to target. To define such a region, pick several residues (one may be sufficient) and generate a list of residues proximal to them.
   (a) Use all residues that have at least one atom within a certain distance (e.g. 10 Å) of at least one atom from the original residues that you picked.
   (b) If you have the UCSF (University of California at San Francisco) version of MS (*Table 4*), use the -i flag to restrict the molecular surface to your site of interest. If you have the Quantum Chemistry Program Exchange version of the program you must flag the atoms individually in the reformatted coordinates file (set the surface request number to 2 for atoms in the site, set all others to 1).
4. Use MS (*Table 4*) and Sphgen (distributed with Dock) to calculate spheres, as described in the Dock manual.
5. Look at the sphere clusters in the Sphgen output file. These represent a small, heuristically chosen subset of the total spheres calculated, the rest are in the large temp files also output by the program. Use Showsphere (distributed with Dock) to generate PDB formatted spheres file and concomitant sphere surface file. Inspect these graphically. If they describe the binding site adequately, stop.
6. If the sphere centres are unsatisfactory, consider the larger set also output by Sphgen.

(a) Convert the temp files to a spheres file with Tosph (distributed with Dock). There will typically be a few thousand spheres in this file.

(b) Using an editor and a graphics program, flag several of the spheres that are in hot-spots on the protein surface.

(c) Use Cluster to calculate a new sphere set that is biased towards the regions of the surface that you have flagged (see Dock manual).

7. For clusters containing more than 100 spheres, delete large radius spheres on the edge of the site. If this does not remove enough spheres, or you have difficulty choosing which spheres to delete, the Cluster program will do this automatically. Often, several subclusters are produced that can be docked to sequentially. Reclustering spheres is essential for protein ligands (20).

3.2.2 Alternative site description methods

High potential energy regions on the surface of a protein can also locate hot-spots. Such high potential regions will complement a ligand atom of opposite atomic properties. Thus an electropositive region of the protein will interact favourably with a negatively charged ligand atom, a hydrogen bond donor will do the same with a hydrogen bond acceptor, and so on. Protein surfaces may be mapped for high energy regions using methods such as those listed in *Table 3*. The Grid program has been used most extensively for this purpose (22). The same rules apply to energy-based descriptors as to spheres: one

Table 3. Programs to determine high potential energy regions on protein surfaces

Energy surface mapped	Program	Corresponding author or distributor
Molecular mechanics	Grid	Peter Goodford, The Laboratory of Molecular Biophysics, The Rex Richards Building, University of Oxford, Oxford OX1 3QU, UK
Molecular mechanics	Chemgrid	Irwin Kuntz, Department of Pharmaceutical Chemistry, School of Pharmacy, University of California, San Francisco, CA 94143, USA
Electrostatic	Delphi	Barry Honig, Department of Biochemistry and Molecular Biophysics, Columbia University, 630 168th Street, New York, NY 10032, USA Also distributed by Biosym Technologies Inc., 9685 Scranton Road, San Diego, CA 92121, USA
Hydrogen bonding	Hsite	Philip Dean, Department of Phamacology, University of Cambridge, Tennis Court Road, Cambridge CB2 1QJ, UK
Hydrophobicity	Hint	Donald Abraham, Department of Medicinal Chemistry, Virginia Commonwealth University, Richmond, VA 23298–0540, USA

needs enough high potential points to map adequately the site, but having too many leads to excessive computation times.

If one knows the structure of a ligand–protein complex, and the ligand is bound in the site you wish to target, then it is a simple matter to use the ligand coordinates themselves as the hot-spots with which to dock other ligands. Assign types to ligand atom descriptors in the same manner as spheres to improve run time and selectivity. Ligand atom descriptors can be usefully combined with computed descriptors to expand the limits of the binding site.

We find it useful to combine the sphere and energy approaches to site description. Spheres are calculated as usual, but are 'labelled' according to chemical potential. A sphere in an electropositive region is 'labelled' positive, one in an electronegative region is 'labelled' negative, and so on. We only allow matches between spheres and atoms with complementary chemical labels. This speeds the docking calculation tenfold and improves selectivity (23).

3.2.3 Experimental constraints

A final site description issue is that of using external constraints in the docking calculations. Experimental data, such as mutagenesis experiments, often suggest that a certain residue is directly involved in ligand binding. Including this information in a docking calculation can dramatically improve selectivity (24) and run time (25). Dock allows one to flag critical protein spheres and ligand atoms and insist that these be involved in any docked complex.

3.3 Scoring docked complexes: energy calculations

Most docking methods calculate an interaction energy to evaluate and rank ligand–protein complexes. This energy usually includes some or all of the terms found in molecular mechanics force fields, such as dispersion, polar, hydrogen bonding, and occasionally hydrophobic interactions (Chapter 10). To calculate the interaction energy (E_{int}) of a given docked complex one needs to know the potential energy of the protein (P) and the atomic properties of the ligand (A) at all ligand atom positions (i):

$$E_{int} = \sum_{i=0}^{n} P_i A_i \qquad [1]$$

The potential P is usually calculated once, and stored on a lattice of points in memory during the docking calculation (see Dock manual for details). Every docked ligand orientation is fit on to this lattice and its energy calculated by summing the product of the lattice point energies and the atomic properties. Lattice energies can be calculated with programs such as Grid, Delphi, and Chemgrid (*Table 3*).

Ligand atomic properties, such as partial charges and van der Waals parameters, can be calculated using a molecular editor such as exist in pro-

Table 4. Some useful computer programs for docking projects

Purpose	Program name	Corresponding author or distributor
Macromolecular graphics	MidasPlus	Tom Ferrin, Department of Pharmaceutical Chemistry, University of California, San Francisco, CA 94143–0446, USA e-mail: `tef@socrates.ucsf.edu`
Macromolecular graphics	Grasp	Anthony Nichols, Department of Biochemistry and Molecular Biophysics, Columbia University, 630 168th Street, New York, NY 10032, USA e-mail: `nicholls@cuhhca.hhmi.columbia.edu`
Molecular surface	MS	Quantum Chemistry Program Exchange, Department of Chemistry, University of Indiana, Bloomington, IN 47405, USA Also distributed with MidasPlus
Molecular mechanics	AMBER	Peter Kollman, Department of Pharmaceutical Chemistry, University of California, San Francisco, CA 94143–0446, USA e-mail: `pak@socrates.ucsf.edu`
Molecular mechanics	CHARMM	Martin Karplus, Department of Chemistry, Harvard University, Cambridge, MA 02138, USA
Molecular modelling, editing and graphics	Sybyl	Tripos Associates, Inc., 16299 S. Hanley Road, Suite 303, St. Louis, MO 63144, USA
Molecular modelling, editing and graphics	MacroModel	Office of Science and Technology Development, Columbia University, New York, NY 10027, USA
3D ligand structures from 2D representations database conversions	Concord	Tripos Associates, Inc., 16299 S. Hanley Road, Suite 303, St. Louis, MO 63144, USA
Multiple conformations for small molecules	Cobra	Oxford Molecular Ltd., Terrapin House, South Parks Road, Oxford OX1 3UB, UK
Ligand solvation energy	Hydren	Alexander Rashin, Biosym Technologies Inc., 1515 Rt 10, Suite 1000, Parsippany, NJ 07054, USA
Small molecule database, substructure searching	Isis	MDL Information Systems, Inc, San Leandro, CA 94577, USA
Substructure searching	Aladdin	Daylight Chemical Information Systems, 18500 Von Karman Avenue 450, Irvine, CA 92715, USA

grams like Sybyl and MacroModel (26) (*Table 4*). One typically needs at least partial atomic charges. If the ligand is a nucleotide or peptide such charges and parameters exist in parameter files distributed with programs such as AMBER, CHARMM, or GROMOS (*Table 4*) or can be found in the literature. Small molecule parameters, especially hydrophobicities, can often also be found in the literature (27), but one must usually calculate charges oneself.

4. Docking a small molecule to a protein

We begin with the structure of a protein and a small molecule (generally less than 50 heavy atoms) and ask: where best does the ligand fit? *Protocol 2* summarizes the preparatory steps we have just considered.

Protocol 2. Preparatory steps before docking

1. Remove appropriate waters and ions.
2. Calculate a potential energy lattice for the protein.
3. Locate descriptor positions on the protein surface (spheres, ligand atoms, points of high potential).
4. Calculate or look up ligand parameters such as partial atomic charge.
5. Calculate alternate conformations for the ligand, if desired (Cobra (28), Sybyl, or MacroModel programs, *Table 4*).

4.1 Controls

Before beginning the docking calculation, perform the following controls, if possible:

(a) Calculate an interaction energy for a known complex of a ligand with your protein. Map the ligand atomic positions from the complex on to the lattice (subprograms to do this are often distributed with the energy potential program) to find the energy (see Equation 1). If the energy is positive (poor fit), you have probably made a mistake in your potential energy calculation or your ligand parameters (*Protocol 3*).

(b) Use Dock to reproduce the complex between a known ligand and the protein, using their bound conformations. This will allow you to see if your docking parameters (below) are sensible. One should be able to reproduce a known complex to better than 1 Å root mean square deviation (rmsd) (*Protocol 4*).

If a complex structure is unavailable, consider doing these controls on another protein–ligand system. At least you will learn whether the general steps you have taken are correct.

Protocol 3. Insuring that a known ligand–protein complex has a negative interaction energy

The purpose is to test your ligand parameterization and lattice energies. Begin with the coordinates of a known ligand–protein complex.

Method

1. Does the ligand stericaly fit the protein site? Use the program Score-opt2 (distributed with Dock) to test this. If there is a close contact between the ligand and the protein, consider reducing the close contact limits and recalculating the contact map. Alternatively, delete poorly determined protein or ligand atoms if these are responsible for

the contact. Insure that the close contact does not involve a water, ion, or ligand inadvertently left in the site.

2. Does the ligand electrostatically fit the protein site? Use Scoreopt2 again, this time with the Delphi phi map (see Dock manual). If the interaction energy is positive, examine the ligand partial atomic charges and the protein potentials at the ligand atom positions. Common problems include:
   (a) Lack of hydrogen atoms on ligand and protein hetero-atoms.
   (b) Mis-parameterized protein residues (check the Delphi output to insure that the overall charge on the protein is sensible and an integer).
   (c) Waters, ions, or ligands included in the site for the electrostatic calculation but not the Distmap calculation.
   (d) Mis-parameterized or non-parameterized, and therefore unrecognized, waters, ions, or ligands.
3. The same approach is appropriate when using force field scoring (Dock manual).

4.2 The docking calculation

Docking programs often involve choosing values for numerous adjustable variables. Dock has nine important ones (*Table 5*); it is unfortunately necessary to play with these numbers in many cases (see the Dock manual). The goal is to sample the site with the ligand densely enough to get the correct answer, but not so densely that the calculation takes forever. This is usually possible with a small molecule. For rigid body docking, expect a calculation time of less than one hour on a Silicon Graphics Indigo 2 or similar computer. Use external constraints, such as those from mutagenesis or kinetic experiments, if you can. Try to generate more configurations of the ligand in the site to see how robust the docking solution is energetically and geometrically. Rigid body energy minimization within Dock can improve the ability of the program to find favourable configurations (29).

Protocol 4. Single ligand docking control

Can you reproduce the experimental configuration of a known ligand–protein complex? Use *Protocol 3* to insure that you have parameterized the ligand and mapped the protein potential correctly.

Method

1. Begin with the parameters in *Table 5* to dock the ligand to the site. If Dock produces configurations within 1 Å rmsd of the bound structure

Protocol 4. *Continued*

the control is successful. If you wish to perform a database search calculation, pay attention to the number of orientations sampled (NMATCH, reported in the OUTDOCK file).

2. To reduce the number of matches, decrease the bin sizes. If you have trouble getting a good correspondence between search time and reproduction of the experimental structure, try lowering *dislim* or increasing *nodlim*. Using rigid body minimization with limited sampling is another sensible strategy (29).
3. If no configurations close to the experimental complex are produced, experiment with the following Dock input parameters:
   (a) Increase the number of matches sampled by increasing the bin sizes.
   (b) If you have had focusing turned off, turn it on.
   (c) Increase the number of allowed close contacts in focusing (*fctbmp, Table 4*). For protein–protein docking, an *fctbmp* of as high as 15 is not unwarranted. For small molecules *fctbmp* should not exceed 3.
   (d) Try increasing *dislim* or decreasing *nodlim*.
4. If the experimental configuration is still not found, look for errors in your site description:
   (a) Do the protein descriptor centres (e.g. sphere centres) overlap your ligand atoms? If fewer than six do so, consider recalculating your spheres.
   (b) If you have labelled your spheres and atoms by chemical type, insure that the types you have chosen are complementary.
   (c) Check the format of your sphere and atom input files. Are the correct numbers of spheres and atoms being read? Insure that the longest distances between sphere and atom points are sensible (OUTDOCK file).
5. Is the ligand, or the ligand sphere set for protein ligands, bigger than the protein sphere set? Consider using fewer ligand atoms or spheres to match with protein descriptors, while continuing to fit the entire set of ligand atoms into the protein site. Create a separate, sphere formatted file using the atom coordinates as descriptor centres. Redock with only a subset of these descriptors (see Dock manual). For protein–protein docking, consider reducing the size of your ligand sphere set. Alternately, consider increasing the number of spheres in the protein cluster.

It is impossible to know whether the docking is correct without experimental verification. When X-ray or NMR experiments are impractical, consider

Table 5. Dock version 3.0 database search: sample input

Variable	Value	Meaning
versn	2	Scoring scheme (2 = electrostatic and contact)
mode	search	Single ligand docking or database search
clufil	file	File containing cluster
nclus	3	Sphere cluster—can search multiple sites
mapnam	file	File containing contact grid
dislim nodlim ratiom **lownod**	**1.5 5** 0. **5**	Distance tolerance on atom-sphere matches Number of matches to orient ligand Ignore Minimum matches to orient ligand
lbinsz lovlap sbinsz sovlap	**.1 .1 .2 .2**	Bins: sphere and atom internal distance histograms, used for matching
ligfil	file	Database filename
outfil	file	Output filename
ictbmp	0	Number of close contacts allowed between a ligand and the protein
natmin natmax nsav	5 100 200	Minimum number of atoms in a ligand to dock Maximum number of atoms in a ligand to dock Number of ligands to save
irestr moltot molsav	0 60000 100	Restart run? Number of molecules to search Save results every molsav ligands
inchyd	Y	Include ligand hydrogens?
expmax (0 or 1)	**1**	Focus—amplifies the number of complementary dockings when value is 1
fctbmp (1)	**1**	Number of close contacts allowed to signal focusing algorithm
phifil (filename)	name	File containing electrostatic grid

Variables in **bold** face affect the amount of ligand orientational sampling.

other tests. For instance, your docking calculations will suggest roles for specific residues in recognizing your ligand, you may often test such roles with site-directed mutagenesis.

4.3 Example: docking phenolphthalein to thymidylate synthase

Some of the possibilities and limitations of a single molecule docking project may be illustrated with an example from our own work (7). We had discovered a new class of micromolar inhibitors, the phenolphthalein family of compounds, of the cancer target thymidylate synthase (TS) and wished to predict an accurate binding mode before the crystallographic solution. We began with:

- the structure of the enzyme in its open conformation
- the location of a lead (but dissimilar) inhibitor in the enzyme site
- the computer-generated structures of a number of analogues that inhibited the enzyme

We used the location of the atoms of the lead inhibitor as descriptors, and calculated a molecular electrostatic potential using the Delphi program. We eliminated all waters from the structure but retained two phosphate ions that were well resolved crystallographically and which occurred, in one form or another, in every crystal structure of TS of which we knew (then over 30). Ligand partial atomic charges were calculated with the Gasteiger–Marsilli (30) algorithm using Sybyl. We assumed that all analogues bound in the same manner, and insisted that a docking of any single analogue must be able to accommodate them all. This reduced the number of possible dockings to two families of highly related orientations, of which one was favoured energetically.

The crystallographic complex, when solved, showed the ligand in the same binding region as the docked complex, but translated by 1.4 Å and rotated by 43° (*Figure 2*). In the crystallographic structure, the inhibitor formed hydrogen bonds with two bound water molecules. One of these waters bound in a

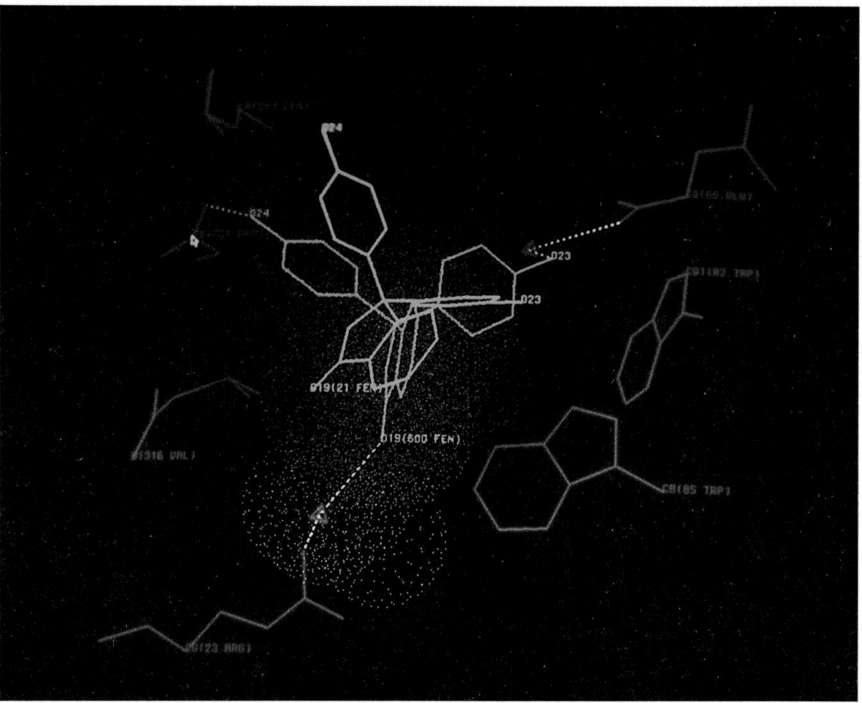

Figure 2. Superposition of experimental (green) and low energy Dock consensus configuration (cyan) positions of phenolphthalein in the target site of TS (graphics made with MIDAS+) (48). Descriptors used in the docking calculations are represented as magenta spheres, enzyme side chains are coloured red. Ligand binding waters are represented as red tetrahedrons, their interactions with TSs residues and phenolphthalein are represented as dashed yellow lines.

novel site in TS, while the second occurred in several other crystal structures of the enzyme. The novel water interacted with an arginine that appeared in a significantly different conformation in the liganded form of the protein than the form that we had used for our modelling. In the unliganded form this arginine was poorly determined crystallographically. There were smaller accommodations involving several other residues, but for the most part the protein structure was unchanged. While we were able to get close with our docking calculations, the precise ligand geometry eluded us because of the roles of specific waters, and that of a residue whose conformation changed, for which we did not account. This is a good example of the sort of problems one can run into from waters, ions, and mobile residues.

5. Docking a protein ligand to a protein receptor

The fundamental problem is that there are many possible interfaces between two proteins, and considering all of them would take a great deal of time (31). It is important, even more so than in the small molecule case, to target regions on the protein surfaces that are likely to interact. One can target more than one region, but consider each separately. This is especially important for descriptor methods.

5.1 Defining interaction regions

Divide both the ligand and the receptor into different possible interaction regions. Interaction regions can be defined using experimental information (16), for instance from mutagenesis or mechanistic considerations. An alternative is to section the surface using structural criteria, such as the sphere method of Kuntz (5,20) (*Figure 1*). When using Dock, use less than 100 descriptors in any given receptor or ligand site.

It is sometimes sensible to eliminate flexible, poorly determined atoms in the binding region from the initial fitting calculations (20). An alternative is to increase the tolerances for close receptor–ligand contacts for such residues, to increase the number of allowed close contacts, or to model these residues as multiple conformations (32). Removing these atoms will reduce the potential complementarity of some complexes but will allow other complexes to be explored (see preparation of input for a rough guide). Use the coordinates of all atoms in the structure to evaluate the complexes once docked, even highly flexible or poorly determined ones.

5.2 Choosing the best docked complexes

The energy of the docked complexes may be evaluated by a number of different criteria, including:

- energy minimization
- solvation free energy (33,34)

- electrostatic complementarity
- packing density (35)
- mechanistic criteria

Of these methods, energy minimization is probably the most reliable (20). Use it before the other methods to remove high energy contacts that can be present in the initial docked structures owing to conformational uncertainties.

The results with test systems in the last few years suggest that it will be possible to dock proteins together using their unbound conformations, generating complexes that resemble the bound forms. In most cases, the invariant side chains and backbone atoms contain enough information to specify to a good approximation, if not uniquely, the binding mode of complex. The catch is 'not uniquely'. We found, in docking protein protease inhibitors into proteases, that Dock would generate not only the native configuration, but non-native complexes as well (20). Such non-native complexes differed from the crystallographic complex by between 8 Å and 22 Å rms—completely different sets of inhibitor residues were docked into the active site of the protease. While some of these non-native complexes were easy to exclude by relatively simple criteria, some others were quite complementary. An example of such a non-native complex, compared to the native, is shown in *Figure 3*. Both complexes show extensive and complementary surface area burial. We could not reliably distinguish the near native complexes from several of the complexes that were far from the native mode using such standard techniques as surface area burial, steric packing, hydrophobicity, mechanistic criteria, electrostatic complementarity, or molecular mechanics interaction energy. The same result, and even some of the same binding modes, was found by Cherfils *et al.* (10) using a grid-based docking approach very different from Dock. An obvious conclusion is that our energy evaluation schemes are still too simple, as is our ability to sample conformation space. A more intriguing hypothesis is that some of these non-native, but complementary dockings, might be sampled in solution, though they will be less favourable than the native (10,20). See Cherfils and Janin (36) for a thoughtful discussion of these issues. Recently, after this chapter was submitted, several docking groups were able to predict a protein–protein complex structure *de novo* (51). This encouraging result suggests that docking programs are increasingly reliable for predicting the structures of protein complexes.

6. Novel inhibitor discovery

There are several strategies towards structure-based inhibitor discovery (1,4,37). Docking programs use a database search approach (*Figure 4*). Rather than try to design a novel inhibitor from scratch, the idea is to scan a database of known, diverse molecules to see which, if any, will complement

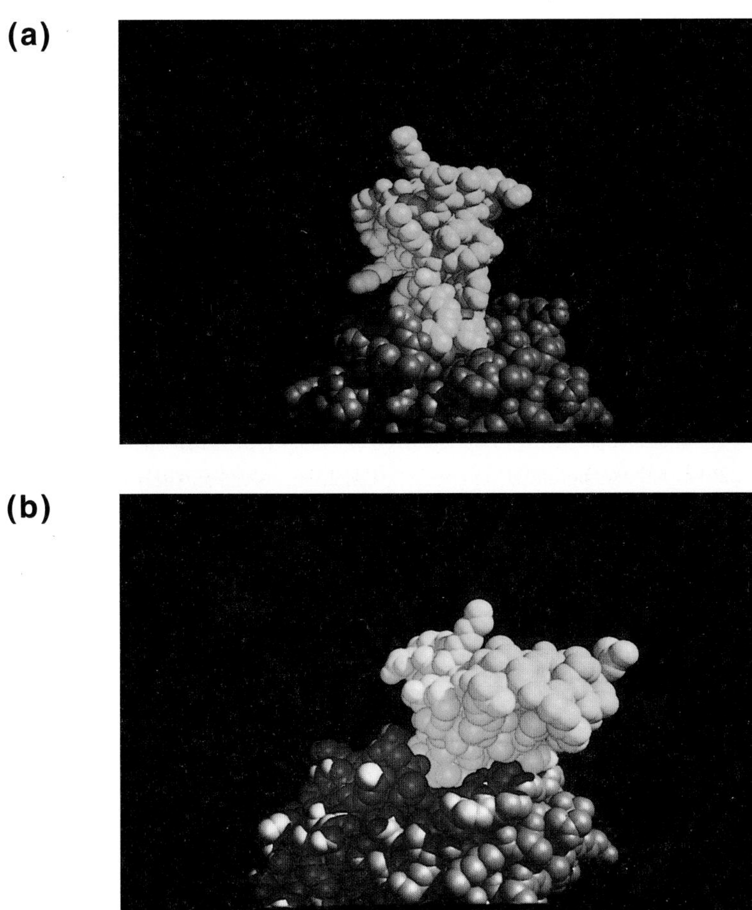

Figure 3. Energy minimized bovine pancreatic trypsin inhibitor (BPTI)/trypsin docking, free conformations, van der Waals representations of the interfaces (graphics made with MIDAS+) (48). (a) The lowest rms docked configuration, 0.52 Å from the crystallographic configuration, BPTI in cyan, trypsin in magenta. (b) A docked configuration of BPTI over 21 Å from the crystal structure of the complex, BPTI in green, trypsin in magenta. Both configurations have extensive, complementary interfaces.

the protein. One typically finds unusual, previously unconsidered molecules as candidate inhibitors. Some of these candidates will not bind to the protein when tested experimentally (false positives). Other molecules, which would inhibit if tested, or are known to inhibit, might not score well in the database search (false negatives). Since the goal is usually to discover *a* novel inhibitor, but not *all* novel inhibitors, such false positives and false negatives are tolerable. Using present technologies, our experience is that one can reasonably hope to find molecules that inhibit in the 1–100 micromolar range.

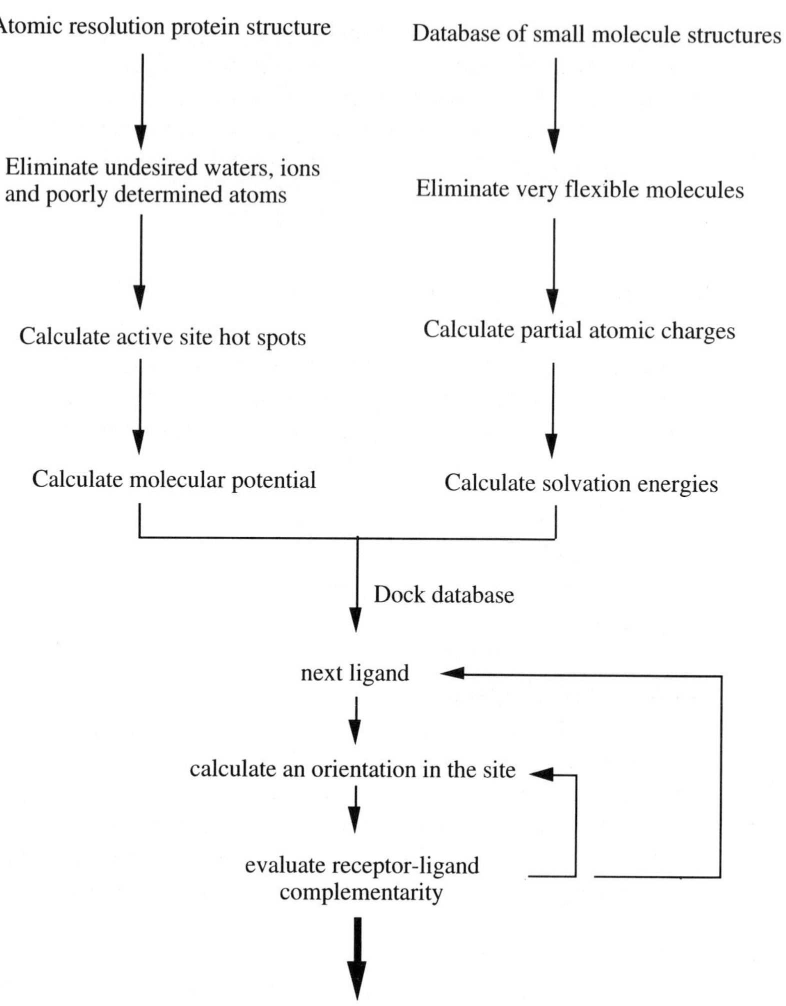

Figure 4. Flow chart for docking a molecular database to a protein.

6.1 Choosing a database

The first practical question is which molecular database to use (*Table 1*). One has two basic choices: using databases of experimentally-determined structures (e.g. the Cambridge Structural Database, CSD) (38), which are often not readily available for testing, or using databases of computer-generated structures (e.g. the Available Chemicals Directory, ACD) that are readily available for testing. I advise using the ACD or a similar database (scientists in the pharmaceutical industry will often have access to databases of proprietary molecules), at least at the beginning of a project. Databases like the ACD allow for short cycle times between a computational hypothesis and an experimental test; you can simply buy any compound that scores well in the docking search. The quality of the structures in an experimentally-derived database, such as the CSD, are unquestionably more reliable than computer-generated. Such compounds are, however, typically difficult to acquire and often must be synthesized.

6.2 Preparing a database for docking

Database docking is single molecule docking magnified. The same input preparation is used (*Protocol 2*), save that one must assign parameters for not one but tens of thousands of molecules. Approximations are called for, as are programs to treat ligands sequentially and automatically. One must often assign the following characteristics to database molecules:

- hydrogen atom positions
- partial atomic charges
- atom types (for van der Waal parameters)
- solvation energies

6.2.1 Ligand atomic charges

Hydrogen atom positions and atomic charges are necessary to evaluate the electrostatic complementarity of the putative ligand–protein complexes. They can be calculated (using a program such as Sybyl or Macromodel) provided the database contains atomic number, hybridization, and connectivity information. Atom types are required for calculating dispersion interactions, for use with Dock they can be assigned with Mol2db (distributed with Dock).

6.2.2 Solvation corrections

Solvation energies are important to include when docking a diverse list of molecules to a protein. Charged molecules, for instance, will usually have a better calculated interaction energy with the protein than uncharged molecules, but they also will be better solvated by water. The proper way to balance the two effects is to calculate the differential desolvation energy of every orientation for every database molecule. This is not currently feasible. As a proxy for this calculation, we have instead calculated the full desolvation energy of every database molecule once, storing the energies in a look up table. This desolvation cost is subtracted from the interaction energy of every orientation of any given ligand on docking. This is a rough treatment a complex phenomenon, but it has significantly improved our ability to distinguish true inhibitors from false positives (23).

Balance each protein–ligand interaction term with a solvation term. Use Born equation methods, such as the Hydren program (39) or Delphi (40), to calculate the electrostatic components of solvation for database molecules. This approach makes most sense when using a Poisson–Boltzmann-type method (41) (Delphi, for instance) to calculate the electrostatic potential of the protein. Set solvent and protein dielectrics to the same values in both the protein potential and ligand solvation calculations. Simple algorithms exist to calculate the hydrophobicity of most small molecules (e.g. CLOGP, available from the Medicinal Chemistry Project, Pomoma College, Claremont, CA 91711, USA) while programs such as Hint (42) (*Table 3*) evaluate hydrophobic contribution to surface area burial in a ligand–protein complex. Dispersion contributions to solvation are harder to explicitly calculate; the methods of Rashin (39) (Hydren program) and Still (43) that treat cavitation energies may be useful in this regard.

6.2.3 Ligand conformations

Small molecule databases for which structures have been calculated, such as the ACD, will often contain large, highly flexible molecules, such as long chain alkanes. Only one conformation of such molecules is typically represented, and this conformation is often unreasonable. Unless you wish to calculate multiple conformations for the molecules you should delete such molecules from the database. There are several programs to calculate multiple ligand conformations based on an initial two- or three-dimensional structure (e.g. Sybyl, MacroModel), we have found the Cobra program (28) useful in this regard (23). The Concord (44) program is widely used for calculating a single three-dimensional structure from a two-dimensional representation of a molecule (*Table 4*).

6.3 Controls

Database searches take a fairly long time. Expect anywhere from 2–14 days on a workstation such as a Silicon Graphics Indigo 2 for about 120 000 ACD molecules, docked in one conformation. There are no formal upper limits on such a calculation. A week of your time might be spent evaluating the results of a search, considerably more testing them. Consider the following controls:

(a) Calculate the interaction energy for a known complex of a ligand with your protein—the energy should be negative (*Protocol 3*).

(b) Reproduce the experimentally-determined structure of a known ligand (*Protocol 4*).

(c) Establish that known ligands of the protein that exist in the database score well when docked into the protein (*Protocol 5*).

It is often possible to find known ligands of your protein in the molecular database. Extract such molecules from the database along with congeners that do not bind to your protein. Dock this mini-database into the protein. Begin with input variables that worked well for your single ligand docking control. You may have to vary several of these to achieve a balance between the length of the calculation and the quality of your results. Molecules known to bind to the protein should score well (negative interaction energies), while the non-binding congeners should score relatively poorly (more positive interaction energies).

Protocol 5. Control for Dock database search

1. Dock the control database. The known ligands should score well (negative scores) and the non-ligand molecules should score less well. If this is true the control has worked.
2. High positive (poor) scores for known ligands often arise because Dock was unable to find the proper configuration for the ligands. Visually inspect the docked complexes to insure that the ligands are bound in the expected geometries. If they are not, the following problems may have arisen:
   (a) Configurational sampling may have been inadequate, check the NMATCH number reported in OUTDOCK. If this is low compared to the single ligand control, increase sampling by varying the Dock search parameters such as bin size, focusing, and so on.
   (b) Ligand parameterization may be incorrect. Inspect partial atomic charges, assigned atom types, atom labels (for matching), and so forth. If you have labelled your protein spheres you must label your ligand atoms in a complementary fashion. If you are correcting interaction energies with solvation energies, insure that you are using the same terms for each. For electrostatic solvation corrections insure that the dielectric of the desolvated phase is the same as the dielectric of the solvent excluded volume of the protein.
   (c) Ligand conformations may be wrong. Focus on ligands with as few 'rotatable' bonds as possible. If you have the resources, consider docking multiple configurations of the ligands.
3. Low negative (good) scores for your negative control molecules often arise because you have not corrected for solvation, because your close contact limits are too loose, or you are expecting the docking program to make too subtle a distinction.

6.3.1 Example database search control

Table 6 shows the results of a database search control calculation, docking against the enzyme thymidylate synthase (TS). Pyrimidine monophosphates should score well, while diphosphates, triphosphates, and higher charged species should score poorly. We chose a subset of 696 ACD molecules that

Table 6. Dock database search control example: sensitivity of results to variable choices

Variable changed	Values	No. of nucleotide monophosphates in top 25 molecules	No. of pyrimidine monophosphates in top 25 molecules	Average orientations per ligand	No. of ligands found	Run time (sec)[a]
None	Below[b]	16	12	390	321	990
dislim	0.75	14	4	11	33	443
dislim	2.0	6 (24/50)	5 (12/50)	1688	380	1984
nodlim	7	24	1	46	48	536
Bin sizes	0.3, 0, 0.8, 0	12	6	400	322	970
Focusing	0	6	1	221	245	800

[a] All calculations performed on a Silicon Graphics PI 25.
[b] Beginning Dock input variables were as in *Table 5*, except that bins were set to 0.15, 0., 1.0, and 0. for lbinsz, lovlap, sbinsz, and sovlap.

contained either a phosphate, a sulfate, a phosphonate, or a sulfonate and fitted them into the dUMP binding site of TS (23). Orientations were scored for electrostatic interaction and steric fit using Delphi and Distmap potential lattices. Ligand electrostatic solvation was corrected for using energies calculated with Hydren (*Table 4*). Generally, the pyrimidine monophosphates docked with large, negative binding energies that were typically better than most other molecules in the database and considerably better than pyrimidine di- and triphosphates, or any molecule bearing more than two negative charges.

We varied several of the Dock input parameters searching for conditions that would:

- maximize the number of high scoring pyrimidine monophosphates
- maximize the number of high scoring nucleotide monophosphates in general

Even though the latter are not recognized by TS, we felt that good scores for these molecules were appropriate given the simplicity of the scoring scheme. Note that different input parameters (*Table 5*) gave different results. Increasing the stringency of the matching criteria, by increasing *nodlim* or decreasing *dislim*, eliminated most non-nucleotide monophosphates. This in turn increased the proportion of nucleotide monophosphates in the final docked list, but at the cost of removing several pyrimidine monophosphates that are known ligands of TS. Decreasing the stringency by increasing *dislim* results in more ligands being found, including more high scoring non-nucleotide monophosphates. More nucleotide and pyrimidine monophosphates are, of course, also found. Using the focusing option (45) considerably improves the results. I favour using this option, but opinion in the Kuntz group is divided as to its utility.

6.4 Choosing molecules to assay for binding

At the end of a Dock database search you will have:

(a) An interaction score for every database molecule docked. One may choose several scoring criteria; Dock reports separate scores for each method.

(b) For each scoring method, a list of the best molecules in the database, ranked by decreasing interaction energy. The atomic coordinates of these molecules are produced in their best scoring configuration in the binding site.

Visually inspect the docked molecules using a molecular graphics program (*Table 4*). Some of the molecules that the docking program suggests may seem inappropriate on graphical inspection. Reasons for discarding a database molecule include:

(a) Complexes dominated by one favourable interaction, with much of the ligand extending out of the binding region into solvent.

(b) Molecules that will be insoluble under the conditions of your assay.

(c) Molecules that will react in a non-specific manner with your protein or components of your assay.

6.5 Strategies for testing the molecules

Many of your candidate inhibitors may be difficult to dissolve in water. Consider DMSO as a general solvent to deliver candidate inhibitors to the assay. You must control for the effect of the DMSO on the reaction, it is usually possible to limit the amount of DMSO delivered to one per cent or less of the total reaction volume.

The absolute interaction energy that a program such as Dock reports for any given database molecule is untrustworthy (1). At best we might hope for a monotonic relationship between relative rank and relative binding free energy, but even here our controls advise considerable caution (23). We are still at a stage where intuition, as well as computed scores, must guide our choices. Individuals will weight the computed scores and their own insight differently; there is no 'right' way to do this. I prefer a two track strategy:

- take several weeks to test the molecules in order of their docked scores
- take several weeks to test the high scoring molecules that you like the best

6.5.1 Specificity controls

When, and if, you find a molecule that binds to the target protein, control for non-specific binding. Generally, a molecule that will bind to most proteins is uninteresting.

(a) Establish that inhibition is reversible, unless one has specifically called for chemical modification of the protein.

(b) Determine whether the molecule binds competitively, uncompetitively, or non-competitively compared to a ligand known to bind in the target site (e.g. the substrate).

(c) Measure the inhibitory potential of the molecule against other proteins.

6.5.2 Finding analogues

Once you have established specific binding, look for analogues of the lead that will also bind to the protein. Such analogues often exist in a database and can be found using similarity searches using programs such as Isis (*Table 4*) or Aladdin (46). To improve the chances of getting useful predictions from the docking program, individually redock the analogues, using greater sampling and multiple conformations.

6.5.3 Establishing binding modes

It is important to know the bound geometry of a lead, especially if one wants to optimize inhibition through chemical elaboration. I have outlined a computational strategy towards geometry elucidation in Section 4.3, ultimately this will depend on the solution of atomic resolution structures. Molecules may bind in multiple configurations and conformations with similar energies, especially in the micromolar inhibition range. The first suggestion of multiple binding modes is often a series of analogues that inhibit but are difficult to fit into a single binding geometry. To test whether a ligand or series of ligands are binding in dissimilar geometries:

(a) Measure the binding of an analogue(s) that dramatically does not fit your favoured geometry (e.g. the analogue contains a bulky group that would intersect a rigid region of the protein). Binding of such an analogue suggests either that your model is wrong or there are multiple binding modes.

(b) Make a residue substitution in the protein that intersects the putative binding region. If all analogues still bind there are either multiple binding modes or the model is wrong. If some analogues still bind but others do not then there are probably multiple binding modes.

Acknowledgements

I thank the UCSF docking group, past and present, especially Tack Kuntz, Elaine Meng, Dale Bodian, Renee DesJarlais, George Seibel, Andrew Leach, Richard Lewis, Dan Gschwend, Cindy Corwin, Diana Roe, Chris Ring, and Kathy Perry. Thanks also to John Irwin, Enoch Baldwin, and Paul Charifson, and to the Cancer Research Fund of the Damon Runyon–Walter Winchell Foundation Fellowship, DRG-1228, for financial support. The Available Chemical Directory and the Isis program were kindly provided by MDL Information Systems, Inc, San Leandro, CA 94577, USA.

References

1. Kuntz, I. D. (1992). *Science*, **257,** 1078.
2. Kuntz, I. D., Meng, E. C., and Shoichet, B. K. (1994). *Acc. Chem. Res.*, **27**, 117.
3. Varney, M. D., Marzoni, G. P., Palmer, C. L., Deal, J. G., Webber, S., Welsh, K. M., *et al.* (1992). *J. Med. Chem.*, **35,** 663.
4. Itzstein, M. V., Wu, W.-Y., Kok, G. B., Pegg, M. S., Dyason, J. C., Jin, B., *et al.* (1993). *Nature*, **363,** 418.
5. Kuntz, I. D., Blaney, J. M., Oatley, S. J., Langridge, R., and Ferrin, T. E. (1982). *J. Mol. Biol.*, **161,** 269.
6. DesJarlais, R., Sheridan, R. P., Seibel, G. L., Dixon, J. S., Kuntz, I. D., and Venkataraghavan, R. (1988). *J. Med. Chem.*, **31,** 722.
7. Shoichet, B. K., Perry, K. M., Santi, D. V., Stroud, R. M., and Kuntz, I. D. (1993). *Science*, **259,** 1445.
8. van Gunsteren, W. F. and Berendsen, H. J. C. (1990). *Angew. Chem. Int. Ed. Engl.*, **29,** 992.
9. Smellie, A. S., Crippen, G. M., and Richards, W. G. (1991). *J. Chem. Inf. Comput. Sci.*, **31,** 386.
10. Cherfils, J., Duquerroy, S., and Janin, J. (1991). *Proteins*, **11,** 271.
11. Jiang, F. and Kim, S. H. (1991). *J. Mol. Biol.*, **201,** 79.
12. Pearlman, D. A. and Murko, M. A. (1993). *J. Comput. Chem.*, **10,** 1184.
13. Bohm, H.-J. (1993). *J. Comput. Aid. Mol. Des.*, **6,** 61.
14. Moon, J. B. and Howe, W. J. (1991). *Proteins: Struct. Funct. Genet.*, **11,** 314.
15. Goodsell, D. S. and Olson, A. J. (1990). *Proteins*, **8,** 195.
16. Stoddard, B. L. and Koshland, D. E. (1992). *Nature*, **358,** 774.
17. Ring, C. S., Sun, E., McKerrow, J. H., Lee, G. K., Rosenthal, P. J., Kuntz, I. D., *et al.* (1993). *Proc. Natl. Acad. Sci. USA*, **90,** 3583.
18. Bax, B., Blaber, M., Ferguson, G., Sternberg, M. J. E., and Walls, P. H. (1993). *Protein Sci.*, **2,** 1229.
19. Rutenber, E., Fauman, E. B., Keenan, R. J., Fong, S., Furth, P. S., Ortiz-de-Montellano, P. R., *et al.* (1993). *J. Biol. Chem.*, **268,** 15343.
20. Shoichet, B. and Kuntz, I. D. (1991). *J. Mol. Biol.*, **221,** 327.
21. Bodian, D. L., Yamasaki, R. B., Buswell, R. L., Stearns, J. F., White, J. M., and Kuntz, I. D. (1993). *Biochemistry*, **32,** 2967.
22. Lawrence, M. C. and Davis, P. C. (1992). *Proteins*, **12,** 31.
23. Shoichet, B. K. and Kuntz, I. D. (1993). *Protein Eng.*, **6,** 723.
24. Wodak, S. J. and Janin, J. (1978). *J. Mol. Biol.*, **124,** 323.
25. DesJarlais, R. L. and Dixon, J. S. (1994). *J. Comput. Aid. Mol. Des.*, **8,** 231.
26. Mohamadi, F., Richards, N. G. J., Guida, W. C., Liskamp, R., Lipton, M., Caufield, C., *et al.* (1990). *J. Comput. Chem.*, **11,** 440.
27. Hansch, C. and Leo, A. J. (1979). *Substituent constants for correlation analysis in chemistry and biology*. John Wiley and Sons, New York.
28. Leach, A. R. and Prout, K. (1990). *J. Comput. Chem.*, **11,** 1193.
29. Meng, E. C., Gschwend, D. C., Blaney, J. M., and Kuntz, I. D. (1993). *Proteins*, **17,** 266.
30. Gasteiger, J. and Marsili, M. (1980). *Tetrahedron*, **36,** 3219.
31. Tilton, R. F. Jr., Singh, U. C., Weiner, S. J., Connolly, M. L., Kuntz, I. D. Jr., Kollman, P. A., *et al.* (1986). *J. Mol. Biol.*, **192,** 443.

32. Wilson, C., Mace, J. E., and Agard, D. A. (1991). *J. Mol. Biol.*, **220,** 495.
33. Eisenberg, D. and McLachlan, A. D. (1986). *Nature*, **319,** 199.
34. Horton, N. and Lewis, M. (1992). *Protein Sci.*, **1,** 169.
35. Gregoret, L. M. and Cohen, F. E. (1990). *J. Mol. Biol.*, **211,** 959.
36. Cherfils, J. and Janin, J. (1993). *Curr. Opin. Struct. Biol.*, **3,** 265.
37. Appelt, K., Bacquet, R. J., Bartlett, C. A., Booth, C. L., Freer, S. T., Fuhry, M. A. *et al.* (1991). *J. Med. Chem.*, **34,** 1925.
38. Allen, F. H., Bellard, S., Brice, M. D., Cartwright, B. A., Doubleday, A., Higgs, H., *et al.* (1979). *Acta Crystallogr. Sect. B*, **B35,** 2331.
39. Rashin, A. A. (1990). *J. Phys. Chem.*, **94,** 1725.
40. Gilson, M. K. and Honig, B. H. (1987). *Nature*, **330,** 84.
41. Warwicker, J. and Watson, H. C. (1982). *J. Mol. Biol.*, **157,** 671.
42. Wireko, F. C., Kellogg, G. E., and Abraham, D. J. (1991). *J. Med. Chem.*, **34,** 758.
43. Still, W. C., Tempczyk, A., Hawley, R. C., and Hendrickson, T. (1990). *J. Am. Chem. Soc.*, **112,** 6127.
44. Rusinko, A., Sheridan, R. P., Nilakatan, R., Haraki, K. S., Bauman, N., and Venkataghavan, R. (1989). *J. Chem. Inf. Comput. Sci.*, **29,** 251.
45. Shoichet, B., Bodian, D. L., and Kuntz, I. D. (1992). *J. Comput. Chem.*, **13,** 380.
46. Van Drie, J. H., Weininger, D., and Martin, Y. C. (1989). *J. Comput. Aid. Mol. Des.*, **3,** 225.
47. Bernstein, F. C., Koetzle, T. F., Williams, G. J. B., Meyer, E. F. Jr., Brice, M. D., Rodgers, J. R., *et al.* (1977). *J. Mol. Biol.*, **112,** 535.
48. Ferrin, T. E., Huang, C. C., Jarvis, L. E., and Langridge, R. (1988). *J. Mol. Graph.*, **6,** 13.
49. Katchalski-Katzir, E., Shariv, I., Eisenstein, M., Friesem, A. A., Aflalo, C. and Vakser, J. A. (1992). *Proc. Natl. Acad. Sci.* (USA), **89**, 2195.
50. Totrov, M., and Abagyan, R. (1994) *Nature Structural Biology*, **1**, 259.
51. Strynadka, N. C., Eisenstein, M., Katchalski-Katzir, E., Shoichet, B. K., Kuntz, I. D., Abagyan, R. *et al.* (1996). *Nature Structural Biology*, **3**, 233–39.

12

The prediction challenge

MANSOOR A. S. SAQI

Genome projects will soon be producing hundreds of kilobases of raw sequence a day. A major effort in biocomputing will be the analysis of sequence data in terms of functional and structural characterization. Homology-based predictions have proved to be useful. However, given the rate of growth of the sequence databases, homology searches are increasingly likely to spot a similarity with another sequence of unknown function. This 'function-homology gap' (1) is likely to increase. There is a need for both statistical analyses to maximize the amount of information that can be used to characterize function and also new algorithms and approaches for secondary and tertiary structure prediction. The level of detail required will vary from establishing function, to getting an idea of the protein fold with a view to prompting site-specific mutagenesis experiments to detailed molecular modelling.

Perhaps the most convincing test of protein structure prediction methods is being able successfully to predict in advance of the X-ray or NMR structure becoming available. The first organized assessment of the various methods for structure prediction was recently initiated in the form of a prediction challenge and the outcome of that challenge was discussed at a meeting at Asilomar in December 1994 (*Meeting on critical assessment of technique for protein structure prediction,* December 4–8, 1994 A special issue of *Proteins: Structure, Function and Genetics* reports the results of the Asilomar Structure Prediction Challenge [13]). Sequences to be predicted fell in to three general categories, namely comparative modelling targets, threading or sequence to fold assignment, and *ab initio* prediction.

Comparative protein modelling (Chapters 6 and 7) is generally considered to be a well-established techinque and success usually correlates with the degree of sequence identity that the query sequence has to the template. A number of totally automated tools for model building have recently become available, e.g. *SwissModel* by Peitsch (2) available on the ExPASy molecular biology server (URL `http://expasy.hcuge.ch/`) and *Modeler* by Sali and Blundell (3). High homology modelling where there are few or no insertions or deletions is really a test of side chain packing optimization and current methods generally perform well. A more difficult problem in high homology modelling is generating loop conformations such as in comparative homology modelling of antibodies. Here conformational search approaches though computationally expensive are promising for small loops. Global free energy optimization using a biased probability Monte Carlo approach has been shown to be a useful methodology (4).

For sequence similarity approaching the twilight zone (about 30% sequence identity between the query sequence and the template) the situation is rather different and such cases present a significant challenge for comparative modelling and also serve to highlight the problems with completely automated approaches. As an example consider the sequence of eosinophil-derived neurotoxin (EDN) a modelling target at the Asilomar challenge. The EDN sequence shares about 33% sequence identity with ribonuclease. The accuracy of predicted models as judged by rmsd on all C^{α} atoms varied from 2.69 Å to 5.22 Å. The main source of error was caused by inaccurate alignments especially with approaches requiring no manual intervention. A *possible* alignment from which to initiate model building is shown in *Figure 1* with 3rn3 (ribonuclease A) and 1onc (pancreatic ribonuclease) as the templates. Other sources of error in the modelling of EDN arose from the modelling of residues between the first two helices (approximate alignment positions 21–31 in *Figure 1*) where the loop is considerably shorter in EDN than in 3rn3, its closest template and particularly the large insertion relative to the templates (alignment positions 121–130).

Loop building is often carried out by searching a structural database and candidate loops for the modelled protein are chosen by consideration of rms fit to the stem residues together with the requirement to have a minimum number of bad contacts generated by the inserted loop fragment. In addition sequence similarity of the candidate loop fragment with the stretch of query

```
     1          11         21         31         41        50
     |          |          |          |          |         |
 1   KPPQFTWAQW FETQHINMTS QQ------CT NAMQVINNYQ RRCKNQNTFL    44     EDN
 1   KETA---AAK FERQHMDSST SAASSSNYCN QMMKSRNLTK DRCKPVNTFV    47    3rn3
 1   PD----WLT- FQKKHITNTR DVD-----CD NIMST-NLF- -HCKDKNTFI    37    1onc

     51         61         71         81         91        100
     |          |          |          |          |         |
45   LTTFANVVNV CGNPNMTCPS NKTRKNCHHS GSQVPLIHCN LTTPSPQNIS    94     EDN
48   HESLADVQAV CSQKNVACKN GQT--NCYQS YSTMSITDCR ETGSS--KYP    93    3rn3
38   YSRPEPVKAI C-KGIIASKN VLTTSEFY-- -----LSDCN VTSRP-----    74    1onc

     101        111        121        131        141
     |          |          |          |          |
95   NCRYAQTPAN MFYIVACDNR DQRRDPPQYP VVPVHLDRII *            135     EDN
94   NCAYKTTQAN KHIIVACEGN ---------P YVPVHFDASV *            125    3rn3
75   -CKYLKKST  NKFCVTCENQ A----PVHFV GVG-----SC *            105    1onc
```

Figure 1. Alignment of EDN (query sequence) with two template sequences, 3rn3 (ribonuclease A) and 1onc (pancreatic ribonuclease).

sequence to be modelled can also be taken into account. Clearly the choice of stem residues will affect the fragment pulled out of the database. Assessment of the accuracy of loop modelling must take into account the temperature factors of the corresponding segments in the crystal structures. Other problems which may influence the model include errors in the template (X-ray) structure. The quality of the model will never be better than the quality of the structure it was based upon.

How accurate do structures generated by comparative modelling need to be? The answer to this depends very much on what is required of the model—is it needed for molecular replacement, for ligand design, or to suggest site-specific mutagenesis experiments. In some cases a C^{α} trace alone may be sufficient whereas for understanding substrate specificity and binding details of the side chain states and loop conformations are important. A measure of the quality of a model is perhaps not best represented by rmsd on all C^{α} but should include other features and there is a need for some standard measures of evaluation.

For sequences with no obvious similarity to sequences of known structure, sequence to fold assignment algorithms are powerful tools (for reviews see refs 5–7 and Chapter 8). The threading method of Jones *et al.* (8) performs well and uses potentials of mean force which are calculated from an analysis of a set of high resolution protein structures. The potential of Jones *et al.* has two components, namely pairwise and solvation potentials. The solvation terms are calculated in a similar manner to the pairwise term but with relative solvent accessibility as the relevant variable.

An important factor in assessing threading predictions is the quality of the alignment of the query sequence with the predicted (known) fold. Cases where predictions are correct but alignments are poor (as judged by the structure-based alignment) are worrying and prompt the question as to what extent are the algorithms recognizing supersecondary structural motifs rather than a global fold?

What is suprising is the ability of essentially simple methods to perform well in sequence to fold assignment. Such methods which take various forms but essentially match predicted secondary structure profiles with the known secondary structure profiles of a non-redundant fold library showed considerable promise at the *Asilomar Challenge*. The method of Bowie *et al.* (9) essentially uses dynamic programming to align the query sequence with structures where the query sequence is represented by a hydrophobicity string and the sequences of known structure by accessibility strings. The approach of Livingstone and Barton (10) makes use of a multiple sequence alignment of the query sequence with its relatives. A variety of secondary structure prediction methods together with an analysis of patterns of residue conservation are employed to obtain a consensus prediction. This is followed by looking for similar patterns of secondary structure in a fold database including any other experimental constraints. An essentially similar approach using secondary structure predictions to identify folds but also including beta-interaction pseudo potentials is descibed by Hubbard (11).

What is highlighted by prediction challenges such as the Asilomar meeting is the need for accurate automatic sequence alignment algorithms. This is essential for automated homology modelling. There is also a need to determine what are the main contributing factors in threading or sequence to fold assignments. Standard methods of assessing models and predicted folds are needed and with threading predictions these must include a measure of alignment accuracy. For model-built structures a step in this direction has been implemented in *SwissModel* where a model B-factor is computed. This is a local coarse grain measure of confidence and takes into account the deviation of the model from the set of template structures. This model B-factor occupies the normal B-factor field in the pdb file and can be visualized by for example using the colour by temperature option in the RasMol molecular rendering program (12). A critical assessment of modelling and prediction methods will serve to prompt further development of techniques in the field.

References

1. Bork, P., Ouzounis, C., and Sander, C. (1994) *Curr. Opin. Sruct. Biol.* **4**, 393.
2. Peitsch, M. C. (1995) *Biotechnology* **13**, 658.
3. Sali, A. and Blundell, T.L. (1993) *J. Mol. Biol.* **234**, 779.
4. Abagyan, R. and Totrov, M. (1994) *J. Mol. Biol.* **235**, 983.
5. Bryant, S.H. and Lawrence, C.E. (1993) *Proteins* **16**, 92.
6. Nishikawa, K. and Matsuo, Y. (1993) *Protein Eng.* **6**, 811.
7. Sippl, M.J. (1993) *Proteins* **17**, 355.
8. Jones, D.T., Taylor, W.R., and Thornton, J.M. (1992) *Nature* **358**, 86.
9. Bowie, J.U., Clarke, N.D., Pabo, C.O., and Sauer, R.T. (1990) *Proteins* **7**, 257.
10. Livingstone, C.D. and Barton, G.J. (1993) *Comput. Appl. Biosci.* **9**, 745.
11. Hubbard, T.J.P. (1994) In *Proceedings of the biotechnology computing track, protein structure prediction ministrack of the 27th HICSS*, pp. 336–54. IEEE Computer Society Press.
12. Sayle, R. A. and Milner-White, E. J. (1995) *Trends Biochem. Sci.* **20**, 37.
13. Lattman, E. E. (ed.) (1995) *Protein structure prediction: a special issue. Proteins* **23**, 295.

Index